Science, Fables and Chimeras

Science, Fables and Chimeras: Cultural Encounters

Edited by

Laurence Roussillon-Constanty
and Philippe Murillo

CAMBRIDGE
SCHOLARS
PUBLISHING

Science, Fables and Chimeras: Cultural Encounters,
Edited by Laurence Roussillon-Constanty and Philippe Murillo

This book first published 2013

Cambridge Scholars Publishing

12 Back Chapman Street, Newcastle upon Tyne, NE6 2XX, UK

British Library Cataloguing in Publication Data
A catalogue record for this book is available from the British Library

ISBN (10): 1-4438-4810-7, ISBN (13): 978-1-4438-4810-7

Quelle chimère est-ce donc que l'homme?
Quelle nouveauté, quel monstre,
quel chaos, quel sujet de contradiction,
quel prodige!
—Blaise Pascal. *Pensées* (1670), 434.

TABLE OF CONTENTS

LIST OF ILLUSTRATIONS

Centerfold illustrations:

ILL. A 3 : Henry Moore, Sculptural Objects, lithograph, 1949 (University of Dundee Museum Services, D'Arcy Thompson Zoology Museum Art Collection. Presented by the Art Fund, 2012. Copyright Henry Moore Foundation). Reproduced by permission of The Henry Moore Foundation.

ILL. A 4 : Bruce Gernand, Coded Chimera—Cat Crocodile Morphs, rapid prototype models with digital prints, 2012 (University of Dundee Museum Services, D'Arcy Thompson Zoology Museum Art Collection. Presented by the Art Fund, 2012). Copyright the artist.

ILL. D 3 : François Miglio, Triptyque EXPLORATION I: Holistic Vision (H 85cm x L 80cm) – Unions of Matrices (H 85cm x L 80cm) – Gemmation (H 85cm x L 80cm).

LIST OF TABLES

ACKNOWLEDGMENTS

This book emerges from an international interdisciplinary conference "*Science, fables et chimères*: *croisements*", which was held at The Natural History Museum in Toulouse and at Université Paul Sabatier, Toulouse 3 (France), from 10-11 June 2011. Although it contains not only some of the papers delivered at that conference but also papers that were submitted to the editorial board at a later stage for publication in this volume, we would like to first acknowledge here the various institutions and partners that allowed us to organize the event and complete the book project. In particular we would like first to thank Paul Sabatier University and its institutional partner, the Natural History Museum which both generously sponsored the event and allowed us to use their venues and resources to run our conference.

To the Natural History Museum, and its director and chief curator, Francis Duranthon and especially to its program coordinator Anne Blanquer-Maumont, we would like to express our gratitude – not only for her professional help, but also for her imagination and encouragement throughout the project. We are also very grateful to our home university and the Paul Sabatier University Science Library for their support and active part in helping us advertise the event and setting up the temporary exhibition "Rêve de chimères" in the Library building. Our warmest thanks go here to Director of the Science University Library (SCD BU Sciences), Pierre Chourreu whose enthusiasm for the project never faltered. We also thank all the members of staff who helped us stage the exhibition, from the selection of books to the actual hanging of Hervé Delamare's drawings on the walls. We should particularly like to thank Christine Launay, Marielle Mouranche, Maryse Vernet and Philippe Viguier for their help with that aspect of the project.

Two research laboratories, the CICADA (Centre Intercritique des Arts et des Discours sur les Arts, EA 1992) based in Pau (Université de Pau et des Pays de l'Adour) and the CAS (Cultures Anglo-Saxonnes, EA 801) based in Toulouse (Université Toulouse 2- Le Mirail) helped us fund the event and we are greatly indebted to them for their financial support. Patrice Simon also supported our initiative and generously helped us sponsor the event. Within the English Department, many people have helped us in various ways at every stage of the project from the early days of organizing the conference to the very last stage of helping us proofread this manuscript. Our warmest thanks and greatest gratitude go to our former Head of Department, Jean-Bernard Hislen and to our colleagues, Serge Arbiol (and Lise), Marie-Agnès Détourbe, Leena Jasani, Claudine Peyre and Jacques Pivato who've all been instrumental in helping us pull this together. From day one, Ralph O'Connor, Olivier Keller, Hervé Delamare, Catherine Gadon and David Midgley responded with great enthusiasm to our initial project and without their encouragements and friendship, this volume wouldn't have come into existence.

The list of others to whom we owe special thanks is long and we can only say that this book is truly the outcome of many fruitful exchanges and encounters with a host of people who share the same appetite for knowledge and curiosity for life in general. Among them, we especially thank the contributors to this volume, most of whom not only proved to be excellent collaborators but also trusted us at every stage of the book project, giving it their full support.

Our warm gratitude also goes to artists Hervé Delamare, Bruce Gernand, François Miglio and Patricia Piccinini who allowed us to publish images of their works free of rights and to Anne Keller for her drawing skills. We also thank the Henry Moore Foundation for granting permission to publish the lithograph of Henry Moore's *Sculptural Objects* of 1949. We also thank Cambridge Scholars Publishing editor, Carol Koulikourdi for her trust and the CSP editorial staff for precious support and help.

Last but not least, we received copious help and advice through the various phases of copy-editing from Marc Constanty whose technical expertise and unfailing support has been invaluable. Any faults in the work remain our own.

INTRODUCTION

The Mayan calendar had predicted a change in December 2012 that was interpreted as the Earth's doomsday. This event – which soon turned out to be a "non event" as the year 2013 began to unfold – is a timely instance of the close relationship between science, belief, and humankind's irrational fear of extinction.

It also illustrates the way in which our ultra modern world is still able to feed on ancient stories, entertain the idea that there might be alternative forms of life and question the ways in which we use or abuse our planet and run our lives. In one word, our contemporary civilizations are still steeped in myth – despite scientific progress and nanotechnology, despite Darwinism and modern art. But is "despite" the right term to use? Or shall we rather say, "because of"? The essays in this volume all address the issue by examining the interplay between 'science, fables and chimeras' – three words that made up the title for a conference that was held at Paul Sabatier University in Toulouse in June 2011.

Deliberately moving away from the way research and expertise have too often been compartmentalised, our concern has been to favour an interdisciplinary dialogue and recreate the dynamics that those cross-boundaries encounters created during the conference. For similar reasons we have decided to publish articles both in French and in English, depending on each contributor's preference, relying on consistent proof-reading and editing to forge a harmonious whole.

The very composition of the volume transgresses established orders and borders as it both uses and subverts the classical 'three-part' structure of conventional French essay-writing: while the first three parts of the volume offer academic contributions to the debate, the fourth part includes contemporary artistic responses that all explore the interaction between science, fables and chimeras in a more intuitive and experimental way. Such an editorial choice is meant first to record the particularly creative atmosphere that enriched the discussions during the conference, spreading to the University Library and further afield to the hall of Toulouse Natural History Museum. As our university library walls were illuminated with the display of Hervé Delamare's works shown alongside a small but beautiful selection of books drawn from our local university Old Books collection, so the conference also greatly benefitted from the presence of the novelist Margaret Drabble who kindly allowed a performed French reading (by actor Phil Wharton) of selected passages from some of her novels. In addition, another contemporary French artist living in California, François Miglio, generously contributed to the conference debates by showing some of his canvasses, one of which so captured some of the themes running through the discussions at stake that we chose to use its three parts as headings to illustrate the first parts of our book.[1]

The union of matrices

The first part of the volume, called "The Union of Matrices" engages with the relation between myth, fable and science. In chapter one, Ralph O'Connor offers a stimulating introduction to the questions underlying the study of "fables" in the broad sense of fictitious stories or narratives, or myths. He retraces the most recent critical theories on myth and suggests that myth should be truly examined and incorporated into the work of science in modern culture. Building on the critical work of the well-established theorist of myth Robert Segal, O'Connor contends that Segal's argument relies on "a sharp dichotomy between science and myth, in which science is assumed to have no function other than that of explanation" (O'Connor, 20). He goes on to suggest that myth can, and should be approached as an integrative part of science with which it shares a number of features – the most prominent of which is what could be called the urge to tell a story.[2] Having thus defined the common ground between myth and science, O'Connor convincingly revisits the cannon of the major intellectual figures who have looked into the definition of myth (such as Popper, Mircea Eliade and Joseph Campbell) to demonstrate how some of their insights could be applied, improved, or prolonged when considering today's science. In the last part of his article, O'Connor thus examines how such contemporary intellectual figures as the entomologist and sociobiologist Edward Wilson and the zoologist Richard Dawkins present the general public with an evolutionary epic that celebrates the progress of science in true mythical mode. What arises from O'Connor's analysis is a clear sense that myth and science go hand in hand, constructing their own definition, uncovering new territories to travel, evolving and building their own narratives. As fields of knowledge,

[1] For the purpose and coherence of the book François Miglio has given us permission to change the order of the panels of his triptych and we are most grateful to him for this.

[2] For a novelist's fine analysis of humankind's urge to tell stories, see Nancy Huston's book: *L'espèce fabulatrice* (Paris: Babel, 2010).

neither can stand alone, and both are clearly steeped in the culture from which they have emerged.

Olivier Keller's subsequent article on mathematics, myth and chimeras similarly stresses the primary role of traditional myths in the elaboration of thought in gerneral and offers a dual perspective on the use of chimeras in the construction of mathematical concepts by making a distinction between fruitful chimeras and sterile ones. In the first part of his article, he describes how the surface of caves was used by the hunters gatherers to symbolise the crossing from the natural world to the world beyond. Arguing that "distorted perspective" (a concept developed by Abbé Breuil) was purposely used so as to embody the ritualistic passage from this world to the next, Keller endeavours to show how the initial surface of the grotto was then used to create key-objects of elementary geometry such as the figure, the line and the dot. His next example takes us to a lesser-known area and the geometrical constructions known as *Sulbasutras* (meaning in Sanskrit "aphorisms of the rope") originating in India in the third century B.C. in which a creation-myth has given birth to an elaborate use of geometrical shapes. By contrast, the second part of Keller's article offers a decidedly provocative analysis on how mathematical knowledge has been and might be used to project modern interpretations on archaic sources – a claim he illustrates through two examples (megaliths and the Ishango bone found in Congo). The result, Keller argues, is purely chimerical and only draws attention to the interpreter's mathematical skills and/or fanciful imagination.

As the next section in part I demonstrates, at the heart of the debate lie three main human faculties: imagination, reasoning and faith, the focus of Alice Lamy's article on medieval thought. The first section of her article firmly anchors the scientific knowledge of medieval philosophers in Aristotle's definition of imagination (or *phantasia*) as a faculty needed to bridge the gap between the mere perception of sensory objects and abstract knowledge. She demonstrates how the English scholastic philosopher and logician, Walter Burley (1275-1344), drew on Aristotle's *Physics* to defend the view that imagination holds a very strong abstracting power. Lamy then retraces Burley's disagreement with William Ockham over Aristotle's categories of substance and quality showing how in his *De formis,* Burley holds that quantity is a form separate from the quantified body. Such an excursion into the complex scholastic debates of the medieval period allows us to become aware of the impact of philosophical thought on scientific theories and the perception of the world. If faith played a major role in the theory of knowledge of the fourteenth century, it also bore a huge weight in the way nineteenth-century natural philosophers tried to reconcile myth with science. As Robert Segal reminds us in the introduction to his massive anthology on myth, "the issue of the relationship between myth and science was the preoccupation of nineteenth-century theorists. Because myth was assumed to be so much like science, the rise of science was assumed to threaten the survival of myth"[3].

The discovery of the fossils of the Burgess Shale by Charles Walcott in 1886 is a case in point and a good example of the collision between myth, science and faith. In her article on the subject, Alexandra Borsari rehearses the terms of the debate and examines the diverging interpretation of the fossils at the time of their discovery and their reassessment by evolutionary biologist Stephen Jay Gould. Looking at Gould's charges against Walcott in his best-selling book *Wonderful Life*, such as how, in Simon Conway Morris's words[4], "Walcott attempted to 'shoehorn' a range of previously unknown creatures into a few familiar categories to fit his preconceptions" on the diversity of living organisms, Borsari shows how the nineteenth-century scientist's knowledge and cultural background could only lead to the conclusions he drew.[5] In the same fashion, she argues, one can probably conjecture that Stephen Jay Gould's analysis of the fossils hinges on his own beliefs and cultural background, an idea also endorsed by Simon Conway Morris himself as he recently argued against Gould, declaring that he "held views on the mechanisms and nature of evolution — and even on humankind's place in the universe — as the 'lessons' to be drawn from the Burgess Shale".

That one should see a miniature replica of the formation and order of the universe in a fossil or a shell, or in William Blake's famous poetic line "to see a world in a grain of sand" is certainly the common feature between such figures as Walcott, Conway Morris, Gould, or D'Arcy Thompson. What is less common though, in the case of the last-named, whose work and achievements are detailed by Matthew Jarron in his article

---

[3] *Myth: Critical Concepts in Literary and Cultural Studies*, ed. Robert Segal (London and New York: Routledge, 2007), 6.

[4] For a glimpse of the current controversy between Simon Conway Morris and Stephen Jay Gould on the fossils of the Burgess Shale, see the following article: Simon Conway Morris and Stephen Jay Gould, "Showdown on the Burgess Shale," *Natural History* magazine, 107 (10): 48-55. http://www.stephenjaygould.org/library/naturalhistory_cambrian.html.

[5] To some extent, Umberto Eco's article, *Ils cherchaient des licornes*, adopts a similar line of argument when he explains that the Europeans travelling abroad all carried with them a whole library of virtual "reference books" that conditioned their approach and understanding of other cultures. See *La licorne et le dragon. Les malentendus dans la recherche de l'universel*, eds.Yue Dayun and Alain Le Pichon (Peking: Peking U.P. 1995).

"Transformations", is the sheer breadth of knowledge covered by the great Scottish naturalist whose studies ranged from foraminifera to geometry and classical literature. Today, one could hardly imagine a university professor giving her students the following advice: "Try also to understand that though the sciences are defined from one another in books, there runs through them all what philosophers used to call the commune vinculum, a golden interweaving link, to their mutual support and interpretation" (quoted in Jarron, 84). And yet, reading about the multifarious and far-reaching repercussions of d'Arcy's observations in such diverse fields as chaos theory, fractals or biomathematics, one can only be struck by the beauty of that "golden interweaving link" running through all things. Among them, perhaps the most unexpected outcome of D'Arcy's interest in geometry is the descendant of his transformation diagrams now used in computerized sculpture form-making – examples of which were recently on display at Dundee.

The final article in part one of the volume also examines the interactions between science and art in the nineteenth century, this time from the perspective of the novel. Isabelle Percebois further explores the blurred frontier between art and science in so-called pseudo-sciences such as phrenology or animal magnetism – all of which became hugely popular throughout Europe at the time. She then goes on to demonstrate how those "soft sciences" such as mesmerism actually derive from real science and technological advances combined with the dream world of the supernatural. Conversely, literature drew from those newly invented pseudo-sciences to imagine new plots and unsolved mysteries. In such novels as Mary Shelley's *Frankenstein* or Villiers de l'Isle-Adam's novel, *Claire Lenoir*, invented notions such as "magnetic fluid" or "vital fluid" are used to substantiate fanciful theories and stories.

Holistic visions

Part two, entitled "Holistic visions", mainly focuses on chimerical perspectives within the exploration of the dialogue between science, spirituality and religion.

The notion of an "inescapable" conflict between religion and science has been the subject of a centuries-old debate. The historiography of the relationship between science and religion highlights the ever-sensitive character of the matter at stake: how have boundaries shifted between the two concepts? The 'Conflict Thesis' firmly substantiated by the Irish apothecary John Rutty in the mid-18th century in *A spiritual diary and soliloqui* (1776), and then reinstated by John William Draper's *History of the conflict between religion and science* (1875) and Andrew Dixon White's *A history of the warfare of science with theology in Christendom* (1896), establishes a formal state of opposition between the cause of science and that of religion. It was then taken over by major historians and scientists such as Galileo Galilei (1564-1642), Charles Darwin (1809-1882) or Richard Dawkins, and prevailed as the commonly accepted theory much into the twentieth century. This approach considered that religion and science are separate but equal, where religion is a source of ethical and religious inspiration, while science deals with objective, quantitative modelling of physical phenomena.

In section I, the three articles offer a broad view of the intertwined relationship between the scientific and the imaginary. Daniel Larangé explores the synthesis of the sciences by engaging the debate between imaginary science and the science of the imaginary. By taking the example of esoteric mysticism, Larangé presents a thorough analysis of the reenchantment mechanism in which faith becomes more rationalized within a search for harmony. Soul, conscience and body are analysed as embodiments of the three categories of sciences. In addition, magnetism, mesmerism, positive thinking and spiritism are auxiliaries of the romantic bridging of science and religion in the nineteenth century.

The next article by Sébastien Grégov stresses the importance of hermeticism in the chimerical encounters of science and religion by unveiling the deeper meaning of Christian Rose-Croix's *Noces Alchimériques.* Back in the seventeenth century the emergence of the Rosicrucian cult laid the foundations for a new pattern designed to show how to heal the world through experientialism and the imaginary. The main objective of hermetic experience is to bridge science, conscience and religion. The author also points out the necessity of a magic thought that results from a combination of religion and the sciences. Through the example of the *Noces*, Gregov shows how important it is to refocus the world onto the human individual himself, away from scientific discoveries. Man is meant to recreate chimeras to re-enchant the world. It is the purpose of a 'new conscience' where hermetic science would include all the other sciences: 'I am in the world, and the world is in myself'.

Egil Asprem's contribution offers an exploration of anti- and non-mechanistic tendencies in early twentieth century science, illustrated by the "vitalistic controversy", that deal with two epistemic aspects of the Weberian disenchantment thesis: first, the idea that the world is completely rationally explicable; second, that there is a strict epistemic divide between the domains of religion, philosophy and science, in which science's vast potential is still restricted to cold facts, whereas it can tell us nothing about "ultimate questions" regarding value, ethics, or metaphysics – the presumed domains of religion and philosophy. The author argues that scientists in Weber's own day were far from universally committed to such a view of science; on the contrary, much groundbreaking scientific work took a very different stance. Through the examination of several historical examples, Asprem suggests that the illusion of disenchantment covers up some intriguing cultural links between empirical branches of science and religion, which also force us to move "the occult" out of the margins of early twentieth century science towards the centre. This holds particularly true for the sciences of life and mind.

In section II, the two articles unfold arguments for new claims regarding the importance of occultism or esotericism in the future of the science/religion dialogue. Over the last three decades, many historians of science have strived to develop a revisionist perspective on the conflict thesis. In the 1970s, John Hedley Brooke was one of the most significant revisionists in this matter, arguing against the partiality of the notions of both science and religion in this thesis, and emphasizing the more 'social' (and complex) implications instead of taking an essentialist approach. Later, David Lindberg and Ronald Numbers demonstrated the complexities and more subtle interactions between science and religion by initiating the 'Complex Thesis'. Science and religion are in agreement when one or the other is nudged, interpreted into conformity. The need for a revision of common approaches to this problematic relationship between science and religion has been supported by many recent discoveries and events, bridging the two categories. One example is the July 2012 announcement of Cern, the particle physics lab near Geneva, concerning their identification of the Higgs boson (aka the God particle), or at least a particle that behaved like it. Indeed, elementary particles, like the quarks and electrons inside atoms, derive their mass from an invisible field that stretches through all of space. Without something to give particles mass, there would be no stars, planets or life as we know it. The recent reappraisals of the categories of 'science' and 'religion' have triggered a new trend in historical and sociological inquiry: a shift from 'beliefs' to 'practices' within the perspective of a deeper questioning of these categories. And as Peter Harrison has suggested, the 'science' category appeared in the nineteenth century, while the ‘religion’ category was introduced during the European Enlightenment.[6]

As Marco Pasi underlines in the first article, modern art is intimately connected to magic and esotericism. He shows how modern art has been influenced by the occult insofar as it promotes a return to the primitive ‘mind’ of magic. Would post-modern art become another religion or New religious Movement? Pasi also questions the equation of magic with modern art, thus suggesting a higher importance of occultism in the evolution of the arts. Should we interpret this phenomenon as a reactionary or anti-establishment stand? Contemporary art must be considered at a distance, as experimentation, a transgression of traditional religious values. Is art a substitute for secularization? Marco Pasi reveals that, with the importance of esoteric initiation and magic, secularization is not so widespread in the arts. With its heritage of resentment towards traditional religion, post-modern art has become an area of experimentation, of constant innovation, challenge, provocation and transgression of aesthetic, as well as political and social conventions. There seems to be an intriguing parallel between occultism and contemporary art.

The article by Philippe Murillo that follows examines the conversion of religion and science into metaphysics to contemplate a state of harmony between the mind, the spirit, the body and the cosmos. To what extent can spirituality and faith be secularized, if at all? Popular metaphysics has always been part of the fostering of American spiritual and religious identity. American metaphysical religion (New Thought) is the epitome of chimerical encounters between practical expectations and a spiritual quest. The author points out a reconceptualization of magical and therapeutic spirituality within a post-secular society. A new way is being proposed to spiritual seekers, like a crossbreed between science and spirituality, where western idealism and eastern mysticism are combined. Moving from a substance perspective to one of ‘process’, New Thought is undergoing a (chimerical) transformation into ‘Process New Thought’. Science and ‘process philosophy’ are used to renew a century-old practical religion. In keeping with new philosophical trends and scientific developments, such as quantum physics, among others, Process New Thought emphasizes that the soul is less a “thing” than a history of experiences.

[6] Peter Harrison, ‘Science’ and ‘religion’: constructing the boundaries, in *Science and Religion, New Historical Perspectives*, Ed. Thomas Dixon, Geoffrey Cantor, Stephen Pumfrey (Cambridge: Cambridge University Press, 2010), chapter 2, 30-31.

Gemmation

Part three of our volume moves away from the issues and historical debates surrounding science and religion to offer a selection of case studies that reflects the fertile cross-pollination between literature, science and the arts. The first section examines how literature and the arts have used the development of scientific knowledge and technology to hold up a mirror in which humankind can contemplate its own limitations and faults. In her article on fables, Sabine Gruffat offers an original account of La Fontaine's famous fables and of the writings of some of his followers, such as Houdar de la Motte and Fénelon. In her analysis she more particularly pays attention to the back-and-forth movement between the scientific exploration of the insect world in the emerging discipline of entomology and their translation into moralistic tales, arguing that the numerous changes in scale (from the very small creatures to be seen through the recently-invented microscope to the infinite and unknown bounds of the universe) illustrates the moralistic excesses of the human mind. She goes on to show how in the 18th century writers like Réaumur or Fénelon used familiar topics such as the history of bees or silk worms to renew the popular genre of the fairy-tale to include what could be called "the marvellous real".

By contrast, Houdar de la Motte – a somewhat second-rate eighteenth-century fabulist – went as far as to advocate the writing of fables combining the "erroneous naturalism" of the past (and its beliefs in fabulous creatures such as the sphinx or mermaids) with "modern truths" (Gruffat, 212). As Gruffat convincingly argues, in those writings, the fable gradually becomes a place of experimentation that allows readers to admire a homely and familiar representation of the marvellous. As such, the fable genre is able to popularize the more unsettling advances of science and to enhance the writings of the encyclopaedists. This is beautifully exemplified in the last part of her article where she analyses l'Abbé Aubert's *New Fables* (1773) and La Fontaine's *Fables* in relation to Buffon's natural history.

In the next article, Céline Cadaureille also offers a focus on supernatural creatures and the sense of wonder that arises from the merging of experimental research in the sciences with the inventiveness of the arts. In her analysis, she shows how Australian artist Patricia Piccinini uses art to show the blurred frontier between the human and the non-human. In her photographic series *Protein Lattice* (1997) and in *So2*, a pseudo-chimerical creature named *Exallocephalla parthenopa* after the mythological figure of Parthenope and inspired by recent developments in synthetic biology, the artist questions the ever-changing definition of humanity. In the Venice Biennale Show (in 2003), her hyper-realistic moulages confront and engage the viewer in an unsettling and possibly disturbing experience. As Cadaureille underlines, the chimera here does not so much refer to mythology as to the potentially unnatural cross-breeding of different species (Cadaureille, 230).[7] In looking at how the artist stages her art works, the author refers to Freud's notion of the uncanny – a particular feeling of strangeness arising from a familiar scene turned alien – and clearly identifies the artist's awareness of the moving frontiers between life and its inanimate counterparts.

The issue of cross-breeding or hybridization is also at stake in Peter Matthiesen's novel *Killing Master Watson* as aptly shown by Marie-Christine Noailles. Here as in most of his other novels, according to Noailles, the author combines scientific observation with lyrical, sophisticated prose that connects the world of art and the world of the natural sciences. As Noailles describes the various protagonists in the book, she underlines the parallel between some of the characters and Charles Darwin and his cousin, Francis Galton (inventor of eugenics and the "nature versus nurture" phrase). She goes on to demonstrate how the main characters in the novel embody underlying ideological positions regarding humankind and some of Darwin's key-notions, like variation under domestication which deals with the breeding of domesticated animals. In *Killing Master Watson*, she argues that hybridization is not limited to natural phenomena but can be seen to operate in all fields and at every level of human existence so that literature and natural science are seen to use and share the same concepts. Even though cross-disciplinary studies tend to even out differences and highlight the common features between science and the humanities as if they both cultivated the same infinite plot of ground, one has to acknowledge that investigating the significance of scientific theories in disciplines other than the science they emerged from may mean travelling long distances through uncharted waters.

As Monique Martinez and Michel Caffarel well illustrate in their jointly authored article on quantum physics, such an intellectual endeavour is inevitably risky and adventurous as it demands both scientific rigour and imagination. In the first part of the article, Martinez details how such

---

[7] Incidently, Piccinini's latest show (and first solo exhibition) in London entitled "Those who dream at night" (December 2012) also addresses the issue of regeneration and fertility in the broadest sense as the artist herself declares in a recent interview. (http://www.rev-ma.blogspot.co.uk/2012/12/patricia-piccinini-haunch-of-venison.html)

theorists as Gregorio Morales have developed "quantum aesthetics", going as far as arguing that quantum theory had first instinctively been discovered by artists rather than scientists. The main point here, though, is to plead for what Edward Wilson called "the unity of knowledge" and become aware that "science needs the power of intuition and metaphor provided by the arts just as the arts need the new stock of knowledge that science can provide (quoted in Martinez, 258)." In the case of quantum physics principles applied to the arts a number of examples are given by Martinez in drama and in avant-garde music (John Gage) which all tend to illustrate the benefit gained from such cross-fertilization. However, in the second part of the article, physicist Michel Caffarel uncovers some of the complexities underlying the over-simplistic view of quantum physics and warns us against any hasty conclusions: the world of quantum physics should not be too readily opposed to classical physics as many aspects of the physical world also challenge our reason and defy our wildest imagination, all things that the science and art of tomorrow will undoubtedly explore.

The final article in the section moves one step further towards the future by discussing the changing nature of humankind through two lenses: science fiction and nanotechnologies, two fields that heavily rely on the power to envision the future through technology and the power of myth. After quickly retracing the major stages of the development of nanotechnology and describing the various actors involved, Mathieu Quet here examines a new kind of bestiary: human hybrids transformed in various ways with the help of technological devices or through medical advances. By choosing to focus on athletes and genetic doping, he demonstrates how numerous modern chimeras get created and designed: part human, part machine, these new beings (cyborgs, genetically-modified creatures) have become the focus of attention because they embody what humankind may turn into. However, as Quet concludes, the discourse on nanotechnology implies that those new chimeras are only part of an overall attempt to supervise as much as to encourage innovation. In this debate hovering between the public sphere of popular science and science fiction and the academic world of fundamental research and "scientific facts", the chimera still looms large in our imagination.

Artistic voices

As mentioned at the beginning of the introduction, the last section in the volume is intended to offer a dialogue between first-hand accounts of how contemporary artists and writers use the raw matter of scientific objects or theories to create their own cosmogony peopled with chimeras and give instances of the critical reception of their practises. In the first contribution in this section Laurence Roussillon-Constanty gives a synthetic and necessarily sketchy account of the representation of chimeras in canonical medical treatises such as Ambroise Paré's *Monstres et Prodiges*. As a very general commentary on the selection of book illustrations displayed as part of the temporary exhibition held in the university library of the science faculty in Toulouse, the article offers a conspectus of the visual representation of well-known chimerical creatures such as the so-called Ravenna Monster or the unicorn. Its main purpose is to show the continuity of motifs that run from the sixteenth century through to contemporary artistic practice, as exemplified by the practice of artist Hervé Delamare.

The interview with the artist reproduced in the volume will hopefully be seen as a lively follow-up to the exhibition and as an insight into Delamare's rich imaginary world and visual vocabulary. In his artistic practice the chimera is translated into various kinds of media (from drawing to sculpture and wax moulages), sometimes getting reproduced at different scales, and ranging from the gigantic (in sculpture) to the smaller, fragile structure (such as the figure that appears on our book cover). In the same way, François Miglio's work stems from close observation of the natural world and its inherent structure which he shares in his contribution on fractals in relation to the chimerical, fanciful and disorderly universe that surrounds us. What emerges from Miglio's personal artistic experience is a deep sense of the harmony underlying the seeming chaos of life and a feeling that the artist has freedom to build on science to create his own cosmogony.

The final chapter in this section takes up the same cue, this time from the point of view of fiction. In her splendid contribution, Margaret Drabble retraces her own experience as a fiction-writer growing up in Britain in the 1960s and writing in the context of the now classic debate about the 'two cultures'. She goes on to illustrate how she herself used motifs and themes from the natural sciences in writing her family saga *The Peppered Moth*, in which this particular insect – said to be the only existing evidence of 'evolution in action' – as a metaphor for her characters' greater or lesser ability to adapt to a new environment. The volume concludes with Claudine Peyre's article on Margaret Drabble's novels, in which she describes how fiction is always intrinsically interlaced with myth, even in the most realistic stories. In Drabble's fiction, the domestic sphere itself resonates with echoes of debased mythical figures and uses science to reflect on chance, free will and unforeseen encounters.

In literature, art or science what ultimately seems to be the thread in all the essays in the volume might just be that profoundly human need to fictionalize, invent new myths or create artistic forms that suggest infinite patterns and "endless forms". As scientists tell us that the universe is for ever-expanding and philosophers argue that "now is the time for a finite world"[8] we find ourselves like the janus-faced chimerical figure in Delamare's sculpture – having to find a new balance in the Moëbius strip, torn between the desire to look forward to our future and the temptation to look back on the mysterious way in which our destinies have so far unfolded.

[8] Albert Jacquard, *Voici le temps du monde fini* (Paris:Seuil, 1993).

PART A:

THE UNION OF MATRICES

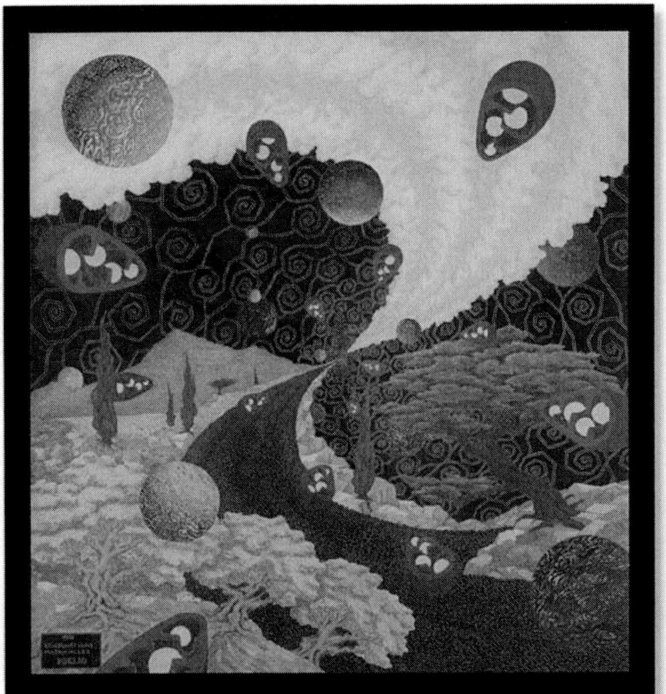

ILL. A-1 : François Miglio, *Unions of Matrices*, detail from Triptyque EXPLORATION I (H 85cm x L 80cm)

SECTION I

The Place of Myth in Modern Science: Some Theoretical Considerations

Ralph O'Connor

The idea that myth has any place in modern science will strike many readers as bizarre. The self-definition of science today is bound up with a rigorous exclusion of fantasy or mythology and focuses (depending on the field) on the pursuit of rationally argued and/or empirically verifiable truths. Examples of science's hostility to myth can be found whether one looks back to the Darwinian revolution, the Enlightenment, the Copernican revolution, the twelfth-century Renaissance and other alleged turning-points in the history of science: it underpins the self-representation both of the modern sciences and of their more capacious predecessor, natural philosophy. If myth has any place in these disciplines, it is surely as the obverse of science, since myth is conventionally accorded no capacity by modern Westerners to embody literally true knowledge about the physical universe. In this view, depending on one's perspective, science may be seen as complementing or displacing myth, but not incorporating it.

In this short chapter I will argue that the incorporation of myth is, on the contrary, an important part of the work of science in modern culture. Rather than attempting a comprehensive survey of how it has done so (which would be the subject for a whole book: see footnote 40), I will simply take a few short steps down two connected avenues of argument: one theoretical, the other empirical. First, I will engage with Robert Segal's influential and lucid analyses of how key modern theorists of myth have responded to science, in order to see how far their theories and Segal's assessments of their theories leave room for an interpretation of science as (in part) mythic. Second, I will briefly discuss a few examples of scientists as mythmakers. It would perhaps have been easier to show the science-myth alliance at work by analysing theistic visions of science, the Gaia hypothesis and its kin, or fantastic treatments of scientific themes by creative artists (themes which are ably developed from different perspectives elsewhere in this book), but such examples risk not being accepted as communicating "real science" by some readers. [1] My main examples will therefore be taken from the works of defiantly atheistic scientists such as Edward Wilson and Richard Dawkins, with a nod towards their Victorian predecessors. The aim is not to treat either the theories or the examples in the detail they deserve, but rather to serve as a springboard for further discussion about the usefulness (or otherwise) of the category "myth" when thinking about science's place in culture.

In an essay published in 1996 Robert Segal stated that "Whether myth has a future depends on its capacity to meet the challenge of modern science".[2] He goes on to explain how theorists of myth have met that challenge. Segal mentions but does not discuss what he sees as the two weakest responses, namely ignoring science and labelling science as mythic. He then briefly discusses the more respectable "surrender" move, epitomized by the Victorian anthropologist Edward Tylor. In this move, myth simply gives up its explanatory function to science and dies. Most of Segal's discussion centres on the twentieth-century theorists of myth whose response to science took the more complex form of what he calls "regrouping", adapting myth so that it becomes compatible with science. This is done either by reconfiguring the content of myth as non-literal, or by reconfiguring the function of myth as other than, or additional to, that of explanation of how things came to be. As Segal shows both here and elsewhere, eminent theorists of myth such as Mircea Eliade, Rudolf Bultmann, C. G. Jung and Joseph Campbell have met "the challenge of science" by combining several of these adaptive strategies in their accounts of what myth is and does, giving myth a number of possible futures within a scientific culture.

Segal's explanations of the relationship between myth and science are extremely lucid and therefore very useful. But part of their clarity comes from their being based on a sharp dichotomy between science and myth, in which science is assumed to have no function other than that of explanation, or at least is treated according to this function alone: "To accept the scientific explanation of the world is to render the mythic one false."[3] Attempts to label science itself as mythic are therefore judged by

---

[1] Robert Segal has recently argued that the Gaia hypothesis does not, in fact, bring myth back into science, but his notion of what constitutes 'myth' is stricter than mine. See Robert Segal, "Myth and Science: Their Varying Relationships," *Religion Compass* 3 (2009): 337-58.

[2] Robert Segal, "Does Myth Have a Future?" in *Myth and Method*, ed. L. L. Patton and W. Dinoger (Charlottesville: University Press of Virginia, 1996), 19.

[3] Ibid.

Segal to be "facile".[4] Segal's analytic summaries of the various theories of myth reveal a rich variety of positions, varying through time according to their authors' personalities and cultural contexts, but science functions in these summaries and in Segal's own commentaries (as indeed in many of the writings of the theorists themselves) as a static and impersonal entity, posing its unquestioned "challenge" to humanity like the extraterrestrial obelisk in Arthur C. Clarke's novel *2001: A Space Odyssey*. In science fiction like Clarke's, of course, the future possibilities of science and of worlds beyond our own are often represented in quasi-mythic form. The obelisk, standing before the awestruck ape-men to bring them into a new phase of evolution, is a secular *deus ex machina*. Science fiction's adoption of many of the functions of myth is well known, and the pages of early science-fiction magazines such as *Amazing* and *Astounding* are littered with thinly disguised Greek and Norse deities and rehashings of their myths.[5] The authors of stories and speculative essays like these, however, are rarely claiming literal truth-value for what they write, so they do not directly affect the point under discussion. Nevertheless, the prevalence of mythic narrative patterns in fictional representations of science does at least prompt us to wonder whether there is something inherent in science which calls out for mythic treatment by storytellers.

Most philosophical and anthropological discussions of science's supposed mythic content have homed in, like Segal, on science's explanatory procedures. One of the most influential exponents of the idea that scientific explanation is in some way mythic was the distinguished philosopher Karl Popper, whose *Conjectures and Refutations* (1963) presented a new version of the old idea that science evolved from myth. As part of his wider refutation of the inductive method, Popper presented scientific theories as myths whose makers invite subsequent audiences to comment on, correct and if necessary reject the "story" they are given. Popper's aim was not to undermine the validity of scientific knowledge (although his views have sometimes been used, in distorted form, to do just this). Rather, his aim was to argue that scientific knowledge has

---

[4] Segal, "Does Myth Have a Future?," 19. See also Robert Segal, "Tylor's Theory of Myth as Primitive Science," in *The Sum of Our Choices: Essays in Honour of Eric J. Sharpe*, ed. A. Sharma (Atlanta: Scholars Press, 1996), 70-84, reprinted in Robert Segal, *Theorizing about Myth* (Amherst: University of Massachusetts Press, 1996), 7; Segal, "Myth and Science," final section.

[5] In *Amazing Stories*, vol. 23 issue 1 (January 1949), Arthur Petticolas's novel "Dinosaur Destroyer" is set in the mythical land of Atlantis and features a race of Amazons; Richard S. Shaver's essay "The Cyclops" conflates extraterrestrial visitors with the Greek Cyclops myth.

always progressed by testing existing theories against the evidence rather than by producing knowledge neutrally from an accumulation of raw data. He saw science as a special form of mythmaking – a kind of mythmaking whose inbuilt critical apparatus enabled it to draw ever closer to the truth.[6]

As theories of knowledge go, this is far from facile. Yet, as Segal has hinted elsewhere, Popper's understanding of myth fell short of his understanding of science. Not only did he limit his discussion to creation-myths alone – understandably, perhaps, given the kinds of science he was discussing; he also used the category "myth" in an ambivalent way, often resorting to it (in the same book) as a label for pernicious falsehoods and thus undermining his more precise and value-neutral use of the term as a precursor of scientific theorizing.[7] Furthermore, when discussing myth in this non-pernicious, pre-scientific sense, Popper treated it as a monolithic and commonsense category whose meaning and significance must be obvious from the context in every case, almost precisely mirroring the way in which some theorists of myth have treated "science". Pre-scientific myth, for Popper, becomes a one-dimensional foil to the more interesting and complex phenomenon of science itself: myth is defined (implicitly) as a narrative of unquestioned truth-value which is handed down, unchanged in all its essentials, from one generation to the next. This commonsense assumption prevented Popper from observing some interesting further common ground between science and myth, which those who study real-life myths in their cultural contexts will immediately think of when reading his account: namely, that the process of handing down a mythological narrative has very often (to a greater or lesser extent) involved a questioning and sometimes a radical alteration of the narrative. For example, the adoption of Christianity in mediaeval Europe did not always result in the simple replacement of one set of myths by another, but often prompted the critical re-evaluation, transformation and adaptation of existing myths. The vernacular literature of mediaeval Ireland and Iceland contains renovated versions of many of these myths in legendary or other quasi-historical forms, and their narrators sometimes *explicitly* invite listeners to correct the story if it seems inaccurate or if they have better information.[8] The narrators of sagas, in fact, are much more often found

[6] Karl Popper, *Conjectures and Refutations*, 3rd ed., new imprint (London: Routledge, 2002), 169-72 (chapter entitled "Towards a Rational Theory of Tradition").

[7] Popper, 185, 211, 256 *et passim*, especially when discussing Baconianism or Nazi ideology. See Segal, *Myth*, 33-34.

[8] On transformations of older myths, see, for example, John Carey's essay "The Baptism of the Gods" in his *A Single Ray of the Sun: Religious Speculation in*

asking for improvements and apologising for inaccuracies than are the impersonal narrators of modern scientific papers, for all that the latter implicitly invite criticism. Nor was this critical attitude towards dogma confined to the later retelling of myths associated with an obsolete religion. The commentary tradition in ancient and mediaeval Christianity, Islam and especially Judaism subjected the received narratives to multiple reinterpretations across time, often in ways which changed the meanings of the original stories in fundamental ways to produce new theologies. These considerations could be used to strengthen Popper's original point that science is akin to myth, but if so, they would require a new way of defining the distinctiveness of scientific discourse as compared to myth.

Such debates about the nature of scientific explanation (which are far from settled today) feed into the question of how far scientific knowledge may be seen as "constructed", in other words, is not value-neutral but has been made by people serving specific social and political interests and reflects those interests in some fundamental way. This question lay at the heart of the "Science Wars" of the mid-1990s, which reduced an important philosophical debate to the level of an academic squabble. During that quarrel, the category "myth" was sometimes used as a shorthand for "falsehood" or "distortion" by partisans on both sides of the debate (although less often in their more considered statements).[9] In its shrillest and simplest expressions, the postmodern-relativist claim that science was a form of modern myth and therefore not objective or trustworthy locked horns with a (simplified) scientific-realist claim that science was objective and therefore not mythic; the assumed opposition between truth and myth was implicit within both claims. It is this kind of facile opposition which Segal rejects when he repudiates the notion that science itself is mythic by linking it with "the postmodern dismissal of the authority of science".[10]

Early Ireland (Aberystwyth: Celtic Studies Publications, 1999), 1-38; Margaret Clunies Ross, *Prolonged Echoes: Old Norse Myths in Medieval Northern Society*, 2 vols (Odense: Odense University Press, 1994-98). On the critical attitude, see Ralph O'Connor, "History or Fiction? Truth-Claims and Defensive Narrators in Icelandic Romance-Sagas," *Mediaeval Scandinavia*, 15 (2005): 101-69.

[9] For contrasting assessments, see *A House Built on Sand: Exposing Postmodernist Myths about Science*, ed. Noretta Koertge (New York: Oxford University Press, 1998); *Beyond the Science Wars: the Missing Discourse about Science and Society*, ed. Ullica Segerstråle (New York: SUNY Press, 2000).

[10] Segal, "Does Myth Have a Future?" 19; Segal, "Tylor's Theory of Myth," 7; Segal, "Myth and Science," final section. Robert Segal has, however, informed me (pers. comm.) that his labelling of science-myth equations as 'facile' in 1996 (in "Does Myth Have a Future?") was not formulated with postmodern philosophers in mind.

There is, however, a different way of conceptualizing the possible mythic component of science which sets to one side the philosophical debates and polemics just mentioned, discards the popular equation of myth with falsehood, and rejects the notion that identifying mythic aspects of scientific thought cheapens or undermines the authority of science. Let us accept for the sake of argument that what we take as "good science" tends to be better at explaining physical phenomena than myths are (when the two are in direct competition on such a question).[11] That point granted, it is nevertheless possible to see science as incorporating myth at other levels. This view requires an appreciation of the wider cultural authority of science as a phenomenon whose procedures and knowledge-claims have moved well beyond scientific institutions and research-communities to become a dominant part of the world-view of most educated moderns, indeed almost a badge of membership of the modern world. The nature and extent of this manifestation of "science" and its correspondence to what scientists actually do and think vary hugely, of course, according to cultures and individuals; but it is uncontroversial to state that science is an inescapable part of modern public discourse, both among scientists and among people in general. Like other components of a worldview, then, we should expect science in this context to perform functions beyond pure explanation or prediction. Whereas Victorian anthropologists such as Frazer and Tylor saw science and mythmaking as kindred but incompatible activities engaged in by people at opposite ends of a cultural spectrum, I propose to examine cases where science has been deliberately made mythic by its own proponents.

Some of these functions of "science in culture" come to mind when considering the views of myth held by some of Segal's foremost examples of the "regrouping" strategy (i.e. preserving myth in the face of science by removing its most important functions from the explanatory level to another level). The American theorist Joseph Campbell suggested several functions for myth which recall the language in which science has been promoted in modern public culture. Campbell's views on the function of myth boil down to four main points: myth creates a sense of awe and mystery before the cosmos (as do the final pages of many popular scientific treatises), provides a symbolic image of the world (his example is the Great Chain of Being, but iconic and idealized images of the Solar System or the evolutionary "tree of life" may be used to serve similar ends), justifies a social order in terms of a cosmic order (as do various

[11] The view, which I accept for the sake of the argument, that science provides explanations (rather than mere predictions, for example) is not universally accepted among philosophers of science.

versions of social Darwinism and more recent scientistic or technocratic ideologies, including calls for new ethical systems based on science)[12] and harmonizes humans with the cosmos, society and themselves (claims often made on behalf of cosmology, evolutionary biology and psychiatry).[13] Campbell himself did not fully acknowledge these resonances between his conception of myth and some of the functions of science in modern culture, despite his own reference to the Great Chain of Being which was so central to early modern science, his many references to space travel as a backdrop for modern myths, or his own insistence that myth and science were compatible with each other. Indeed, in *The Power of Myth* he suggests that the new metaphors brought into being by science will not become truly "mythological" until they have had time to become generally accepted and stable.[14] Campbell's avoidance of the notion that science itself is mythic does, however, make sense in light of his desire to prove myth's independence from science.

For the Romanian-born historian of religion Mircea Eliade, myth does have an explanatory function, but it also has the power to regenerate the individual by enabling them to make a journey back to a time beyond time, in recitations or re-enactments of stories of primordial origin, and in this way reconciles the individual with the conditions of existence by bringing them face to face with the numinous moment at which those conditions came into being.[15] Eliade saw this regenerative function of myth surviving in modern life in the act of being "carried" into another time by reading a novel or watching a film. Segal has criticized Eliade on this last point for ignoring the fact that such "journeys" are nowadays only imaginative in nature;[16] but it is unlikely that pre-moderns really travelled

[12] For some examples, see Mary Midgley, *Science as Salvation: A Modern Myth and Its Meaning* (London: Routledge, 1992).

[13] This summary (without the scientific bits) is drawn from Robert Segal, *Joseph Campbell: An Introduction* (New York: Mentor, 1987), 119-20.

[14] Joseph Campbell, *The Power of Myth* (New York: Doubleday, 1988), 18. Elsewhere in this series of interviews, Campbell treats science as a separate domain from myth (see *The Power of Myth*, 31, 132, 139). Here science is represented as knowledge and myth as mystery: compatible, but serving different functions. Yet Campbell also treats machines like computers as modern gods (e.g. *The Power of Myth*, 18-20), without registering any dissonance.

[15] Mircea Eliade, *The Sacred and the Profane*, trans. Willard R. Trask (New York: Harvest, 1968), 82, discussed in Segal, "Does Myth Have a Future?", 22. For a different application of Eliade's ideas about myth to modern scientistic ideology, see Lessl, "Science and the Sacred Cosmos: The Ideological Rhetoric of Carl Sagan," *Quarterly Journal of Speech*, 71 (1985): 175-87.

[16] Segal, "Does Myth Have a Future?" 23.

back in time either. For moderns as for pre-moderns, the experiential intensity and "reality" of the imaginary journey depends on the strength of the individual and/or collective imagination making the journey, and on the skill of those staging the re-enactment of past time, whether in words, images or actions. Segal has also faulted Eliade for ignoring the fact that the alternate times into which novels and films transport their audiences are not particularly distant and in any case not primordial. Scientific texts and films, however, bear out Eliade's scheme better than fiction or history. Imaginary journeys back to the origins of life or of the cosmos, of the kind we are familiar with from our television screens or science books for general readerships, fit his criteria for modern myths better than most novels or films, because the past they envisage is very distant and in some cases genuinely primordial. The appeal of contemplating such a past, aided by the spectacular time-travel devices of CGI imagery or verbal theatrics, surely goes beyond the mere desire for explanation.[17]

My examples of science's possible mythic functions given above are all taken from the world of public or popular science, whose cultural products mediate scientific knowledge to wider audiences. This is where the science-myth interface is most obvious and most easily documented, especially if we focus on how the science is presented to its audiences (this, however, being only part of the story). It is primarily when scientists address wider publics that they present the cosmological and evolutionary origin-narratives revealed by modern science as a form of "myth". Of course many scientists would say that science communication, with its relatively free use of metaphor and the language of value, is not "proper" science but only "popularized" science; but this distinction has been called into question by recent sociological and historical studies of science's relationship with its many publics, and can no longer be taken as a given.[18] Granted, specialized scientific papers constitute a very different form of

---

[17] For a particularly numinous example of such scientific contemplation, see Richard Jefferies, *The Story of My Heart: My Autobiography* (London: Longmans, 1883), 17-18. On time-travel voyages into the geological past, see Ralph O'Connor, *The Earth on Show: Fossils and the Poetics of Popular Science, 1802-1856* (Chicago: University of Chicago Press, 2007), 251-54 and 366-433.

[18] Stephen Hilgartner, "The Dominant View of Popularization: Conceptual Problems, Political Uses", *Social Studies of Science*, 20 (1990): 519-39. Roger Cooter and Stephen Pumfrey, "Separate Spheres and Public Places: Reflections on the History of Science Popularization and Science in Popular Culture", *History of Science*, 32 (1994): 237-67. Jonathan Topham, "Rethinking the History of Science Popularization / Popular Science", in *Popularizing Science and Technology in the European Periphery, 1800-2000*, ed. Faidra Papanelopoulou, Agusti Nieto-Galan and Enrique Perdiguero (Aldershot: Ashgate, 2009), 1-10.

communication (with different rules and epistemological status) from works for general readers; but works aimed at general readers have always played an important role in communicating knowledge about the physical universe (and the meanings of that knowledge) not just from experts to non-experts, but also between different expert groups and within the discipline itself. In other words, so-called "science communication" or "popular science" sometimes feeds back into the practice of science.[19] The two figures discussed below are prime examples of this phenomenon.

Not all scientists or science communicators are happy to risk using the category "myth" in this context, but a number of eminent scientists, when speaking as public intellectuals, promote the (quasi-Popperian) view that the grand narrative of cosmic and evolutionary history is a myth which happens to be true, and is for that reason far superior to all preceding and alternative creation-myths. This mythic dimension informs the choice of the goddess-name "Gaia" in James Lovelock's well-known hypothesis of the earth as a self-regulating system, and it moves up to a quasi-religious level in his numinous representation of the earth as a secular "place for worship, with all life as its congregation".[20] Other environmentalists have taken this religious dimension of scientific narrative still further, as documented in Jon Turney's important study of contemporary cosmological writing. But (as Turney notes) it is not just Gaians or pantheists who paint science as the true creation-myth for modern times, the "cosmic creation story", but also materialists such as the entomologist and sociobiologist Edward Wilson and the zoologist Richard Dawkins.[21]

Wilson, in his seminal book *On Human Nature* (1978), prophesies a "new age" in the near future when what he calls "the evolutionary epic" – a materialist vision of the lawlike evolution of matter, life and mind – will become the basis of a new and optimistic worldview. Whereas Popper's

---

[19] Topham, "Rethinking the History of Science Popularization."

[20] James Lovelock, *Healing Gaia: Practical Medicine for the Planet* (New York: Harmony, 1991), 17, discussed in Turney, 240; compare Thomas Lessl, "The Priestly Voice," *Quarterly Journal of Speech*, 75 (1989): 183-97; Segal, 'Myth and Science', last section.

[21] Turney, 227-32; Martin Eger, "Hermeneutics and the New Epic of Science," in *Perspectives on Popular Science Writing*, ed. Murdo William McRae (Athens, Georgia: University of Georgia Press, 1993),188-200. Bernard Lightman, *Victorian Popularizers of Science: Designing Nature for New Audiences*, (Chicago: University of Chicago Press, 2007), 500-02. For 'cosmic creation story', see the overtly environmentalist proposal by Brian Swimme, "The Cosmic Creation Story", in *The Reenchantment of Science: Postmodern Proposals*, ed. David Ray Griffin, Albany (New York: State University of New York Press, 1988), 47-56.

account of scientific knowledge saw individual theories as self-correcting myths, Wilson's myth consists of many different hypotheses and theories stitched together into a grand narrative of origin, and its scale is reflected in his collapsing of the terms "myth" and "epic". As he has put it, "the evolutionary epic is probably the best myth we will ever have", and "scientific materialism is the only mythology that can manufacture great goals from the sustained pursuit of pure knowledge". It is not only truer but also more spectacular than existing or previous mythologies: "the origin of the universe in the big bang ... is far more awesome than the first chapter of Genesis or the Ninevite epic of Gilgamesh".[22] Similar claims, albeit serving a very different agenda, pepper the evolutionary writings of Dawkins, who urges his readers to approach the phenomenon of life with wonder and amazement: religious accounts of the creation "miserably fail to do justice to the sublime grandeur of the real world", unlike their materialist evolutionary counterparts.[23] Imaginary voyages back to the primordial past become, in this context, more than merely imaginary. Dawkins presents the history of life in *The Ancestor's Tale* (2004) as an "epic pilgrimage" to "the dawn of life": this is a metaphorical pilgrimage only, consisting of a contemplation of evolutionary history in reverse order, but it is meant to be "deeply satisfying", using wonder to reconcile readers with the conditions of their existence and to fill them with "true reverence" for the mere fact of life's existence.[24] Dawkins ends his book with the words, "I feel I have returned from a true pilgrimage", while Wilson ends his, with characteristic candour, by extolling the capacity of the materialist evolutionary epic to sustain "blind hopes that the journey on which we are now embarked will be farther and better than the one just completed".[25]

These reflections bring us close to the regenerative and justificatory function of myth as envisaged by Eliade. They also call to mind Campbell's looser definition of myth as "the study of man's greatest

---

[22] Edward O. Wilson, *On Human Nature* (Cambridge, Massachusetts: Harvard University Press, 1978) 201, 207, 202.

[23] Richard Dawkins, *The Ancestor's Tale: A Pilgrimage to the Dawn of Life* (London: Weidenfield & Nicolson 2004), 13 and 509. This is the main theme of Dawkins's earlier book, *Unweaving the Rainbow Science, Delusion and the Appetite for Wonder* (London: Allen Lane, 1998). On the very different ideologies at work in the writings of Wilson and Dawkins, see Michael Ruse, *Monad to Man: The Concept of Progress in Evolutionary Biology* (Cambridge, Massachusetts: Harvard University Press, 1996), 467-70 and 507-25.

[24] Dawkins, *The Ancestor's Tale*, 509; compare Dawkins, *Unweaving the Rainbow*, x.

[25] Dawkins, *The Ancestor's Tale*, 509; Wilson, *On Human Nature*, 209.

story", a story in which "we have come forth from the one ground of being as manifestations in the field of time".[26] The self-reflexive aspect of Campbell's phrasing here resonates with a characteristic trope of modern popular science, in which the thinking subject is seen as matter made conscious. Carl Sagan put this trope in its most characteristic modern form when he announced in his 1980 television series *Cosmos*, "We have begun to contemplate our origins; starstuff pondering the stars". The same trope, in simplified form, introduces all six episodes of the 2011 television series *Wonders of the Universe*, presented by the physicist Brian Cox: "we are part of the universe, so its story is our story".[27] And its latent justificatory function (in Eliade's sense) is made explicit in the astronomer Eric Chaisson's remark, in a book narrating the rise of life, that "[human] life's purpose and meaning" is "to act as an animated conduit for the Universe's self-reflection".[28]

The claim which Wilson and Dawkins make explicitly, and which these other examples follow through in their different ways, is that science has the power of replacing traditional myth. To return to Segal's analysis of myth and science, this is a more comprehensive and also more risky strategy than the straightforward "surrender" scenario, exemplified by Tylor and his followers, in which "myth is conceded to be an outdated and incorrect explanation of the world", science takes over that role and myth falls dead to the ground.[29] We may see a version of that simpler strategy in nineteenth-century naturalists' appropriation of mythical or legendary terminology for newly discovered creatures: giving the mermaid/siren-name *Sirenia* as a Linnaean binomial for manatees and dugongs or (less successfully) using the vernacular "dragon" to denote extinct reptiles. These two examples embody a suggestion that such animals are "the real" mermaids or dragons, and are no less wonderful than their mythological equivalents.[30] But Wilson and Dawkins want more than this: they want

---

[26] Campbell, *The Power of Myth*, 54.

[27] Carl Sagan, *Cosmos* (New York: Random House, 1980), 345 (issued simultaneously with the TV series *Cosmos: A Personal Voyage*, directed by David Kennard *et al.*, in which this phrase comes in the introductory episode). Sagan's mythic aims are discussed by Lessl, "Science and the Sacred Cosmos," and Turney, 239; Brian Cox's narration in *Wonders of the Universe*, directed by Chris Holt (BBC Discovery Channel, 2011).

[28] Eric Chaisson, *The Life Era* (New York: Norton, 1987), 229, quoted in Eger, 198.

[29] Segal, "Does Myth Have a Future?", 19.

[30] On dinosaur terminology, see Ralph O'Connor, "Victorian Saurians: The Linguistic Prehistory of the Modern Dinosaur," *Journal of Victorian Culture*, 17 (2012): 492-504. I am currently preparing a fuller analysis of the dragon analogy,

science to take over the functions of myth not only at the level of explanation, but existentially, ideologically, even spiritually, thus removing the need for traditional myth at all levels. Science replaces myth by becoming mythic, by addressing itself (as Wilson puts it) "with precise and deliberately affective appeal to the deepest needs of human nature".[31] Its communicators achieve this both by alluding to traditional mythologies and by using a range of verbal, visual and aural techniques, some of them drawn from the rhetorical armoury of the literary genres associated with mythology itself, from epic and drama to romance and fairytale.[32]

This creative engagement with myth is not new to the late twentieth century, although some of the most perceptive studies of it have tended to treat it in this way. It has taken place wherever sacred narratives have been secularized. Today this process is associated primarily (albeit not exclusively) with the rise of a materialist vision of scientific progress. In the mid-nineteenth century, a closely analogous process was going on in the use of geological evidence to rewrite Genesis 1, both within and without the churches, drawing on similar epic and mythic tropes to those found in the works of Dawkins and Wilson.[33] Later Victorian naturalists and writers in general did much the same with evolutionary theory, with Robert Chambers's would-be "romance" *Vestiges of the Natural History of Creation* (1844) and Charles Darwin's *On the Origin of Species* (1859) as key reference-points.[34] As Gillian Beer has shown, Darwin grappled with the deep mythic structures of existing creation-narratives (especially

which was the subject of the paper I gave at the conference from which this volume emerged; see also O'Connor, *The Earth on Show*, 37, 97, 157-58, 341-45, 355 and 443, and Heather Brink-Roby, "*Siren canora*: The Mermaid and the Mythical in Late Nineteenth-Century Science," *Archives of Natural History*, 35 (2008): 1-14 (with much discussion of mermaids besides).

[31] Wilson, 200.

[32] On epic, see further David Amigoni and Michael Elwick, eds., *The Evolutionary Epic*, vol. 4 of *Victorian Science and Literature*, ed. Gowan Dawson and Bernard Lightman, 8 vols (London: Pickering and Chatto, 2011-12). On romance and fairytale, see Melanie Keene, "Object Lessons: Sensory Science Education 1830-1870," unpublished PhD dissertation, University of Cambridge, 2008; O'Connor, *Science as Romance*.

[33] O'Connor, "From the Epic of Earth History"; O'Connor, *The Earth on Show*.

[34] James A. Secord, *Victorian Sensation: The Extraordinary Publication, Reception, and Secret Authorship of* Vestiges of the Natural History of Creation (Chicago: University of Chicago Press, 2000), 87-108; Lightman, 219-94; Amigoni and Elwick.

that of Milton's *Paradise Lost*) to shape his own argument,[35] and aspects of this argument were then rendered into more overtly narrative forms by later writers such as Edward Clodd in his prose treatise *The Story of Creation* (1881) and Henry Knipe in his verse epic *Nebula to Man* (1905), the latter illustrated by spectacular full-page colour restorations of prehistoric scenes. The human mind, especially its personification as "science", was often the *de facto* hero or culminating point of such stories (whether creationist or evolutionary), recalling Wilson's "Every epic needs a hero; the mind will do";[36] and the related self-reflexive trope of matter becoming mind to contemplate its starry origins, discussed above in its more recent variants, emerges in patriarchal Edwardian guise as the iconic frontispiece to Knipe's poem (Fig. 1). The primeval alpha male, with his faithful bride leaning on his arm, gazes up to the stars from which he was made, while an ape is dimly visible, lurking in the shadows which they have left behind for ever.

If, however, science communicators have long shared a common desire to make science mythic and/or epic, they have not been compelled to do so in the same way. To return to Popper's insight, myth does not preclude innovation. Martin Eger has suggested that the common structure and central claims of almost all the texts he examines (these being from the 1970s, 1980s and 1990s) suggest a level of consensus which seems "overwhelming, perhaps somewhat boring";[37] but in the manner of presentation there is considerable variation, and the manner shapes the matter rather than merely adorning it. In narrative terms no less than ideology, Knipe's poem in rhyming couplets is worlds away from Clodd's prose treatise, even though both relate to the same underlying story. Watching Cox's *Wonders of the Universe* is a very different experience from reading Wilson's *Consilience*, just as Aeschylus's tragedy *Prometheus Bound* offers a very different experience from the prose account of the same myth in Hesiod's *Theogony*: one is a drama, the other a narrative poem. The narrative voice, for example, can vary greatly, from the priestly solemnity of Sagan and the sometimes Olympian tone of Wilson to the chatty and argumentative style of Dawkins and the mercurial voice of Cox, who combines several of these styles.[38] As in the ancient Greek example, these differences affect the meaning of the story, especially when

---

[35] Gillian Beer, *Darwin's Plots: Evolutionary Narrative in Darwin, George Eliot and Nineteenth-Century Fiction* (London: Routledge & Kegan Paul, 1983).

[36] Wilson, 203.

[37] Eger, 199.

[38] On the importance of the mode of voice as well as its narrative stance, see Lessl, "The Priestly Voice."

combined with the even greater differences of narrative framework and genre. Cox pays explicit homage to Sagan's *Cosmos* in the third episode of his television series, yet Cox's road-movie framework, with its rock-and-roll soundtrack and zapping segues from one scene to the next, contrasts in both its pace and its earthbound reference-points with the galactic-voyage device which frames Sagan's series: a spaceship steadily approaching the earth from outer space, accompanied by Vangelis's slow, hypnotic soundtrack and selected pieces of well-known classical music. When the narrators come into view, Cox is usually moving around; Sagan is typically standing still. These differences reflect not only the changing tastes of television audiences and differing uses of optical technology between the early 1980s and the mid-2000s, but also reflect different notions of what scientific knowledge is and how it should manifest itself. Sagan's history of the cosmos is communicated as a vision from on high, whereas Cox mixes visionary communication with hands-on pedagogy, demonstrating fundamental cosmic principles using ordinary objects such as pebbles and empty bottles. Interestingly, neither series proceeds in strictly narrative order: the mythically resonant notion of the grand narrative does not structure the presentation of cosmic history, but is an outer framing device. It is alluded to at key points but does not move the show forward in itself. This is a longstanding paradox of many so-called evolutionary epics, setting them apart from epics in the commonsense understanding of the term and raising doubts as to the existence of "the evolutionary epic" as a distinct narrative genre.[39]

In their attempt to harness the narrative power of myth and win over the hearts and minds of their audiences to the (or their) scientific cause, the techniques and tropes I have outlined here are a necessary complement to the rhetoric of persuasion in argument which has hitherto been the primary focus of attention to the literary nature of science writing. The study of narrative as conventionally understood (i.e. stories) therefore has a potentially powerful role to play in helping us to understand science's place in culture and the role of myth in this relationship. The communication, and perhaps also the self-image, of science as a producer of grand narrative may be felt to be increasingly necessary to the survival

[39] I have discussed this non-narrative aspect of (Victorian) evolutionary epics in papers delivered at the University of Glasgow on 3 September 2010 and the University of Cambridge on 9 April 2011, and am currently preparing a fuller study for publication. For treatment of the evolutionary epic as its own genre, see Eger; Lightman, 500; Amigoni and Elwick.Contrast Buckland, "Losing the Plot: The Geological Anti-Narrative," *19: Interdisciplinary Studies in the Long Nineteenth Century*, 11 (2010).

of sciences like cosmology and evolutionary biology, given their increasing dependence on diminishing pools of private funding. One could even say, reversing Segal's claim, that whether these sciences have a future depends (in part) on their capacity to harness the imaginative power of myth.[40]

Works Cited

Primary Sources

Campbell, Joseph. *The Power of Myth*. New York: Doubleday, 1988.

Chaisson, Eric. *The Life Era*. New York: Norton, 1987.

Cox, Brian. *The Wonders of the Universe*, directed by Chris Holt. BBC Discovery Channel, 2011 [television series].

Clodd, Edward. *The Story of Creation*. London: Longman, 1990.

Dawkins, Richard. *The Ancestor's Tale: A Pilgrimage to the Dawn of Life*. London: Weidenfield & Nicolson, 2004.

Dawkins, Richard. *Unweaving the Rainbow: Science, Delusion and the Appetite for Wonder*. London: Allen Lane, 1998.

Eliade, Mircea. *The Sacred and the Profane*, translated by Willard R. Trask. New York: Harvest, 1968.

Jefferies, Richard. *The Story of My Heart: My Autobiography*. London: Longmans, 1883.

Knipe, Henry. *Nebula to Man*. London: Dent, 1905.

Lovelock, James. *Healing Gaia: Practical Medicine for the Planet*. New York: Harmony, 1991.

Popper, Karl. *Conjectures and Refutations*, 3rd ed., new imprint. London: Routledge, 2002.

[40] I wish to record my gratitude to Laurence Roussillon-Constanty and Philippe Murillo for their editorial advice, and for inviting me to speak on this topic at Toulouse in June 2011 (albeit this essay has taken a different direction to the original paper). I wish to thank them and all the participants in the *Chimères* conference for their comments and discussion. I am also very grateful to Robert Segal for inviting me to speak on the same topic at the Centre for the Study of Myth at the University of Aberdeen in June 2011, for many stimulating discussions with him and other members of that audience, and for his helpful comments on the present chapter. I wish to thank my wife Clémence for her valuable suggestions at every stage. Finally, I wish to acknowledge an important recent book I heard about too late to incorporate into my essay: Gregory's Schrempp's *The Ancient Mythology of Modern Science: A Mythologist Looks (Seriously) at Popular Science Writing* (Montreal: McGill-Queen's'UP, 2012). My thanks to Joseph Falaky Nagy for alerting me to this stimulating book.

Sagan, Carl. *Cosmos*. New York: Random House, 1980.

—. *A Personal Voyage*, directed by David Kennard *et al.* Public Broadcasting Service, 1980 [television series].

Swimme, Brian. "The Cosmic Creation Story". In *The Reenchantment of Science: Postmodern Proposals*, edited by David Ray Griffin, 47-56. Albany, New York: State University of New York Press, 1988.

Wilson, Edward O. *On Human Nature*. Cambridge, Massachusetts: Harvard University Press, 1978.

Secondary Sources

Amigoni, David, and Michael Elwick, eds. *The Evolutionary Epic*. Vol. 4 of *Victorian Science and Literature*, edited by Gowan Dawson and Bernard Lightman, 8 vols. London: Pickering and Chatto, 2011-12.

Beer, Gillian. *Darwin's Plots: Evolutionary Narrative in Darwin, George Eliot and Nineteenth-Century Fiction*. London: Routledge & Kegan Paul, 1983.

Brink-Roby, Heather. "*Siren canora*: The Mermaid and the Mythical in Late Nineteenth-Century Science". *Archives of Natural History*, 35 (2008): 1-14

Buckland, Adelene. "Losing the Plot: The Geological Anti-Narrative", *19: Interdisciplinary Studies in the Long Nineteenth Century*, 11 (2010), www.19.bbk.ac.uk.

Carey, John. *A Single Ray of the Sun: Religious Speculation in Early Ireland*. Aberystwyth: Celtic Studies Publications, 1999.

Clunies Ross, Margaret. *Prolonged Echoes: Old Norse Myths in Medieval Northern Society*, 2 vols. Odense: Odense University Press, 1994-98.

Cooter, Roger, and Stephen Pumfrey. "Separate Spheres and Public places: Reflections on the History of Science Popularization and Science in Popular Culture". *History of Science*, 32 (1994): 237-67.

Eger, Martin. "Hermeneutics and the New Epic of Science". In *Perspectives on Popular Science Writing*, edited by Murdo William McRae, 186-209. Athens, Georgia: University of Georgia Press, 1993.

Hilgartner, Stephen. "The Dominant View of Popularization: Conceptual Problems, Political Uses". *Social Studies of Science*, 20 (1990): 519-39.

Keene, Melanie. "Object Lessons: Sensory Science Education 1830-1870". unpublished PhD dissertation, University of Cambridge, 2008.

Koertge, Noretta, ed. *A House Built on Sand: Exposing Postmodernist Myths about Science*. New York: Oxford University Press, 1998.

Lessl, Thomas. "The Priestly Voice". *Quarterly Journal of Speech*, 75 (1989): 183-97.

—. "Science and the Sacred Cosmos: The Ideological Rhetoric of Carl Sagan". *Quarterly Journal of Speech*, 71 (1985): 175-87.

Lightman, Bernard. *Victorian Popularizers of Science: Designing Nature for New Audiences*. Chicago: University of Chicago Press, 2007.

Midgley, Mary. *Science as Salvation: A Modern Myth and Its Meaning*. London: Routledge, 1992.

O'Connor, Ralph. *The Earth on Show: Fossils and the Poetics of Popular Science, 1802-1856*. Chicago: University of Chicago Press, 2007.

—. "From the Epic of Earth History to the Evolutionary Epic in Nineteenth-Century Britain". *Journal of Victorian Culture* 14 (2009): 207-23.

—. "History or Fiction? Truth-Claims and Defensive Narrators in Icelandic Romance-Sagas". *Mediaeval Scandinavia*, 15 (2005): 101-69

—, ed. *Science as Romance*. Vol. 7 of *Victorian Science and Literature*, edited by Gowan Dawson and Bernard Lightman, 8 vols. London: Pickering and Chatto, 2012.

—. "Victorian Saurians: The Linguistic Prehistory of the Modern Dinosaur". *Journal of Victorian Culture*, 17 (2012): 492-504.

Ruse, Michael. *Monad to Man: The Concept of Progress in Evolutionary Biology*. Cambridge, Massachusetts: Harvard University Press, 1996.

Secord, James A. *Victorian Sensation: The Extraordinary Publication, Reception, and Secret Authorship of* Vestiges of the Natural History of Creation. Chicago: University of Chicago Press, 2000.

Segal, Robert. "Does Myth Have a Future?" In *Myth and Method*, edited by L. L. Patton and W. Dinoger, 82-106. Charlottesville: University Press of Virginia, 1996. Reprinted in Robert Segal, *Theorizing about Myth*, 19-35. Amherst: University of Massachusetts Press, 1996.

—. *Joseph Campbell: An Introduction*. New York: Mentor, 1987.

—. *Myth: A Very Short Introduction*. Oxford: Oxford University Press, 2004.

—. "Tylor's Theory of Myth as Primitive Science". In *The Sum of Our Choices: Essays in Honour of Eric J. Sharpe*, edited by A. Sharma, 70-84. Atlanta: Scholars Press, 1996. Reprinted in Robert Segal, *Theorizing about Myth*, 7-18. Amherst: University of Massachusetts Press, 1996.

Segerstråle, Ullica, ed. *Beyond the Science Wars: The Missing Discourse about Science and Society*. New York: SUNY Press, 2000.

Topham, Jonathan. "Rethinking the History of Science Popularization / Popular Science", in *Popularizing Science and Technology in the European Periphery, 1800-2000*, edited by Faidra Papanelopoulou, Agusti Nieto-Galan and Enrique Perdiguero, 1-10. Aldershot: Ashgate, 2009.

Turney, Jon. "Telling the Facts of Life: Cosmology and the Epic of Evolution". *Science as Culture*, 10 (2001): 225-47.

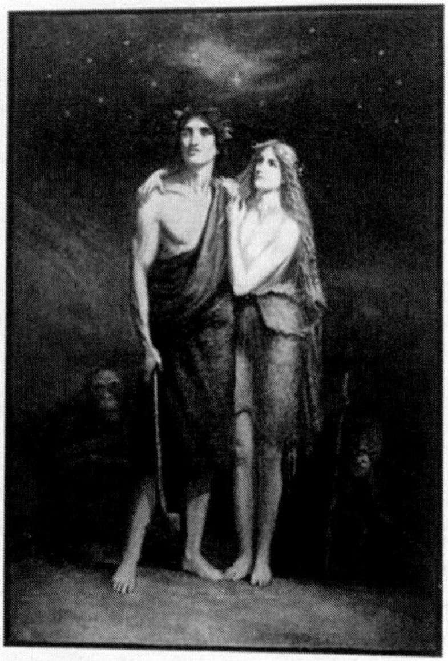

ILL. A-2 : Frontispiece to Henry R. Knipe's evolutionary epic poem, *Nebula to Man*, 1905

MATHEMATIQUES, MYTHOLOGIES, CHIMERES

OLIVIER KELLER

Il est bien connu que les mathématiques non seulement ne craignent pas les chimères mais qu'elles s'en nourrissent. Les qualificatifs d'« irrationnels », « imaginaires », « négatifs », toujours utilisés dans le domaine des nombres, trahissent un ancien statut d'êtres impossibles, tolérés comme intermédiaires de calcul mais sans droit à l'existence dans le résultat. Plus récemment, les nombres « transfinis » inventés par Cantor à la fin du 19ème siècle sont encore l'objet de disputes parmi les mathématiciens, entre ceux qui les acceptent et ceux qui les tiennent pour de pures chimères.

Ce n'est pas ce genre-là de chimères, finalement bien domestiquées, qui nous occupera ici, mais des chimères les unes très anciennes, qui relèvent de la mythologie et du rituel, les autres très récentes, qui relèvent de l'affabulation. Les premières, fécondes, peuvent créer des conditions favorables à l'apparition et au développement d'embryons de mathématiques ; les secondes, stériles, se nourrissent au contraire de mathématiques.

Chimères fécondes

Le monde surnaturel inventé probablement par les chasseurs-cueilleurs du Paléolithique supérieur, et que nous pouvons comprendre grâce à l'ethnologie des sociétés traditionnelles[1], peut à bon droit être qualifié de chimère, quel que soit le sens que l'on donne à ce terme. A l'image de la bête composite malfaisante mentionnée pour la première fois par Homère, le surnaturel peut être un monstre redoutable ; mais il peut être à l'inverse la douce illusion d'un monde d'ancêtres bienfaisants.

Cette vision contemporaine est cependant, pour une part, une vision d'héritiers ingrats. Car nous sommes bel et bien les enfants du surnaturel

---

[1] Le lecteur intéressé par cette question et par les controverses qu'elle soulève peut consulter par exemple le premier chapitre de : Olivier Keller, *Aux origines de la géométrie. Le Paléolithique et le monde des chasseurs-cueilleurs* (Paris:Vuibert, 2004).

et des mythologies qui le constituent. Nous leur devons en effet les premiers pas de la pensée, une pensée qui, avant de pouvoir s'occuper directement d'elle-même en philosophant, a dû se poser comme un objet autre sous l'aspect d'un second monde, celui des pouvoirs, monde « autre » mais étroitement lié au monde des hommes et auquel ceux-ci ont accès. L'échange entre ces deux mondes est le fondement de la pensée des peuples traditionnels et le sens de toute activité, des rituels les plus compliqués aux gestes les plus banals. Des formes nombreuses et variées de l'échange que la documentation ethnographique nous révèle, une seule nous occupera ici, celle qui se manifeste dans l'art pariétal. Et si l'on accepte de rapprocher les sources archéologiques des sources ethnographiques, c'est-à-dire les chasseurs-cueilleurs de la préhistoire des chasseurs-cueilleurs récents, on évite beaucoup d'interprétations fantaisistes et le sens profond de l'art pariétal préhistorique se découvre. Il s'avère que nous lui devons, entre autres, des supports des futurs concepts géométriques[2] : surface de représentation, figure, ligne et point.

Dans les sociétés traditionnelles, l'art est avant tout un acte rituel, une pratique qui résulte de l'invention d'un second monde :

> D'après les anciens, le paysage fut formé par les actions des Êtres Ancestraux [...] A plusieurs endroits les Êtres Ancestraux pénétrèrent directement le paysage après leurs pérégrinations et leurs activités créatrices en laissant derrière eux leur image sur la surface rocheuse [...] En terre d'Arnhem de l'Ouest, les aborigènes établissaient des campements saisonniers dans des abris situés au bas d'escarpements rocheux et ornaient les murs et les plafonds de milliers de mains négatives, ou d'avant-bras et de mains négatifs, ainsi que de peintures monochromes et polychromes. Au moyen de ces marques et de ces peintures ils se liaient étroitement à des sites spécifiques et accédaient directement au pouvoir des êtres ancestraux représentés sous forme animale, humaine ou mythique[3].

---

[2] Ces supports ne sont pas les premiers. Ils font suite aux outils taillés depuis plus de deux millions d'années, dont l'étude révèle des antécédents matériels du concept d'espace et de sa structuration. Il est même possible de déceler un lien entre les outils-figures et les lignes-tranchants d'une part, les figures et leurs contours dessinés ou gravés dans l'art pariétal d'autre part. Pour une étude détaillée de la géométrie embryonnaire que révèle la taille des outils, voir l'ouvrage mentionné dans la note précédente.

[3] Paul S.C. Taçon, "The Power of Stone; Symbolic Aspects of Stone Use and Tool Development in Western Arnhem Land, Australia," Antiquity 65 (1991).

Cette origine rituelle incontestable dans le cas de certains aborigènes d'Australie est attribuée également, par le préhistorien Michel Lorblanchet entre autres, à l'art pariétal en général :

> Ce sont des images latentes, des créatures endormies dans la roche qu'éveille, en un geste fécondateur, l'artiste-prêtre des temps glaciaires. Dans la grotte, ventre de la terre, il présidait sans doute à la naissance du monde et à une communication avec le monde des esprits. Cette idée [...] n'a pas encore donné lieu au développement et à la recherche méthodique qu'elle mérite[4].

La paroi est donc une *limite* entre deux espaces, celui du monde naturel et celui du monde des « êtres ancestraux » et des « esprits ». Elle doit être *sans épaisseur* pour que le contact direct soit possible, avec *une étendue* pour contenir la trace du passage d'un monde dans l'autre ou sa reproduction rituelle. La paroi *fonctionne* donc comme une surface, même si elle n'est pas conceptualisée comme telle ; pensée comme objet sans épaisseur mais avec étendue, elle est donc un modèle, un objet de référence, un « matériau » pour ce qui sera plus tard conceptualisé comme « surface ».

Avec la trace du passage des êtres ancestraux dans l'autre monde, c'est rien de moins que l'acte final de la création du monde que l'« artiste-prêtre » *re-produit*, en même temps que leur retour et la diffusion de leur pouvoir grâce à son geste « fécondant ». Tel est le thème fondamental, qui a dû connaître des variations infinies. On le voit, c'est une affaire vitale, une *re-création* et non une vague commémoration, et le geste graphique, chargé de lui donner à sa façon un maximum de vérité, ne peut en réalité se contenter d'une simple trace. Il lui faut évoquer, sur un espace de dimension deux, le passage à travers cet espace d'un objet réel, d'un animal par exemple. Avec le trompe-l'œil, l'animal est parfaitement évoqué, mais pas le passage. Avec un profil absolu, l'accent est sur le passage, mais l'animal est incomplet. Selon notre hypothèse, la solution fut ce que l'abbé Breuil appelait la « perspective tordue », technique qui consiste en rabattement de certaines parties (cornes par exemple) sur la surface, de telle sorte que le maximum du corps de l'animal s'y trouve (Figure 2-1). Le corps est ainsi reconstruit indépendamment de la vraisemblance visuelle, mais il est tout entier dans le lieu de passage, la paroi : la vraisemblance rituelle est respectée, et cela seul compte.

---

[4] Michel Lorblanchet, *Les grottes ornées de la préhistoire. Nouveaux regards* (Errance, 1995), 173.

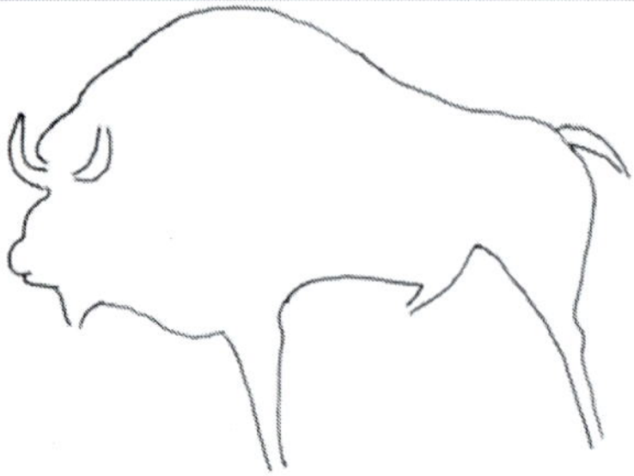	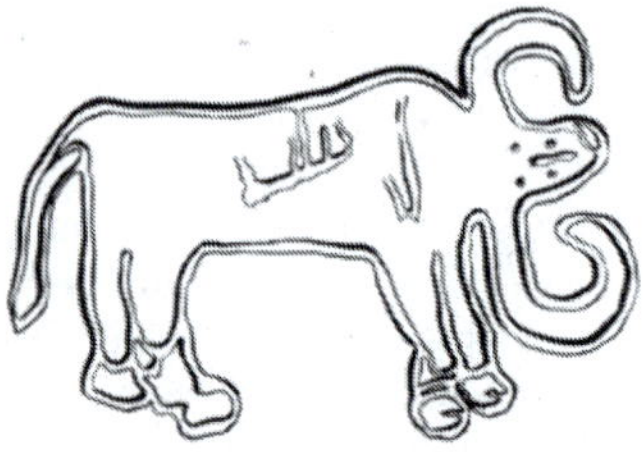
Bison gravé. La Grèze (Dordogne), vers -18000. Profil absolu, sauf les cornes représentées en « perspective tordue ».	Buffle gravé sur bloc rocheux. Yemen, date incertaine. Exemple extrême de « perspective tordue ».

Figure 2-1 : Exemples de « perspective tordue ». Dessins Anne Spanek.

Du coup, la surface n'est plus seulement une frontière abstraite entre deux mondes, mais elle prend géométriquement consistance en tant que lieu de construction. Et plus tard, à partir de l'ère post-glaciaire, elle changera de caractère : elle ne sera plus seulement lieu de passage, mais lieu de séjour puisque des scènes pourront s'y dérouler.

L'invention de la surface de représentation est en même temps l'invention des codes de représentation, avec les futurs objets fondamentaux de la géométrie ordinaire : figure, ligne et point[5]. Voici comment. Lorsque nous avons affaire à un sujet entièrement peint, surtout s'il est représenté en trompe-l'œil, on pourrait défendre l'idée qu'il s'agit d'une reproduction de l'impression visuelle ; il y a bien dans ce cas une figure et son extérieur, mais la limite entre les deux n'a pas de signe spécifique, elle n'est au plus qu'une idée. Il en est de même avec les milliers de mains négatives dans les grottes ornées partout dans le monde ; c'est cette fois-ci la peinture de son extérieur qui crée la figure par la technique du pochoir, mais là encore la limite entre les deux n'a pas son signe propre.

Au contraire, avec la figure réduite à son contour gravé ou peint, marqueur de frontière entre un intérieur et un extérieur, nous sommes certains d'avoir affaire à un signe de ligne ; non pas à une vraie ligne, on le sait, mais à son signe spécifique. En outre, tout se passe comme si la

[5] Il s'agit bien de codes inventés, et non de reproduction de sensations visuelles. Nous ne pouvons voir ni point, ni ligne, ni figure (en tant que « ce qui est contenu par quelque ou quelques frontières » selon la définition d'Euclide) ; nous ne pouvons que les penser.

contradiction entre la ligne représentée d'une part, épaisse en cas de dessin, épaisse et profonde en cas de gravure, et l'idée qu'elle objectivise d'autre part, inclinait le réalisateur à rapprocher l'objet de son idée, avec un tracé partiel ou discrétisé : la fraction de contour évoque le contour entier, et le tracé en pointillés (et non pas en tirets) réduit la ligne à son expression minimale. La série de taches à quoi se réduit le contour de tel bouquetin de la grotte de Lascaux est faite pour inciter l'œil à les suivre, et donc à créer en idée la ligne comme limite de figure : si l'on admet notre raisonnement on conviendra que les taches sont dans ce cas des signes de points sans dimension, de même que les traits continus sont des signes de lignes à une dimension.

Si les objets de base de la géométrie furent donc inventés entre 40.000 et 10.000 ans avant notre ère[6] en tant qu'outils nécessaires à l'actualisation d'une conception mythique du monde, on pourrait penser que la vraie mise en forme et en acte mathématiques de ces objets ne peut surgir que de besoins pratiques ; ce fut en effet probablement le cas des corpus des débuts du deuxième millénaire, en Egypte et en Mésopotamie. Il existe pourtant un autre corpus beaucoup moins connu et moins commenté que les précédents, à usage des brahmanes védiques, et dont l'unique motivation est la construction géométrique d'autels. Il s'agit des *Sulbasutras*, mot sanscrit qui signifie « aphorismes de la corde »[7]. Les textes en notre possession, composés probablement deux ou trois siècles avant notre ère, sont la transcription de traditions rituelles transmises oralement, probablement depuis un millénaire au moins. Malheureusement, on ne sait rien d'éventuels antécédents dans la riche civilisation de l'Indus (-2400 à -1700 environ) ou ailleurs.

D'après le mythe védique, le monde fut créé par l'auto-sacrifice du démiurge Prajapati, c'est-à-dire son démembrement et sa dispersion dans l'espace. Le monde ainsi conçu, en tant qu'objet composite, est une authentique chimère, métaphoriquement aussi cruelle que la bête de l'Iliade, ce « fléau pour tant d'hommes », puisqu'elle fournit la

[6] De notre point de vue, fondé sur l'analyse des frises de l'art mobilier, des figures particulières comme le rectangle et le cercle furent également inventées à cette époque. Au Néolithique, un élargissement de la théorie des deux mondes et l'intégration dans le mythe et dans le rituel des mouvements apparents du soleil furent sources d'autres inventions à caractère géométrique. Voir Olivier Keller, *Une archéologie de la géométrie, peuples paysans sans écriture et premières civilisations* (Paris:Vuibert, 2006).

[7] S.N. Sen and A.K. Bag, *The Sulbasutras with Text, English Translation and Commentary,* Indian National Science Academy, 1983. Voir aussi Olivier Keller, *Une archéologie de la géométrie,* chapitre 6, et : http://www.univirem.fr/commissions/reperes/consulter/40keller.pdf (consulté le 30 janvier 2013).

justification du sacrifice humain. Classiquement, à travers le sacrifice des denrées ou de l'animal de substitut, c'est la naissance du monde qui est reproduite, même si son commanditaire n'a en vue que des buts particuliers. Mais l'édification elle-même de l'autel mime la création[8] en donnant forme et mesure à l'énergie créatrice et à son accroissement, par le biais de la construction géométrique, métaphore de l'engendrement :

- par analogie avec le corps de Prajapati comme substance du monde à créer, le corps du sacrifiant fournit la mesure de base, le *purusa* (« homme »), qui est sa hauteur les bras levés. Les seuls instruments autorisés sont la corde (éventuellement graduée) et les piquets, toute construction doit donc être faite à l'aide de droites (ou éventuellement de segments mesurés) et de cercles. On commence par installer les repères du monde, et donc de l'autel, en construisant les lignes est-ouest et nord-sud, puis le carré de côté 1 *purusa*.
- l'énergie créatrice est symbolisée par un carré d'aire 7,5 *purusas*-carrés qui se détermine en autels de différentes formes (triangulaire, losangique, circulaire) suivant les besoins du commanditaire, et qu'il s'agit de construire rigoureusement. Sauf dans le cas du cercle, les constructions sont en effet rigoureuses.
- la création-extension se réalise avec la construction de l'autel en forme d'oiseau (Figure 2-2), d'aire 7,5 *purusas*-carrés, suivie de son agrandissement homothétique jusqu'à 101,5 *purusas*-carrés[9], unité par unité ! Il s'agissait probablement d'un processus théorique, peut-être réalisé seulement dans sa première étape. En tout cas, celle-ci est décrite dans les *Sulbasutras* de plusieurs manières, toutes parfaitement rigoureuses.

---

[8] L'autel est assimilé à Prajapati lui-même.

[9] 7 et 101 interviennent pour des raisons d'ordre numérologique. Le *Satapatha Brahmana*, traité des observances rituelles, dit que Prajapati est « sept » par les saisons (six saisons et l'année elle-même font sept) et par l'espace (les quatre directions cardinales, le haut, le bas et l'espace lui-même font sept). Par l'année, Prajapati est « cent-un » grâce à un bricolage numérique associant mois, jours, demi-mois et saisons. *Satapatha Brahmana* 10-2-6, en ligne sur le site www.sacred-texts.com.

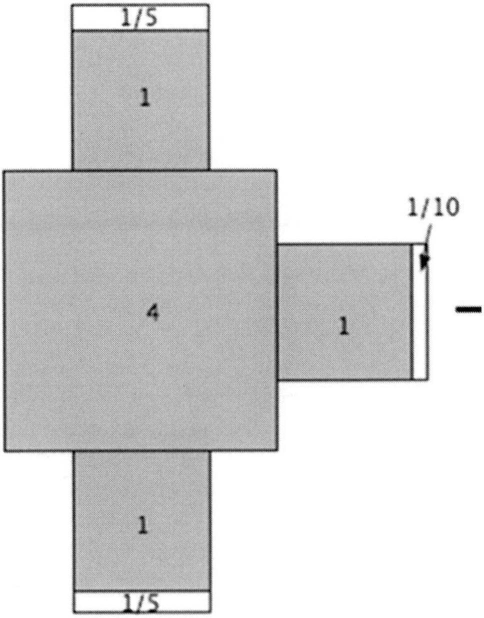

Figure 2-2 : Autel en forme d'oiseau d'aire totale 7,5 *purusas*-carrés.

Nous ne pouvons décrire ici en détail cette dernière construction réellement savante. Nous dirons seulement qu'elle exige de savoir construire « à la corde et aux piquets », donc « à la règle et au compas » : un carré de côté donné, un carré équivalent en aire à deux carrés donnés et un carré équivalent en aire à un rectangle donné. Les deux derniers cas supposent la connaissance de la propriété de la diagonale du rectangle[10], notre « théorème de Pythagore ». Nous avons là un savoir géométrique considérable[11] qui était peut-être partagé par d'autres, en Mésopotamie des

[10] « Les aires produites séparément par la longueur et la largeur d'un rectangle valent ensemble l'aire produite par la diagonale », peut-on lire dans les *Sulbasutras* de Baudhayana. S.N. Sen and A.K. Bag, *The Sulbasutras with Text*, 78.

[11] Autre trait remarquable, mais que nous ne pouvons développer dans le cadre de cet article : les deux problèmes posés et partiellement résolus dans les *Sulbasutras*, la construction d'un carré égal en aire à une figure rectiligne donnée, puis l'agrandissement de la figure en gardant sa forme, sont également les deux problèmes fondamentaux résolus dans les livres II et VI des *Eléments* d'Euclide.

débuts du deuxième millénaire notamment, mais les *Sulbasutras* en sont la seule trace incontestable ; et tout pousse à croire qu'en Inde de l'époque védique ce savoir fut, sinon créé de toutes pièces, du moins développé et systématisé pour des raisons exclusivement rituelles, découlant elles-mêmes de « chimères » mythologiques.

Chimères stériles

Nous ne nous intéresserons pas aux chimères au moins trois fois millénaires comme l'astrologie et la numérologie, mais à certaines affabulations modernes produites par des mathématiciens. Le principe de la fable mathématicienne est le suivant : prenez d'un côté un document de la préhistoire, muet par lui-même mais que vous avez décidé de faire parler, et de l'autre mobilisez votre savoir géométrique, arithmétique et statistique. Avec de l'habileté, de la patience et un sens aigu de l'approximation, vous réussirez *nécessairement* à fabriquer une interprétation mathématique de votre document parce que les mathématiques, en tant que formes abstraites, peuvent être plaquées sur tout ce que l'on veut[12]. Puis vous déciderez sans autre forme de procès que le contenu mathématique de votre interprétation fut également ce qui a motivé et guidé le créateur du document dans les temps préhistoriques ; vous serez alors persuadé d'avoir fait œuvre de science et d'histoire, en démontrant par exemple que les bâtisseurs des mégalithes de Grande-Bretagne disposaient de la science des triplets pythagoriciens[13]. C'est ainsi qu'anachronisme et formalisation arbitraire réunis donnent naissance à un animal composite, un peu de préhistoire et un peu de mathématiques, qui mérite bien le qualificatif de chimère. En voici deux exemples typiques, dont la stérilité scientifique et historique apparaîtra d'elle-même.

Mêmes problèmes et analogie d'instruments, la corde et les piquets dans un cas, la règle et le compas dans l'autre.

[12] On se souvient de l'expérience de Jean-Pierre Adam, qui s'est amusé avec les diverses dimensions d'une guérite de marchande de billets de loterie, avenue de Wagram à Paris ; il lui a fait avouer, entre autres, la distance terre-soleil, le nombre π, le cycle de Meton. Cf. Jean-Pierre Adam, *Le passé recomposé, chroniques d'archéologie fantastique* (Paris : Seuil, 1988).

[13] Mesures en nombres entiers des côtés d'un triangle rectangle, comme (3, 4, 5), (12, 5, 13) etc.

Géométrie des mégalithes

La géométrie des mégalithes a connu un regain d'intérêt dans les années 70 du siècle dernier, avec les travaux de deux ingénieurs anglais, Alexander Thom[14] et son fils Archie. Après avoir fait une grande quantité de relevés de mégalithes en Bretagne et en Angleterre, ils ont avancé[15] :

- que les constructeurs employaient une unité, le yard mégalithique, valant 0,829 mètres,
- que beaucoup de cercles de pierre seraient en réalité des "oeufs" de deux types, faits d'arcs de cercles raccordés dont les centres sont les sommets de triangles rectangles,
- et que les côtés de ces triangles, exprimés en yards mégalithiques, sont des triplets pythagoriciens. Par exemple à Carnac, on aurait un « oeuf mégalithique de type I », avec pour base un triangle rectangle de côtés 37,5-50-62,5, ce qui donne bien un triplet pythagoricien puisque $37{,}5^2 + 50^2 = 62{,}5^2$ qui est lui-même, à une homothétie près de rapport 12,5, le fameux triangle 3-4-5.

[14] Aucun lien avec le mathématicien René Thom.

[15] Alexander Thom, *La géométrie des alignements de Carnac* (Rennes: Université de Rennes, 1977). Alexander Thom et Archibald S. Thom et Aubrey Burl, *Stone Rows and Standing Stones, Britain, Ireland and Brittany* (Oxford: BAR, International Series 560 (1), 1990).

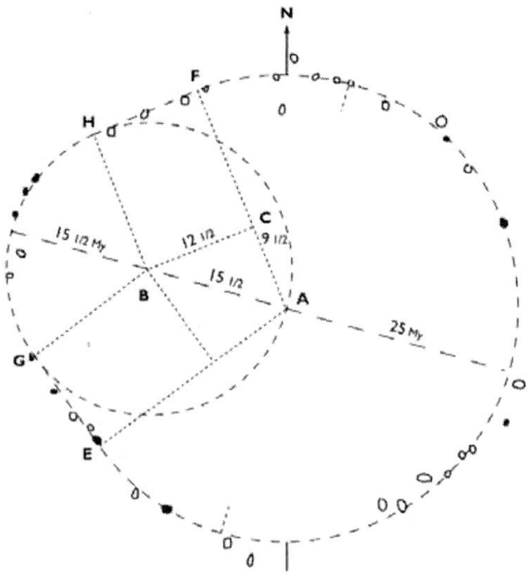

Figure 2-3 : « Œuf mégalithique » de Borrowstone Rig[16] (Ecosse).

Le lecteur verra sur la figure ci-dessus comment on peut fabriquer un tel triangle rectangle à partir d'un relevé de terrain. Il saluera peut-être l'ingéniosité des Thom, mais l'aspect « fabriqué », justement, lui sautera sûrement aux yeux. Il est certain que la position des blocs a été plusieurs fois modifiée au cours des siècles, et comme le dit Jean-Pierre Mohen :

[16] D'après Thom, *Stone Rows and Standing Stones*,op.cit. Seules figurent sur le terrain les pierres indiquées par de petites taches ou de petits ronds. A et B sont les centres des cercles supposés être les supports des arcs FE et HG, reliés par les tangentes HF et GE ; le point Cest la projection orthogonale de B sur AF. Les longueurs des côtés du triangle ABC sont 12,25, 9,5 et 15,5 yards mégalithiques, ce qui donne à très peu près un triplet pythagoricien puisque $12{,}25^2 + 9{,}5^2 = 240{,}3125$ et $15{,}5^2 = 240{,}25$.

> L'emplacement d'origine ne peut être toujours connu avec la précision du centimètre qu'adopte Thom. De plus, elles [les pierres] ont un volume et il est aléatoire de choisir quel est le point représentatif de leur positionnement, même s'il correspond au milieu des extrémités du bloc[17].

De plus, toute la documentation ethnographique s'inscrit en faux contre la vraisemblance, entre des petites communautés paysannes du Néolithique dans une région aussi vaste que les deux Bretagnes, d'une unité commune du type « yard mégalithique »[18]. Mêmes les bâtisseurs d'autels védiques n'avaient pas d'unité de longueur fixe puisqu'elle dépendait de la taille de l'officiant. Outre cela, que l'on pense seulement à la France d'avant la révolution de 1789, avec ses unités de mesure qui pouvaient varier d'un bourg au bourg voisin !

On le voit, la géométrie des mégalithes version « yard mégalithique » et triplets pythagoriciens, construite avec beaucoup de talent par les Thom, n'est néanmoins qu'une chimère. Et si elle fut accueillie favorablement par des mathématiciens, et non des moindres[19], c'est que parmi les motifs d'adhésion à telle ou telle interprétation, une cohérence théorique, même factice, l'emportera aisément sur toute autre considération.

L'os d'Ishango

L'os d'Ishango[20] a été trouvé dans la localité du même nom, en République Démocratique du Congo. Un fragment de quartz encastré à l'une des extrémités montre qu'il s'agit d'un manche d'outil, daté d'environ -20.000. Il présente dans le sens de sa longueur (10 cm) trois rangées d'encoches à peu près parallèles, regroupées en paquets inégaux (Figure 2-4) ; plusieurs sont effacées ou à peine visibles, ce qui rend déjà suspecte *a priori* toute interprétation fondée sur leur dénombrement. Plusieurs chercheurs prirent néanmoins ce risque, à commencer par son découvreur, Jean de Heinzelin, qui expliqua dans le *Scientific American* de juin 1962 comment on pouvait y lire « une sorte de jeu arithmétique

---

[17] Jean Pierre Mohen, *Le monde des mégalithes* (Paris: Castermann, 1989), 38. Cet auteur est bien sûr loin d'être le seul, parmi les préhistoriens, à avoir critiqué les Thom.

[18] Ici comme dans beaucoup d'autre cas, on voit que le comparatisme ethnographique est au moins un bon garde-fou contre les affabulations.

[19] C'est le cas de Bartel L. van der Waerden, par exemple dans son ouvrage *Geometry and Algebra in Ancient Civilizations* (Berlin : Springer, 1983).

[20] Il y en a en réalité deux. Nous ne parlerons que de celui qui a connu la célébrité dans les médias.

inventé par une peuplade possédant un système numéral basé sur 10 ainsi qu'une connaissance de la duplication et des nombres premiers ». Plus tard, Alexander Marshack[21] détecta deux lunaisons dans chaque colonne comportant 60 encoches, et un peu plus d'une lunaison et demi dans la colonne de 48 encoches.

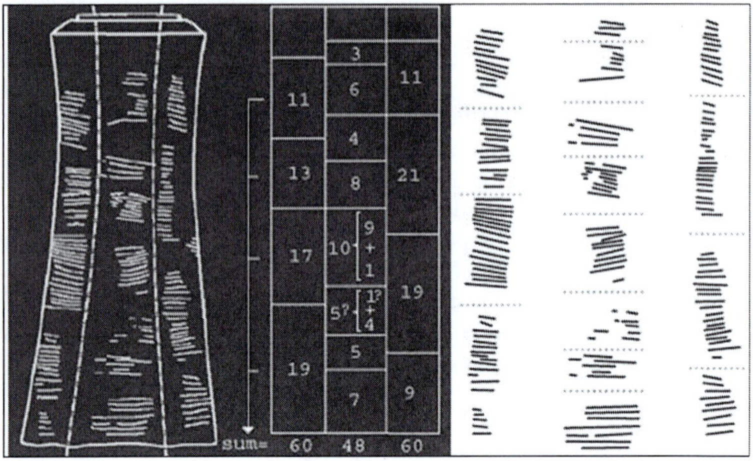

Figure 2-4 : Vues étalées du premier os d'Ishango, avec ses trois colonnes, dites 'de gauche', 'du milieu' et 'de droite'. Source : Dirk Huylebrouck, 'L'Afrique, berceau des mathématiques', dans *Mathématiques exotiques*, Dossier Pour la Science, avril/juin 2005.

Mais c'est seulement grâce au dévouement et aux nouveaux calculs de deux scientifiques belges, le mathématicien Dirk Huylebrouck et Vladimir Pletser, de l'Agence Spatiale Européenne, que l'os d'Ishango acquit sa célébrité dans le grand public. S'ils rejettent les conclusions particulières de De Heinzelin, ils en acceptent les prémisses, à savoir une interprétation purement arithmétique. Ils vont même plus loin que leur prédécesseur en affirmant que les nombres des trois colonnes sont liés de telle sorte que leur ensemble forme une règle à calcul : en prenant les groupes d'encoches (toujours assimilés à des nombres) de la colonne du milieu tantôt par deux, tantôt par trois, ils les additionnent et lisent la somme tantôt dans la colonne de gauche, tantôt dans la colonne de droite. Voici le détail résumé

[21] Alexander Marshack, *Les racines de la civilisation* (Paris: Plon, 1972).

dans un tableau que nous avons confectionné d'après la variante la plus récente, exposée en 2007[22], des calculs des auteurs :

Colonne de gauche	Additions sur la colonne du milieu	Colonne de droite
	3+6 (+2)	11
11	6+4 (+1)	
13	4+8 (+1)	
	4+8+10 (-1)	21
17	8+10 (-1)	
	10+5+5 (-1)	19
19	5+5+7 (+2)	
	7 (+2)	9

La première ligne du tableau veut dire que l'on distingue sur l'os, en haut de la colonne du milieu, un groupe de trois puis de six encoches, et à la même hauteur, sur la colonne de droite, un groupe de onze encoches ; à condition d'ajouter deux encoches, mises entre parenthèses, le total des nombres de la colonne du milieu se trouve sur la colonne de droite. Les lignes suivantes se lisent selon le même principe. On voit que toutes les opérations sont fausses, mais cela ne trouble pas trop les auteurs :

> Pourquoi l'addition supplémentaire de 2 ? Aucune raison ne peut être proposée mais il semble que les positions relatives de ces trois groupes ne sont pas une coïncidence et reflètent une intention inconnue[23].

Dans une version antérieure, les mêmes auteurs avaient lu 9 plutôt que 10 dans le cinquième groupe d'encoches de la colonne du milieu,

---

[22] Vladimir Plester and Dirk Huylebrouck, "An Interpretation of the Ishango Rods", in *Ishango, 22000 and 50 years later : the cradle of mathematics ?*, ed. Dirk Huylebrouck, (Brussel : Koninklijke Vlaamse Academie van België voor Wettenschappen en Kunsten, 2008), 139-170.

[23] Ibid.

moyennant quoi quatre des « additions » étaient exactes ; s'ils ont finalement opté pour 10, bien que ce choix ruine complètement la table, c'est qu'ainsi le total des nombres de la colonne du milieu est 48, qui est un multiple de 12, comme le total 60 des colonnes de droite et de gauche. D'où des spéculations possibles sur une base 12.

Il est bien clair que tout cela n'est guère défendable. En outre, dans cet exemple comme dans le précédent, la documentation ethnographique vient en appui de notre opinion : non seulement il n'y a pas la moindre trace de telles « tables », mais beaucoup de documents dont il serait facile d'inventer une interprétation arithmétique ont en réalité une signification qui lui est complètement étrangère[24]. Malgré tout, l'os d'Ishango est encore l'hôte d'honneur de tout article de vulgarisation sur l'origine des nombres et leur histoire. Il rejoint ainsi d'autres hôtes de marque comme les mégalithes et les pyramides qui, eux aussi, alimentent et continueront à alimenter de nombreuses fables. Et c'est précisément cela, leur caractère de fable, qui garantit leur succès : les chimères sont sources de plaisir. On aime bien imaginer une confrérie de bâtisseurs de mégalithes gardant jalousement leur secret de triplets pythagoriciens, et si les additions d'Ishango sont fausses « pour une raison inconnue », cela renforce encore le caractère énigmatique de l'os et par conséquent l'attraction qu'il exerce : s'il y a tant d' « à peu près », tant d'invraisemblances dans la production des fabulistes, c'est qu'il y a encore tant de mystères à percer, et par suite tant d'autres chimères possibles ! Pas si stériles que cela, en fin de compte !

Œuvres citées

Adam, Jean-Pierre. *Le passé recomposé, chroniques d'archéologie fantastique*. Paris: Seuil, 1988

Eggeling, Julius (translator). "The Satapatha Brahmana, Part IV." In *Sacred Books of the East*, edited by Max Müller, vol 43. Oxford: Oxford University Press, 1897. En ligne à l'adresse http://www.sacred-texts.com/hin/sbr/sbe43/index.htm.

Huylebrouck, Dirk. « L'Afrique, berceau des mathématiques.» *Dossier Pour la Science* 47 (2005) : 46-50.

[24] On trouvera des exemples et la critique complète des os d'Ishango dans un article publié sur le site Bibnum, à l'adresse : http://www.bibnum.education.fr/files/Ishango-analyse.pdf (consulté le 1er février 2013).

Keller, Olivier. *Aux origines de la géométrie. Le Paléolithique et le monde des chasseurs-cueilleurs*. Paris: Vuibert, 2004.

—. *La figure et le monde. Une archéologie de la géométrie. Peuples paysans sans écritures et premières civilisations*. Paris: Vuibert, 2006.

—. « Les fables d'Ishango, ou l'irrésisitible tentation de la mathématique-fiction », http://www.bibnum.education.fr/mathématiques/algébre/l'os-d'ishango.

Lorblanchet, Michel. *Les grottes ornées de la préhistoire*. Nouveaux regards, Paris: Errance, 1995.

Marshack, Alexander. *Les racines de la civilisation*. Paris: Plon, 1972.

Mohen Jean Pierre. *Le monde des mégalithes*. Paris: Castermann, 1989.

Plester, Vladimir and Dirk Huylebrouck. "An Interpretation of the Ishango Rods". In *Ishango, 22000 and 50 years later: the cradle of mathematics?* edited by Dirk Huylebrouck, 139-170. Brussel: Koninklijke Vlaamse Academie van België voor Wettenschappen en Kunsten, 2008.

Sen, S.N. et A.K. Bag. *The Sulbasutras with Text, English Translation and Commentary*. New Dehi: Indian National Science Academy, 1983.

Taçon, Paul S.C. "The Power of Stone; Symbolic Aspects of Stone Use and Tool Development in Western Arnhem Land,Australia"". *Antiquity* 65 (1991) : 192-207.

Thom, Alexander. *La géométrie des alignements de Carnac*. Rennes : Université de Rennes, 1977.

Thom, Alexander et Archibald S. Thom et Aubrey Burl. *Stone Rows and Standing Stones, Britain, Ireland and Brittany*. Oxford : BAR, International Series 560 (1), 1990.

Van der Waerden B.L. *Geometry and Algebra in Ancient Civilizations*. Berlin, Heidelberg, New York, Tolyo: Springer, 1983.

SECTION II

LES RAISONNEMENTS IMAGINAIRES DANS LES COMMENTAIRES MEDIEVAUX DE LA PHYSIQUE

ALICE LAMY

Introduction

Fidèles aux convictions d'Aristote, les commentateurs médiévaux des XIIIème et XIVème siècles en particulier reconnaissent l'importance de la perception et de l'observation de la nature. Pourtant, s'ils se font les héritiers de ces principes épistémologiques, bien peu d'entre eux ont repris ou approfondi les nombreuses données empiriques parfois peu fiables du Stagirite.Délaissant l'observation propre à expliquer certains fonctionnements de phénomènes, les maîtres de la scolastique ont plutôt tenté de percer le mystère du pourquoi des êtres et des créatures par l'élaboration d'explications théoriques fondées sur l'imagination, tant en philosophie naturelle, dans l'étude des mouvements et des changements que dans la science mathématique.

Cette étude voudrait caractériser le rôle ambivalent de l'imagination dans la construction des connaissances et des raisonnements scientifiques, à une époque où la philosophie scolastique exalte les prouesses de la raison et l'excellence intellectuelle de la démonstration logique.

Nous rappellerons tout d'abord les principes épistémologiques de l'imagination depuis Aristote jusqu'aux médiévaux dans la philosophie de la connaissance, avant d'étudier ses trois principales valeurs. Premièrement, l'imagination présente des qualités cognitives qui guident la raison dans la difficile construction de modèles et de schèmes conceptuels destinés à comprendre le monde. Elle requiert des instruments appartenant à la logique médiévale. Deuxièmement, l'imagination détient un pouvoir irrationnel qui désoriente étrangement l'acuité des esprits et les contraint au tâtonnement pour les conduire souvent vers des théories prometteuses encore célèbres aujourd'hui. Troisièmement, l'imagination est dotée d'une dimension verticale qui souligne l'omniprésence de la toute-puissance divine dans la connaissance de l'objet de science. Tout au long de cet article, nous évoquerons les contextes doctrinaux les plus célèbres du XIVème siècle tout en citant ponctuellement des extraits d'œuvres écrites par Walter Burley. Personnalité intellectuelle renommée pour sa grande érudition philosophique, ce maître de la scolastique est représentatif des ébullitions scientifiques de son époque et exerce ses activités de philosophe et de théologien à Oxford et à Paris dans les années 1310-1350.

Les principes épistémologiques de l'imagination, d'Aristote aux commentateurs médiévaux

La place de l'empirisme et de l'imagination dans la science aristotélicienne

Pour Aristote, la science constitue un idéal de perfection cognitive du monde qui nous entoure et s'oppose à l'opinion qui peut être vraie ou fausse. Si la science est un processus de connaissance infaillible, son objet ne peut être contingent. Par conséquent, la perception du réel changeant s'inscrit dans un long processus cognitif, de l'observation du singulier à l'élaboration de lois universelles sur les fonctionnements de la nature. Toute appréhension des phénomènes perçus dans le monde (le réchauffement d'un corps, le lever du soleil ou l'anatomie des animaux), s'effectue de façon empirique à partir de la sensation puis s'érige à une science totale et immuable. L'imagination ou *phantasia* est la faculté incontournable favorisant la transition de la simple perception de l'objet sensible (grandeur, couleur, contour, nombre, mouvement) à la connaissance. Pour Aristote, la *phantasia* n'est donc pas une manifestation erronée de la réalité ou la preuve que la sensation est une source d'erreur pour appréhender la vérité. Ce n'est pas parce que le soleil m'apparaît (*phainetai*) avoir un pied de grandeur que je vais croire qu'il est plus petit que la terre habitée[1] . La *phantasia* ne peut pas se former sans la sensation et sans elle, il ne peut y avoir de jugement (*upolêpsis*)[2]. Dans la mesure où elle ne requiert pas d'objet sensible, elle est dite immatérielle[3] .

Dès lors, de la sensation, la connaissance s'élabore à l'aide de l'imagination, elle est aidée de la mémoire, qui garantit la répétition de l'expérience. L'ascension vers l'Universel semble donc s'effectuer par la persistance et la superposition des images sensibles. De cette façon, la

---

[1] Aristote, *De Anima,* III, 3, 428a 24-b 10, traduction et notes par J. Tricot (Paris : Vrin, 1995).

[2] Aristote, *De Anima,* 427 b 14-16.

[3] Artistote, III, 8, 432 a 9-10.

sensation, par l'imagination, nous ménage un accès à la connaissance scientifique[4].

La faculté d'imagination au XIVe siècle dans l'accès à la connaissance, l'exemple de Walter Burley

En 1330, les Universités de Paris et d'Oxford, les deux pôles européens les plus prestigieux, sont en pleine notoriété et ont déjà 200 ans d'âge. Elles se sont dotées progressivement d'un enseignement philosophique exigeant et ambitieux transmis par des maîtres de la scolastique renommés comme Jean Duns Scot (autour des années 1300-1350), Walter Burley, Jean de Jandun, Thomas Wylton, Jean Buridan, Guillaume d'Ockham, à la faculté des arts, où les étudiants effectuent leur premier cycle d'études et à la faculté de théologie, ultime finalité du cursus universitaire pour les plus brillants d'entre eux. Tous les futurs diplômés de la faculté des arts produisent dans leur carrière au moins deux commentaires sur la physique et la logique d'Aristote, étayés et diffusés par les apports nouveaux de la philosophie d'Averroès. Ces commentaires écrits répondent à une structure souvent stéréotypée mais constituent l'occasion de performances intellectuelles déterminantes pour la distinction et la reconnaissance de chaque maître parmi ses pairs. Les analyses de la philosophie d'Aristote se développent sous forme d'une batterie de questions traditionnelles, expliquées par chaque maître puis interprétées de façon plus personnelle et plus libre.

Ainsi, par leurs lectures approfondies de la *Physique*, les médiévaux sont marqués par les valeurs empiristes et sensualistes inscrites dans la construction des connaissances scientifiques. Albert le Grand au XIIIème siècle, Roger Bacon puis Jean Buridan exaltent les valeurs des données empiriques et de l'induction. Walter Burley souligne les opérations mentales de l'imagination capable d'intégrer les qualités accidentelles de chaque être naturel puis, en cas de destruction ou de disparition de l'objet observé, de retenir le principe plus général de l'espèce, grâce auquel les êtres peuvent s'ordonner les uns par rapport aux autres.

---

[4] Aristote, *De Anima*, (II, 19 100 a 12-b 5).

> Mais l'imagination grâce à laquelle on saisit l'espèce de la chose sans matière n'a pas besoin que la forme dont elle est l'espèce soit en dehors de la matière. Tant que l'espèce est fixée dans l'imagination, que la chose soit présente, absente ou totalement détruite dans le monde, l'imagination est capable de l'appréhender par l'intermédiaire de son espèce. Ainsi, l'imagination est dotée d'une forme dont la perfection est plus abstraite que le sens extérieur. [...] L'imagination n'imagine rien sans la quantité et la qualité accidentelles déterminées et ces accidents proviennent de la matière[5].

Conformément à Aristote, l'imagination selon Burley est dotée d'une forte capacité d'abstraction. Bien plus, l'imagination conditionne l'acte d'intellection et permet à l'homme de s'élever vers la conception de l'Universel à partir de son expérience. Ainsi, la production imaginative exerce une impression sur le monde.

> [...] L'intellect agent n'imprime pas totalement la forme abstraite de la chose car autrement, nous n'aurions pas besoin des sens et de l'imagination pour accéder à ce que nous voulons connaître. [...] L'intellect agent produit des intentions de connaissance imaginées [...] et exercer ses capacités d'abstraction n'est rien d'autre que produire ces intentions imaginées. [...] L'espèce qui existe dans la capacité de fantaisie imaginative permet de faire naître l'espèce dans l'intellect. [...] L'intellect agent est agent principal et le fantasme, l'agent instrumental[6].

L'imagination marque ainsi une modalité de réception (*intentiones*) qui appréhende déjà le principe de classification de l'espèce. Cependant, pour s'élever à la connaissance en acte de cette espèce, l'imagination, trop encombrée par les données contingentes de l'être, doit attendre la lumière intelligible de l'intellect agent. Burley utilise alors des termes plus péjoratifs pour l'imagination, très typiques au Moyen-Âge, qui soulignent le caractère changeant et sensuel de l'image (*virtus fantastica*). Au § 127, il marque la fonction secondaire et instrumentale de l'imagination par le terme *fantasma* qui ne fait qu'introduire la possibilité d'une connaissance (*in intellectum possibilem*) par rapport à la fonction éclairante et rationnelle de l'esprit, qui seul peut garantir la valeur actuelle et universelle des savoirs.

---

[5] Walter Burley, *De potentiis animae*, M.J. Kitchel, "The *De Potentiis animae* of Walter Burley,"*Medieval Studies* 33 (1971): 85-113, § 113-114, 106. Traductions françaises des textes latins, A. Lamy.

[6] Walter Burley, *De potentiis animae*, § 125-127, p. 109-110.

Au XIVème siècle, l'imagination inscrit ses valeurs noétiques[7] dans l'ambivalence de nos capacités cognitives et abstraites. Conscients que notre esprit est limité dans ses conceptions pour connaître l'infinité des savoirs, les commentateurs accordent une place grandissante à l'imagination, aux côtés de la science et de ses instruments d'élaboration.

Première valeur de l'imagination : Les fondements rationnels de l'imagination dans l'élaboration scientifique médiévale

La place centrale de la logique dans l'activité de la philosophie scolastique

La faculté imaginative est renforcée et prolongée par la logique instrumentale. Au IIIème siècle après J.-C., à l'époque tardo-antique, le philosophe Porphyre examine les traités de logique d'Aristote réunis sous le nom d'*Organon* qui étudient la fonction, la hiérarchie et le rôle des termes utilisés dans le langage pour désigner et connaître le réel. Les dix catégories d'Aristote proposent une classification de l'être désignée dans le langage, selon ses différentes caractéristiques accidentelles – qualité, quantité, temps, lieu, relation etc., dominées par la plus absolue et la plus essentielle de toutes, la substance, qui fait que l'être est nécessairement ce qu'il est. Ces catégories, mais aussi les genres et les espèces des êtres sont-ils des choses réelles ou des conceptions de l'esprit lorsqu'on les utilise dans le langage ? Telle est la question centrale introduite par Porphyre. Quels sont les degrés d'abstraction qu'il faut reconnaître pour désigner puis étudier avec justesse tout ce qui nous entoure ? Ces premières interrogations sur l'expression de la nature commune de choses singulières, réelle ou simplement intelligible, (tel homme Socrate et l'humanité, tel cercle et la circularité) donnent lieu chez les maîtres de la scolastique à de célèbres débats logiques durant toute la période médiévale entre nominalistes et réalistes appelés « querelle des universaux ». Pour les réalistes, les catégories sur lesquelles s'appuie notre langage représentent des aspects réels de choses absolues. Ce qui est prédiqué à partir de ces catégories est complètement extérieur au sujet lui-même. Pour les nominalistes, comme Guillaume d'Ockham, le célèbre adversaire de

[7] Le champ noétique concerne l'étude de tous les procédés et de toutes les démarches présents dans l'âme et dans sa partie la plus haute, l'intellect, pour construire et assimiler les connaissances du monde.

Walter Burley sur ces sujets, les catégories indiquent avant tout une division des sens dans lesquels l'être se dit, considérée comme une distinction de modes de signification exercée par notre intelligence. Ces catégories sont d'abord mentales.

La logique instrumentale dans les investigations scientifiques

Ces débats ont des répercussions centrales dans la construction médiévale des connaissances scientifiques. En effet, l'acte de savoir et de croire va-t-il être jugé comme vrai, fiable, universel, aux termes de raisonnements issus de l'observation et de l'examen des choses du monde en elles-mêmes ou aux termes de raisonnements exceptionnels emportant l'assentiment de tous, par la force d'une démonstration purement immergée dans l'analyse des propositions du langage ? Ce débat reste omniprésent dans les méthodes scientifiques médiévales *in re* ou *in mente*. Par exemple, pour Walter Chatton, l'acte de connaître doit s'adresser aux choses elles-mêmes et non aux seules propositions. L'assentiment, c'est-à-dire l'acte par lequel l'esprit reconnaît quelque chose comme vrai, ne porte pas d'abord sur la proposition mais sur la chose signifiée par la proposition. Assentir à la vérité d'une proposition, c'est appréhender qu'il en est ainsi du côté des choses. Pour Guillaume d'Ockham, son émule franciscain, au contraire, seul l'assentiment à une proposition vraie, motivé par une démonstration, peut être constitutif d'un savoir scientifique au sens strict.

De façon dominante au Moyen-Âge, la science porte sur certaines propositions (surtout lorsqu'il s'agit d'un syllogisme proprement scientifique) qui sont appréhendées et auxquelles on reconnaît une vérité, pourvu que le raisonnement soit solide et rationnel. Comme Aristote, les médiévaux considèrent qu'une démonstration scientifique tire de prémisses vraies la raison pour laquelle une proposition est nécessairement vraie. Si l'on pose que les hommes sont des animaux vertébrés et que tous les animaux vertébrés sont mortels, on peut conclure que les hommes sont mortels. Démontrer par syllogisme, c'est admettre que la science est une connaissance causale et admettre la nécessité de tel ou tel phénomène présent dans le monde. Les commentateurs ont multiplié et diversifié à leur époque les raisonnements logiques pour développer les connaissances scientifiques. Une proposition se démontre par la composition syllogistique des prémisses. La conclusion obtenue par voie syllogistique est l'objet propre de la science.

Ainsi, l'exercice de la logique instrumentale renvoie à l'exercice d'une imagination apparentée à la raison et à l'habileté à formuler des hypothèses. Faire preuve d'imagination, c'est exercer ses capacités rationnelles. La logique instrumentale devient un critère distinctif pour délimiter deux domaines d'objets d'études, d'un côté, la philosophie naturelle, de l'autre, le modèle reconnu de science hypothétique, à savoir la science mathématique. Cette dernière envisage des entités sans existence réelle à partir de raisonnements purement hypothétiques. Par conséquent, lorsque les termes désignant ces entités sont pris pour sujets de propositions existentielles, les propositions obtenues sont fausses.

Deuxième valeur de l'imagination : La puissance chimérique de l'imagination dans quelques theories scientifiques médiévales célèbres

L'imagination polémique ou conformiste : se couper du réel pour contester ou approuver la philosophie aristotélicienne. La chute des corps dans le vide et la composition du feu.

La foi et l'imagination sont souvent réunies pour contester Aristote, qui nie l'existence du vide. Cette position déjà discutée par Philopon au Vème siècle après J.-C, est reprise par les médiévaux de façon originale. En accord avec Aristote sur le fait que le vide n'existe pas, ils intègrent pourtant progressivement à leur commentaire sur la *Physique* un ensemble de questionnements sur la possibilité ou non de l'existence du vide, la chute des corps et la vitesse de ces corps en mouvement dans le vide. Ils avancent ainsi que Dieu, dans son infinie puissance, pourrait créer un tel espace et considèrent sérieusement des conclusions originales, totalement imaginaires, pour aboutir à des résultats proches de leurs successeurs du XVIIème siècle. Ainsi, le mouvement des corps dans le vide est impossible mais on peut élaborer toutes sortes de théories si l'on fait preuve d'imagination et si l'on admet des conditions surnaturelles. Dès lors, toutes les données empiriques des corps en mouvement, tout ce que l'on peut observer d'eux dans le monde ordinaire, il suffit mentalement de les reproduire dans le milieu imaginaire du vide et de construire des raisonnements rationnels étayés d'une observation minimale.

C'est ainsi que, contrairement à Aristote qui soutient que la vitesse des corps chutant dans le vide est proportionnelle à leur poids, Thomas Bradwardine ou Albert de Saxe défendent, comme Galilée le confirme plus tard, leurs premières intuitions selon lesquelles deux corps en chute libre, bien que différents en poids et en taille, tombent avec une accélération au même taux constant de vélocité (la force motrice de résistance interne est égale dans tous les corps, bien qu'ils ne présentent pas tous les mêmes unités de matière).

Dans d'autres cas au contraire, les positions d'Aristote sont automatiquement reprises, sans être discutées et l'imaginaire relaie une croyance aveugle basée sur une fausse évidence qui s'apparente à un axiome utilisé dans de nombreux raisonnements. Bien des médiévaux comme Nicole Oresme affirment sans discuter que dans l'altération, un corps chaud peut se refroidir et inversement, et que le feu, par ce même phénomène, peut se transformer en air ou l'air en feu. La puissance chimérique de l'imagination se trouve au service de théories finalement constructives et est également présente dans les raisonnements mathématiques.

L'imagination mathématique : le cas de la latitude des formes

L'intensification pour Aristote est une altération limitée qui ne fait pas changer d'espèce (*Physique* V, 2, 226^{b} 1-2). Par exemple, un corps ne blanchit pas de plus en plus parce que des formes qualitatives de la blancheur s'ajoutent dans le sujet les unes aux autres. Une forme qualitative contraire à celle qui vient d'être introduite disparaît progressivement, laissant la place à une qualité plus déterminée et plus actualisée que la précédente. Le constat d'une blancheur plus éclatante n'est donc pas dû à la variation de la forme communiquée dans l'altération mais à l'élimination de son contraire.

Au XIVème siècle, les qualités sont considérées comme des grandeurs étendues pouvant quantitativement augmenter ou diminuer dans le corps qui subit une intensification ou une rémission. Thomas Bradwardine imagine que la rougeur d'une pomme est susceptible de s'accroître dans la durée de maturation du fruit comme l'intensité de la vélocité d'un corps en train de chuter accroît sa vitesse. Loin du contexte original aristotélicien et d'une observation de la latitude des formes, les maîtres du collège de Merton se lancent dans des calculs mathématiques et des schémas géométriques, échafaudant de pures fictions mentales. Pourtant, au terme de leurs raisonnements coupés des fondements du réel, ils sont revenus sur des principes importants relatifs à la vitesse et ont contribué à finaliser plusieurs théorèmes proches de la loi galiléenne de l'accélération uniforme de la vitesse des corps qui tombent en chute libre précédemment évoquée.

Un dernier type d'imagination est présent dans l'œuvre de Walter Burley : ce sont les arguments *De potentia divina*.

Troisième valeur de l'imagination : La nature spirituelle de l'imagination et la main de Dieu

Une imagination qui décuple et exalte les capacités abstractives de la raison.

Les commentateurs médiévaux consacrent une attention toute particulière aux questionnements cosmiques : la possibilité d'un infini au-delà de notre monde, la pluralité des mondes, le mouvement de la terre, sa sphéricité, l'existence d'un vide infini intra ou extra-cosmique. Pour traiter de telles interrogations, les philosophes ne disposent pas de moyens empiriques et recourent à leur imagination comme capacité d'abstraction permettant de se substituer à l'observation. Fidèles à Aristote, les médiévaux, dans leur grande majorité, rejettent l'idée d'un vide extra-cosmique infini puisqu'il est impossible qu'il existe quelque créature que ce soit, univers, espaces vides ou pleins au-delà de notre monde sublunaire et de ses voûtes célestes.

Toutefois, Thomas d'Aquin soutient que le terme « au-delà de », la notion d'extra-cosmique ouvrent à notre esprit imaginatif des images plausibles et des perspectives permettant d'envisager une place pour d'autres mondes, d'autres corps et d'étendre les dimensions spatiales actuelles à des sphères et des cosmos inconnus. Si ces hypothèses *secundum imaginationem* ne sont pas réelles, elles font partie intégrante de la pensée vivante et scientifique. Bien plus, Nicole Oresme conçoit au contraire au-delà de notre cosmos une immensité incorporelle, infinie et indivisible. Il soutient que, dans la mesure où notre imagination nous permet d'envisager un tel au-delà et prouve que nous n'avons aucun obstacle à accepter la dimension extra-cosmique, il n'y a aucune raison de croire que cette immensité n'existe pas à moins d'apporter de sérieuses preuves contraires. La dimension surnaturelle de l'imagination nous conduit à découvrir la vérité. Elle agit comme une intuition ou une inspiration qu'il convient de préserver et d'écouter, tel l'homme soucieux de faire le bien autour de lui prêterait une oreille attentive à la dimension verticale de ses actions vertueuses et aux messages divins. Avant Nicole Oresme et contre Thomas d'Aquin, Thomas Bradwardine, le « profond Docteur » du réputé collège de Merton, admet, comme Jean Duns Scot, que cette conception est imaginaire pour l'homme. Pour Dieu, en revanche, cet espace vide existe bel et bien.

Une imagination justifiée par la toute-puissance divine

Dans le courant du XIVème siècle, les innovations scientifiques s'accélèrent avec les avancées majeures de Jean Buridan ou Nicole Oresme sur le concept d'*impetus*, les arguments de la toute-puissance divine se multiplient et s'appliquent à des raisonnements géométriques ou mathématiques. Pratiqués dès le début du XIIIème siècle, ces arguments connaissent un regain d'intérêt particulier au moment de la condamnation de 1277 qui entend lutter contre l'usage tendancieux et païen des maîtres-es-arts des commentaires traditionnels d'Aristote et d'Averroès pour expliquer le cosmos. Ainsi, deux règles doivent être préservées : premièrement, Dieu peut réaliser tout ce qui n'entraîne pas contradiction, deuxièmement, Dieu n'agit jamais inutilement, et dans cette mesure, il ne peut pas avoir la volonté de tromper sa créature. Au XIVème siècle, ces deux règles sont articulées à la définition complexe de l'impossible selon Averroès. Ce dernier distingue l'impossible absolu qui entraîne une contradiction logique et l'impossible relatif qui n'entraîne pas contradiction et qui est donc réalisable par Dieu. Dieu peut réaliser tout ce qui n'entraîne pas contradiction et effectuer toute sorte de séparation, par exemple entre la matière et la forme, entre le sujet et l'accident. De la sorte, une nouvelle physique se développe et s'émancipe des contenus traditionnels aristotéliciens à partir d'hypothèses fausses selon la nature mais rendues réalisables et analysables grâce au pouvoir de Dieu. Ainsi, ces arguments imaginaires assument une fonction heuristique et analytique importante et sont mis en valeur dans le cadre instrumental de la logique. De plus en plus d'auteurs comme Walter Burley, cherchent à intégrer la puissance divine dans des contenus de philosophie diversifiés.

Les raisonnements imaginaires de Walter Burley contre Guillaume d'Ockham dans le débat sur le statut de la quantité

Walter Burley remet en cause la conception de Guillaume d'Ockham à propos de la logique et physique de la catégorie aristotélicienne de quantité. Si Burley privilégie et préserve la portée concrète hors de notre esprit (*extra mentem*) de la quantité dans le réel, son adversaire au contraire, qui appartient au courant nominaliste, retient principalement des catégories une distinction de modes de signification exercée par notre intelligence (*in mente*) pour comprendre le monde. De la sorte, la quantité ne contribue pas à définir l'être de la chose, elle ne fait que signifier une valeur particulière des termes substantiels et qualitatifs : la division et la position des parties de la substance. Ainsi, d'un point de vue ontologique, Ockham considère que la substance est en elle-même dotée de parties.

Les arguments imaginaires de Burley viennent réfuter ceux d'Ockham dans son *Traité de la quantité* et dans son *Traité du corps du Christ* en particulier. Burley présente d'abord littéralement l'argumentaire de son adversaire :

> Dieu peut enlever la quantité de la substance corporelle, alors que la substance demeure dans la même position et conserve toutes les parties qu'elle avait avant et si c'est ce que l'on suppose, cette substance corporelle, puisqu'elle occupe une position et un lieu, aura des parties distinctes [...] et par conséquent la substance sera de soi quantifiée et il n'y a pas besoin de poser autre chose dans la substance pour qu'elle soit étendue. Ainsi, elle est étendue par elle-même et quantifiée sans rien d'autre[8].

Ces arguments agissent comme une technique persuasive, un tour dramatique et tangible à une forme d'hypothèse intuitive : si Dieu peut séparer la quantité de la substance, la substance demeure étendue par elle-même, comme avant la séparation, puisque la quantité ne se distingue pas de la substance. En réponse à Ockham, Burley soutient que si Dieu peut séparer la quantité de la substance, la substance va demeurer sans extension, ni parties quantitatives, car la quantité est distincte de la substance: « Il faut dire que si toute quantité était éliminée de la substance corporelle, elle n'aurait pas de parties distinctes. Seule la quantité confère à la substance ses parties distinctes.[9]»

Il cherche donc à conceptualiser des situations retenues comme absurdes et impensables auparavant (la substance est séparée de son accident, la quantité) en dissipant l'équivoque entre les impossibilités physiques et les impossibilités logiques. Substance et quantité ne peuvent être séparées *naturaliter*, dans le cadre des capacités de la nature. En revanche, d'un point de vue surnaturel et divin, cet état de fait est possible. Dans les deux cas, c'est le principe logique de non-contradiction qui détermine ce qui est possible au sens absolu (*simpliciter*). Cette fiabilité du raisonnement imaginaire permet d'apporter la preuve pour Burley que la

---

[8] Walter Burley, *Tractatus de formis*, *Walter Burley's treatise De formis,* ed. F. Scot (München: Bayerische Akademie der Wissenschaft, Bd 4, 1970), *pars posterior*, 51.

[9] Walter Burley, *Gualterii Burlei super Artem veterem Porphyrii et Aristotelis expositio. Venetiis*, 1493, f°. 25ra-28ra (chapitre sur la quantité), *Expositio in Aristotelis praedicaementa*, f°. 28ra.

substance ne peut être étendue sans la quantité accidentelle et que cette catégorie a une utilité particulière.

L'opposition des deux philosophes porte aussi sur le domaine géométrique et mathématique. Burley soutient l'existence des points, des lignes et des surfaces, comme limites du corps, tandis qu'Ockham les considèrent inutiles dans la structure des corps. Dans le passage suivant, Burley souligne que Dieu ne peut pas créer un corps sans une surface qui le limiterait, sous peine de produire un phénomène contradictoire et impossible : l'infini en acte, rejeté par tous les médiévaux depuis Aristote.

> De là, le corps est terminé par la surface comme par son sujet proche et la surface est terminée par la ligne et la ligne par le point ; pour cette raison, de même que le point n'est pas la partie de la ligne, de même la ligne n'est pas la partie de la surface, ni la surface, la partie du corps. Alors, la surface est ainsi le terme essentiel du corps, parce que Dieu ne peut faire le corps du genre de la quantité sans surface qui le limite, parce que si la surface ne terminait pas le corps, le corps n'aurait pas un terme ni une limite, et par conséquent, le corps serait infini en acte, et, si Dieu ne pouvait faire le corps en acte infini, comme le disent les philosophes, il s'ensuivrait que Dieu ne pourrait faire le corps du genre de la quantité sans surface qui le termine, etc[10].

Au nom du même principe de non-contradiction, Burley affirme que Dieu ne peut compromettre la structure de l'infini et a donc créé une ligne auquel seul le point peut mettre un terme.

> Et ainsi j'affirme que Dieu ne peut pas produire de ligne sans point, ligne à laquelle il sert de terme bien qu'il ne puisse produire la ligne sans ce point-ci limitatif et sans ce point-là qui coupe la ligne. Et de la même manière, j'avance un autre argument théologique. Dieu ne peut pas produire une ligne sans point limitatif bien qu'il puisse produire une ligne sans ce point et sans cet autre qui divise la ligne, pourtant il ne peut pas produire la ligne sans ces points ensemble[11].

[10] Walter Burley, *Traité des formes*, 63.

[11] Walter Burley, *Gualteri Burlei in physicam Aristotelis; expositio et questiones ac etiam quaestio de primo et ultimo instanti denuo revisa ac mendis purgata et accuratissima quantum ars perficere potest impressa*, Venezia, 1501, dubia 13, 14 et 15. *Exposition et questions sur la Physique d'Aristote*, livre I, qu. 14, f°. 14vb.

Conclusion

A première vue, la faculté de l'imagination au XIVème siècle et ses embardées inattendues semble à la fois arrimée à la raison et capable, à l'horizon d'un point de fuite invisible, d'engager les esprits à concevoir les hypothèses les plus délirantes. Elle ne vient donc pas modifier les représentations traditionnelles du Moyen-Âge : période-gouffre dans la frise chronologique de l'évolution scientifique, sortie de route malheureuse ou jachère de savoirs entassés et inégaux coupés de l'expérience, jonchés de naïvetés et d'erreurs grossières.

A y regarder de plus près cependant, l'imagination médiévale, dans toute son ambivalence, a permis des découvertes significatives et des théories intéressantes qui rendent hommage aux principes bachelardiens d'une science chaotique, nourrie de trajectoires diverses, irrégulières et spectaculaires, profondément vivante du fait de son lot d'itinéraires inattendus. Ainsi, pour les médiévistes, il reste toujours difficile de tisser des liens d'évolution, de Thomas d'Aquin, qui aborde la chute des corps et la notion de masse de matière interne au corps, à Albert de Saxe ou Guillaume d'Ockham, chez lesquels on retrouve deux siècles plus tard les concepts de densité, de vitesse et d'inertie, jusqu'à Descartes et Newton. Les théories médiévales restent discrètes, agissent indirectement les unes sur les autres, s'adjoignent ou se repoussent, de façon lointaine.

Enfin, l'imagination médiévale nous rappelle avec pertinence jusqu'à aujourd'hui que toute connaissance est fondée sur des croyances. Le principe même de cette imagination repose sur la conviction que les expériences envisagées dans l'intellect produiront une vérité si les conditions hypothétiques décrites pour l'émergence de ces phénomènes sont compatibles avec le réel. A condition de croire en un Dieu puissant qui de la main fait surgir l'ordre du monde qu'il souhaite, en l'homme et ses prouesses rationnelles ou techniques, tout résultat probant pourra voir le jour. La science doit donc à l'imaginaire sa dynamique et son audace, la liberté et la créativité de ses fondements, l'imagination doit à la science l'intensité du désir humain de produire des visions du monde toujours renouvelées.

Œuvres citées

Aristote. *De Anima,* III, 3, 428a 24-b 10, traduction et notes par J. Tricot. ParisVrin, 1995.

Kitchel, M.J. "The *De Potentiis animae* of Walter Burley." *Medieval Studies* 33 (1971): 85-113.

Bianchi, Luca. *Vérités dissonantes: Aristote à la fin du Moyen Âge*, Paris, Cerf. 1993.

—. *L'Errore di Aristotele : la polemica contro l'eternità del mondo nel XIII secolo*, Firenze, la Nuova Italia, 1984.

Biard, Joël et Celeyrette, J. *De la théologie aux mathématiques, l'infini au XIVe siècle*, Paris, Les Belles Lettres. 2005.

Burley, Walter. *Tractatus de formis*, *Walter Burley's treatise De formis,* edited by F. Scot. Müchen: Bayerische Akademie der Wissenschaft, Bd 4, 1970.

—. *Gualterii Burlei super Artem veterem Porphyrii et Aristotelis expositio*. *Venetiis*, 1493, f°. 25ra-28ra

—. *Gualteri Burlei in physicam Aristotelis; expositio et questiones ac etiam quaestio de primo et ultimo instanti denuo revisa ac mendis purgata et accuratissima quantum ars perficere potest impressa*, Venezia, 1501.

Funkenstein, Amos. *Théologie et imagination scientifique*, *Du Moyen Âge au XVIIème siècle*, trad. J.-P. Rotschild, Paris, PUF, 1995.

Grant, Edward. "Empirism whithout observation", in C. Leijenhorst, C. Lüthy, J. M. M. H Thijssen (eds.), *The dynamics of aristotelian natural philosophy from antiquity to the seventeenth century*, ("Medieval and early Modern Science", 5), Leiden, Boston, Köln, Brill, 2002: 141-168.

—. "Scientific Imagination in the Middle Ages". *Perspectives on Science* 12/4 (2004) : 394-423.

Grellard, Christophe (ed.). *Méthodes et statut des sciences à la fin du Moyen Âge*, Actes des tables rondes de Tours et Lyon, Avril 2000-2001, Villeneuve d'Ascq, Presses Universitaires du Septentrion, 2004.

Hugonnard-Roche, H. "Analyse sémantique et analyse *secundum imaginationem* dans la physique parisienne au XIVe siècle". *Studies in Medieval Natural Philosophy*, edited by S. Caroti, ("Biblioteca di Nuncius. Studi e Testi", 1), Florence, L. S. Olschki,1989 : 133-153.

LES FOSSILES DE BURGESS ET LES INTERPRETATIONS MILITANTES DE L'EVOLUTION

ALEXANDRA BORSARI

L' « East side story » largement reprise dans les enseignements racontait comment la bipédie serait venue à l'être humain à la faveur d'un changement climatique survenu avec la formation du rift est africain. Les ancêtres de l'homme auraient été obligés de se redresser pour voir au dessus des hautes herbes de la savane tandis que leurs « cousins » les grands singes auraient conservé un mode de vie arboricole.

Or, différentes découvertes, parmi lesquelles celle de « Toumaï » au Tchad, vieux de 7 MA – ou celle « d'Ardi » - *Ardipithecus ramidus*, vieux de 4,4 MA, bipède en milieu couvert – ont depuis largement déconstruit ce roman présenté comme un savoir universel. La paléoanthropologie et plus généralement la science se sont construites par une succession de remises en question. A ce propos, la littérature concernant l'influence des contextes sociaux sur la construction des savoirs est déjà abondante et met en garde contre la relecture historicisante des conditions de succès d'une théorie sur une autre, ce que Dominique Pestre appelle l'« historicité »[1]. Toutefois, il existe un point aveugle qui est rarement étudié et qui consiste à interroger l'imaginaire des scientifiques dans leurs contextes sociaux.

Cet article s'intéresse ainsi, sous cet angle, à un épisode bien connu de l'histoire de la paléontologie dans lequel convictions religieuses et politiques ont joué un grand rôle. Il s'agit ici d'une première approche, depuis le point de vue des théories de l'imaginaire, de la polémique qui opposa Stephen Jay Gould à la célèbre figure de Charles Doolittle Walcott lorsque le premier entrepris de dénoncer les « erreurs » du second concernant les fossiles découverts dans des dépôts de schiste à Burgess

---

[1] Dominique Pestre, *Introduction aux science studies* (Paris: Editions La Découverte, coll. Repère, 2006).

dans son livre *Wonderful Life*[2], paru en français sous le titre *La vie est belle.*[3]

Après une présentation succincte du site, de la personnalité et du parcours de Walcott et de Gould, les arguments mobilisés par ce dernier contre Walcott dans *Wonderful Life* permettront d'approcher le problème de la place des convictions personnelles et le rôle probablement joué par l'imaginaire dans le cadre de cette polémique.

Le site

Le Schiste de Burgess est un ensemble géologique situé en Colombie britannique, au Canada. Il s'agit de dépôts de schiste noir situés à proximité du mont Burgess, non loin de la ville de Field dans les Rocheuses. Le site, classé au patrimoine mondial par l'UNESCO depuis 1981, fait aujourd'hui partie intégrante du Parc national de Yoho.

Ces dépôts sont exceptionnels dans la mesure où ils recèlent des fossiles de corps mous vieux de 505 millions d'années. Il s'agit de fossiles du Cambrien moyen, période charnière dans l'histoire de la vie, au cours de laquelle les écosystèmes marins sont devenus particulièrement complexes.[4] Un grand nombre des phyla[5] marins représentés dans les schistes de Burgess est aujourd'hui éteint. Le Cambrien a ainsi pu être considéré comme une période d'essais évolutifs suivie d'une radiation importante. Cette radiation supposée a été largement commentée par Stephen Jay Gould. Aujourd'hui, elle est remise en question et l'hypothèse privilégiée est celle d'une lacune de fossilisation[6].

---

[2] Stephen Jay Gould, *Wonderful life, The Burgess Shale and the Nature of History* (New York: WW Norton & Company, 1989).

[3] Stephen Jay Gould, *La vie est belle, Les surprises de l'évolution*, traduction par Marcel Blanc (Paris: Seuil, Collection Points Science, 1991).

[4] Il n'y a sans doute pas encore de végétation sur les terres émergées. Les premières traces considérées comme fiables de végétaux terrestres datent de l'Ordovicien moyen, il y a env. 468 Ma. Cf. Philippe Steemans *et al.*, "Origin and Radiation of the Earliest Vascular Land Plants," Science *April 17, 2009, Vol. 324, no. 5925, 353 DOI: 10.1126/science.1169659.*

[5] Phylum : embranchement

[6] Cyril Langlois, « De Burgess au Gabon : les plus anciennes traces fossiles de pluricellulaires », http://planet-terre.ens-lyon.fr/planetterre/XML/db/planetterre/metadata/LOM-fossiles-pluricellulaires.xml (consulté le 29 janvier 2013) : « L'hypothèse concurrente étant une lacune de fossilisation, certains de ces taxa cambriens n'ayant pas de squelette minéralisé (éponges, vers, cnidaires). La présentation, au printemps 2010,

Le regain d'intérêt pour ce site date des années 1960, lorsqu'Harry Blackmore Whittington, paléontologue britannique, décida de réexaminer les données accumulées. Parmi les étudiants qui s'attelèrent avec lui à cette tâche, figurent Derek Briggs et Simon Conway Morris. Ce dernier est sans doute le plus connu du grand public, du fait de l'hommage que lui a rendu Stephen Jay Gould dans *Wonderful Life*. Malgré les critiques que Simon Conway Morris a pu émettre par la suite sur le livre de Gould et ses prises de position parfois déroutantes[7], il reste l'un des principaux artisans de ce nouveau regard porté sur les fossiles de Burgess, vus traditionnellement comme des formes trop primitives pour être aussi complexes que les espèces actuelles[8].

Gould *vs* Walcott

Même en faisant abstraction de la distance temporelle qui les sépare, il semble difficile de trouver deux personnalités plus éloignées. Charles Doolittle Walcott (1850-1927) s'est intéressé aux fossiles dès son plus

---

de fossiles ordoviciens rappelant ceux de Burgess, et provenant d'un nouveau gisement marocain (les argiles et les silts de la formation Fezouata), vient confirmer cette hypothèse. Un certain nombre d'organismes de morphologie «basale », « cambrienne » — par rapport aux fossiles typiquement ordoviciens —, rappelant *Marella* ou *Anomalocaris*, par exemple, perduraient dans ces écosystèmes ordoviciens. La transition entre les faunes cambriennes et ordoviciennes aurait donc été plus progressive qu'on ne le croyait auparavant. »

[7] Dans ses écrits, Simon Conway Morris n'hésite pas à dénoncer une sorte de relativisme qui serait caractéristique de la postmodernité: une perte des valeurs au nom d'une pseudo rigueur scientifique. Cette posture le rapproche des mouvements spiritualistes voulant réintroduire la métaphysique dans les sciences et tout particulièrement en biologie évolutive. Il est ainsi surprenant de le voir diriger avec des théologiens un ouvrage collectif présenté comme un ouvrage de discussion scientifique de la théorie de la convergence : Simon Conway Morris (ed.), *The deep structure of biology, Is convergence sufficiently ubiquitous to give a directional signal* (West Conshohocken, Pennsylvanie: Templeton Foundation Press, 2008). Il faut souligner que cette maison d'édition a pour objectif affiché d'influencer les leaders d'opinion en livrant des lectures chrétiennes de sujets scientifiques. Concernant plus précisément le Schiste de Burgess, cf. Simon Conway Morris, *The crucible of creation, The Burgess Shale and the Rise of Animals,* New York : Oxford University Press, 1998, traduit en français sous le titre : le creuset de la création.

[8] Cf. Harry Blackmore Whittington et Simon Conway Morris, *Extraordinary Fossil Biotas: Their Ecological and Evolutionary Significance* (London: Royal Society, 1985).

jeune âge. Issu d'un milieu très modeste, il n'a pas fait d'études mais a continué à collectionner les fossiles qu'il commercialisait, notamment à Yale. C'est à cette époque qu'il a fait la connaissance de Louis Agassiz, alors en poste à Harvard, célèbre pour son hostilité aux thèses de Darwin, qui l'encouragea à poursuivre ses recherches. Il fut ensuite recruté en 1879 en tant qu'assistant en géologie au sein d'un organisme alors tout récent : le *US Geological Survey* dont il devint le directeur en 1894. Cette institution n'est pas la seule dans laquelle Walcott était destiné à jouer un rôle : en 1907, il devint également secrétaire de la *Smithsonian Institution* puis, en 1923, président de *l'American Association for the Advancement of Science*[9]. Walcott découvrit le site de Burgess en 1909 et y retourna régulièrement jusqu'en 1924. Au cours de ses différentes campagnes de fouilles, il collecta jusqu'à 65 000 spécimens. Nulle surprise à ce que figurent parmi ses préoccupations principales la conservation des sites et des collections.

Stephen Jay Gould, quant à lui, est né en 1941, presque un siècle après Walcott, et il est décédé en 2002. Issu du monde universitaire qu'il n'a jamais quitté, il fit toute sa carrière à Harvard. Son intérêt pour la paléontologie dépasse de loin les frontières de sa discipline puisqu'il s'intéressa au cours de sa carrière non seulement à l'histoire du vivant mais également à l'histoire des sciences. Malgré de nombreuses récompenses, et contrairement à Walcott, Gould ne dirigea aucun institut prestigieux. Alors que Walcott conforta petit à petit sa position au sein de la classe dirigeante, Gould fut plus enclin à mener un travail de vulgarisateur et de scientifique engagé dans la cité. Sa théorie la plus célèbre est celle des équilibres ponctués[10] qui semble aujourd'hui dépassée en biologie évolutive. Cette théorie porte sur les rythmes des changements morphologiques : de longues périodes de stabilité (ou stases) seraient entrecoupées par des épisodes de variations (ou ponctuations), contredisant la vision d'une évolution plus graduelle[11].

---

[9] Pour une biographie détaillée, cf. le travail consacré sur plusieurs décennies à la figure de Walcott par Ellis Leon Yochelson (références en fin d'article).

[10] Cf. son ouvrage de synthèse: Stephen Jay Gould, *The structure of evolutionary theory* (Cambridge, Massachussetts and Londres: The Belknap Press of Harvard University Press, 2002), 745-1022.

[11] Patrice David et Marie-Laure Parmentier, « Evolution et développement » in *Biologie évolutive*, ed. Frédéric Thomas, Thierry Lefèvre et Michel Raymond (Bruxelles: De Boeck, 2010), chapitre 8, 292 et glossaire, 802. L'ouvrage *Les mondes darwiniens* est moins catégorique sur le caractère dépassé de cette théorie et insiste davantage sur sa dimension controversée : Henri Cap, « Comportement et évolution : regards croisés », in *Les mondes darwiniens. L'évolution de l'évolution,*

Parmi ses plus grands combats figurent ceux qu'il a menés contre toutes les formes de déterminisme: il a ainsi combattu la sociobiologie et la psychologie évolutive, disciplines dans lesquelles il voyait le danger de limiter l'explication des comportements au strict héritage évolutif. Ses prises de position révèlent son attachement aux libertés fondamentales et sa dénonciation depuis ses années d'études, des différentes formes d'inégalité.

Son engagement se traduisit par une lutte contre les intégrismes et les créationnismes. Il déboucha sur la théorie du non recouvrement des magistères (Non-Overlapping Magisteria ou NOMA), qui considère que les champs du religieux et des sciences peuvent coexister mais ne doivent pas se rencontrer[12].

La critique de Gould vis à vis de Walcott

Dans *Wonderful Life*, Gould poursuit le même but que dans nombre de ses écrits : dénoncer et démonter les préjugés qui biaisent le regard porté sur les découvertes scientifiques, permettent de les instrumentaliser pour les mettre au service d'intégrismes de tout type, ou de toute entreprise destinée à légitimer les inégalités.

Gould se livre ainsi à la critique de l'idéologie d'un progrès linéaire en soulignant à quel point la conception d'une évolution sous la forme d'une marche vers le progrès, ou d'un arbre se ramifiant vers toujours plus de complexité a pu influencer des générations de scientifiques, entraînant des interprétations complètement erronées à partir d'observations qui auraient dû, si elles avaient été analysées avec moins de préjugés, donner lieu à une vision beaucoup moins anthropocentrique de l'évolution.[13] Pour Gould, Walcott restait soumis à cette conception du vivant et voulait à tout prix faire correspondre ce qu'il découvrait avec sa vision de l'histoire du vivant.[14] Selon lui, Walcott ne pouvait donc pas découvrir quoi que ce soit de nouveau avec les animaux du schiste de Burgess : il était condamné à y chercher des preuves qui appuieraient sa vision d'un monde ordonné par la

ed. Thomas Heams and others (Paris: Editions matériologiques, 2011), chapitre 23, 808.

[12] Stephen Jay Gould, *Rocks of Ages: Science and Religion in the Fullness of Life*, New York, Ballantine Publishing Group, 1999.

[13] Stephen Jay Gould, *Wonderful Life, op. cit*, chapter one, « The Iconography of an Expectation », 23- 52.

[14] Stephen Jay Gould, *La vie est belle, Les surprises de l'évolution*, *op.cit.*, 316.

volonté divine, allant des créatures simples vers des formes plus complexes pour aboutir enfin à l'être humain.

Pour appuyer son interprétation de l'œuvre de Walcott, Stephen Jay Gould exploita dans son livre un passage d'une déclaration publiée en 1923. Dans cette déclaration, Walcott tentait de contrer les thèses séparationnistes et de défendre une vision fondamentaliste refusant de séparer science et religion. Gould s'en est servi pour insister sur le caractère idéologique des recherches de Walcott.[15] Peu importait en effet pour Walcott de s'affranchir de sa subjectivité : les faits « bruts » n'ayant pas d'autre intérêt que de servir les croyances et d'œuvrer à la paix civile.[16]

Pour s'affranchir du conditionnement opéré par sa propre culture et par son groupe social, Stephen Jay Gould rappelle ainsi qu'il est indispensable d'avoir conscience que le fonctionnement du monde naturel n'est pas fait pour répondre aux angoisses de l'être humain. La vision que l'homme se construit, est toujours une vision *a posteriori*, une explication faisant sens à partir d'événements qui n'en ont pas forcément. Pour Gould, l'histoire de la vie est d'abord faite de hasards. Si les bouleversements successifs qui ont agité le vivant depuis ses débuts ont abouti au monde naturel actuel, ce n'est certainement pas en suivant un plan préétabli, mais bien parce qu'à chaque changement climatique, à chaque extinction, il s'est trouvé un type d'être vivant pour survivre, non pas parce qu'il était « meilleur » dans l'absolu mais parce que le hasard a voulu que son mode de fonctionnement - ou un avantage particulier - lui permette de s'en sortir. Ainsi, la variété considérable d'espèces fossilisées sur le site de Burgess ne permettait pas de deviner quelle serait l'évolution future du monde vivant. Si le phylum des chordés[17] ne s'était pas développé, l'évolution aurait pu aboutir à un monde où non seulement l'homme mais l'ensemble des vertébrés ne seraient jamais apparus.[18]

---

[15] Ibid., 340-341.

[16] Pour donner une meilleure idée du contexte de l'époque, il faut rappeler que deux ans plus tard s'est tenu le Procès du Singe : affaire dans laquelle, en juillet 1925, un instituteur de Dayton, John Thomas Scopes, a été condamné à payer une amende de 100 dollars pour avoir enseigné à ses élèves la théorie de l'évolution appliquée à l'être humain. A l'époque, dans le Tennessee, même s'il n'était pas interdit d'enseigner la théorie darwinienne de l'évolution en la restreignant aux plantes et aux animaux, il était interdit d'aller contre la Bible et d'enseigner autre chose que l'origine divine de l'être humain. Le *Butler Act* n'a été abrogé qu'en 1967.

[17] Le phylum des chordés se divise en trois branches parmi lesquelles celle des vertébrés dont le système nerveux et la corde sont dorsaux.

[18] Stephen Jay Gould, *La vie est belle, Les surprises de l'évolution*, *op.cit.*, 429-431. La place de Pikaia comme premier ancêtre connu des vertébrés est aujourd'hui

La notion de hasard est très débattue en biologie évolutive et Stephen Jay Gould a bien conscience que l'organisation du vivant suit certaines règles, notamment celles que lui dictent les contraintes physiques de l'environnement terrestre. Cependant, ces contraintes ne définissent que des cadres généraux dans lesquels la vie peut se déployer de multiples façons[19]. La contingence donnerait donc au monde vivant son caractère profondément imprévisible même si, au niveau des espèces, les lois de la sélection naturelle opèreraient. Les contraintes imposées par les conditions physiques terrestres – pesanteur, magnétisme ou composition de l'atmosphère, par exemple.- joueraient donc un grand rôle mais seulement sur les « options de base » du vivant.

Les soubassements imaginaires de la vision traditionaliste de Charles Doolittle Walcott

Les interprétations successives des fossiles du Schiste de Burgess soulèvent la question des interactions entre convictions personnelles et recherche. Charles Doolittle Walcott et Stephen Jay Gould sont deux scientifiques expérimentés et reconnus par leurs pairs. Pourtant, chacun d'entre eux semble s'intéresser en priorité à l'aspect du réel qui appuie le mieux ses convictions profondes. Sans oublier que le discours des chercheurs est toujours une construction, un discours sur le réel dont la complexité empêche qu'il soit saisi entièrement[20], et que non seulement les faits, les données du terrain, ne parlent pas d'eux-mêmes[21], mais que le discours qui s'élabore autour de ces faits n'est jamais parfaitement

remise en cause, notamment par des chercheurs français : Jean Vannier du Laboratoire Paléoenvironnements et paléobiosphère de l'Université Claude-Bernard Lyon-1 et Philippe Janvier du Laboratoire Paléodiversité : histoire et dynamique du Muséum National d'Histoire Naturelle. Haikouichthys, un poisson de 530 millions d'années découvert en Chine en 1999, serait ainsi le plus ancien vertébré connu.

[19] Stephen Jay Gould, *La vie est belle*, *op. cit.*, 380-381.

[20] Ce regard porté sur le réel est aussi le résultat d'une histoire des représentations et des pratiques scientifiques qui place la méthode expérimentale au cœur de la démarche des sciences à l'Age classique. Cf. à ce sujet, la mise au point effectuée par Jean-Marc Lévy-Leblond pour la Chaire Modélisations des Imaginaires de Telecom ParisTech et Rennes 2 : Jean-Marc Levy-Leblond, *Le grand écart, La science entre technique et culture* (Paris : Editions Manucius, 2013), 18-19.

[21] Cf. l'analyse, à partir d'un cas emblématique de l'histoire de la physique, de Dominique Pestre et Michel Atten, *Heinrich Hertz, l'administration de la preuve* (Paris : Presses universitaires de France, 2002).

dépouillé d'*a priori*[22]. Que peut apporter une approche par l'imaginaire à la compréhension de ce phénomène ? L'exemple de Walcott est en effet emblématique de cette influence de l'environnement social et culturel mais aussi de celle de l'imaginaire le plus fondamental sur l'intelligence – influence d'autant plus forte lorsque le sujet pense, à tort, être parfaitement maître de sa pensée.

L'imaginaire n'est pas un domaine annexe de la pensée. Il est bien au contraire le cadre dans lequel se déploie toute pensée, comme l'a montré Gilbert Durand tout au long de son œuvre[23]. Il existe plusieurs niveaux au sein de l'ensemble des manifestations de l'imaginaire. Le plus profond et le plus ancien peut se définir comme un ensemble de traits ou de plis mentaux hérités de l'évolution : le repli sur le groupe d'origine par exemple, la maîtrise des ressources ou encore l'investissement paternel et les stratégies destinées à prévenir l'infanticide, la coopération, etc[24].

Toutefois, l'imaginaire ne se limite pas seulement à l'héritage de l'évolution. Il est aussi constitué par les éléments fondateurs d'une culture, d'une civilisation, c'est-à-dire par les orientations privilégiées pour expliquer le monde. Il en va ainsi de l'imaginaire typiquement néolithique qui continue sans doute de marquer et de définir notre rapport au monde. L'interprétation de la place de l'homme au sommet de l'échelle du vivant en est peut-être l'un des aspects les plus déterminants et le moteur de certaines interprétations biaisées de l'histoire du vivant, à l'image de celle de Walcott.

Cette vision de l'homme comme créature suprême pourrait ainsi être l'un des résultats étonnants de la sédentarisation progressive qui s'observe à la surface du globe avec la néolithisation. D'un point de vue symbolique, la néolithisation se caractérise par l'installation de l'homme au centre du monde. Olivier Keller a proposé une explication du phénomène avec l'observation de la course du soleil. Celle-ci place tout observateur sur un

---

[22] Le nécessaire travail de réflexivité qui est exigé du chercheur sur ses propres opinions de même que la confrontation avec les pairs doit permettre de limiter l'influence des préjugés. Cf. notamment Baudouin Jurdant, « Communication scientifique et réflexivité », *Espaces réflexifs* [carnet de recherche], http://reflexivites.hypotheses.org/695 (consulté le 17 octobre 2012).

[23] « Il n'y a pas de coupure entre le rationnel et l'imaginaire [...] l'on peut assimiler la totalité du psychisme, dès qu'il décolle de l'immédiate sensation, à l'Imaginaire, et la pensée en sa totalité se trouve intégrée à la fonction symbolique. » *In* Gilbert Durand, *L'imagination symbolique* (Paris: Presses Universitaires de France, coll. « Quadrige Grands Textes », [1964], 2008), 88. Cf. également Gilbert Durand, *Les structures anthropologiques de l'imaginaire* (Paris: Dunod, [1969], 1992).

[24] Cf. *Biologie évolutive*, ed. Frédéric Thomas, Thierry Lefèvre et Michel Raymond, *op. cit.*

centre géométrique et mythique assimilé progressivement au centre du monde[25]. Ce positionnement central de l'homme s'accompagne d'un exil des divinités, exil qui va permettre à l'être humain de se penser en gardien de la création.

Tout le début de *Wonderful Life* est d'ailleurs consacré aux représentations erronées de l'évolution sous forme d'arbres partant du plus simple au plus complexe, de cônes ou d'échelles menant inévitablement à l'être humain et reposant sur le présupposé que ce qui est en bas est forcément ancien et correspondrait donc nécessairement à des formes de vie plus élémentaires[26]. Cette manière de représenter l'évolution semble ainsi une autre manifestation de cette conception de l'homme comme super créature qui serait non seulement maître du monde, mais aussi en charge de sa bonne marche. Il est alors tentant de voir dans l'attitude de Walcott une expression de cette vision de l'homme au sommet de la création qui, croyant exercer sa liberté, perpétue peut-être un comportement résultant d'un long parcours anthropologique entamé avec le Néolithique. Walcott appartient au genre humain, qu'il juge supérieur à toute autre espèce vivante et qu'il estime être en charge de l'ordre du monde. De surcroît, il est blanc et c'est un homme : il est donc bien, pour son époque, au sommet de la création[27].

Pour Walcott, tout semblait aller de soi : il était satisfait de sa contribution au savoir et à la société car ses choix le confortaient dans l'idée que l'être humain est au sommet de la création. Walcott était sans doute un grand scientifique mais son esprit n'était peut-être pas préparé à aller au-delà des dogmes qu'il avait adoptés et qui avaient contribué à sa réussite. Toutefois, plus que sa réussite sociale, il semble que sa foi et sa volonté de mettre la science au service de la religion aient été les éléments les plus déterminants dans son parcours. Ainsi, fidèle à ses croyances, il n'était pas prêt à interpréter autrement les fossiles extraordinaires qu'il avait sous les yeux. Pour lui, ces animaux étaient des témoignages de formes anciennes et donc forcément très primitives selon l'équation

---

[25] Cf. les travaux novateurs d'Olivier Keller qui viennent confirmer les intuitions de certains préhistoriens (Jacques Cauvin notamment), en particulier : Olivier Keller, *Une archéologie de la géométrie, Peuples paysans sans écriture et premières civilisations* (Paris: Vuibert, 2006).

[26] *Wonderful Life, op. cit*, chapitre 1 [mentionné plus haut en note 13].

[27] Toute la fin du XIXème siècle est traversée par l'obsession de la quête du chaînon manquant. De même, jusqu'à la Seconde Guerre Mondiale, les discours raciaux qui classent les être humains en fonction de leur couleur mais aussi les discours eugénistes qui s'inscrivent dans le cadre d'une recherche de « la race pure » sont légion. La position de Walcott est donc conforme à l'idéologie alors dominante en Occident.

traditionnelle établie entre ancienneté et simplicité, entendue comme pauvreté des formes[28].

Stephen Jay Gould entre vulgarisation et militantisme

Le même processus est-il à l'œuvre chez Stephen Jay Gould? En effet, la jubilation perceptible dans *Wonderful Life* n'est sans doute pas due au seul caractère scientifique de son travail. Stephen Jay Gould était un militant progressiste, très engagé contre les mouvements réactionnaires et intégristes. Prendre en défaut un ancien représentant de l'*Establishment* puritain et une figure très respectée du « panthéon » scientifique américain représentait certainement pour lui une grande source de plaisir. Ainsi, et sans remettre en question le sérieux de Gould, il est possible, à partir de son travail sur le Schiste de Burgess et de ses engagements personnels, de s'interroger également sur l'influence de son engagement dans la cité. Stephen Jay Gould, en effet, était révolté par les injustices sociales et d'une certaine manière, une grande partie de son travail, notamment son travail de vulgarisation, consiste à démontrer que l'état actuel des sociétés n'est pas le résultat inéluctable de l'évolution ni le prix à payer pour vivre en être humain[29]. Il voulait redonner à l'homme sa place dans la nature en tant qu'espèce parmi les autres dont l'apparition est plus liée à la contingence qu'à une quelconque « nécessité » ou projet divin.

Il y a donc clairement un parti pris dans sa démarche. De par sa volonté de contrer la vision traditionnelle de l'évolution, il a par exemple surestimé la disparité et la diversité au Cambrien. Il est ainsi probable que s'il avait abordé le problème en ayant moins le souci de discréditer la vision traditionnelle de l'évolution, il aurait sans doute été plus objectif. Très clairement, Stephen Jay Gould a instruit un dossier à charge contre Walcott et s'est surtout intéressé à ce qui confortait ses vues. Cependant, *Wonderful Life* est un ouvrage de vulgarisation, ce qui peut expliquer l'exagération des partis pris. Mais dans la mesure où Stephen Jay Gould défendait l'idée qu'il était possible d'écrire des livres de vulgarisation qui puissent intéresser à la fois le grand public et la communauté

[28] Cf. Charles Doolittle Walcott, "Evidence of primitive life: Annual Report of the Smithsonian Institution for 1915" (Washington, 1916), 235-255.

[29] Pour une synthèse de la pensée de Stephen Jay Gould, cf. Richard York and Brett Clark, *The science and humanism of Stephen Jay Gould* (New York: Monthly Review Press, 2011).

scientifique[30], la frontière entre science et engagement militant devient plus ténue.

Au-delà des contextes particuliers du début du XXème siècle et des années 1970 et 1980 aux Etats-Unis, l'exemple de la polémique de Burgess met en lumière l'influence des trajectoires personnelles dans les parcours et productions scientifiques de Walcott et Gould. La volonté de ce dernier de corriger son aîné, de surcroît une figure encore très respectée de la communauté scientifique aux Etats-Unis, relevait ainsi sans doute tout autant de l'engagement militant de Gould que de sa carrière de chercheur.

En se concentrant sur les critiques formulées par Gould à l'encontre de Walcott, cet article a cherché également à illustrer une application possible d'une approche par l'imaginaire pour cerner, dans ce cas précis, les éléments qui ont pu être décisifs, en dehors de toute considération strictement scientifique. La proposition qui est faite ici concerne ainsi la mobilisation d'éléments imaginaires agissant sur le temps long – l'installation au centre du monde et au sommet de la « création » avec la sédentarisation – et venant se combiner à un contexte social et à un parcours personnel particuliers.

Œuvres citées

Conway Morris, Simon. *The crucible of creation, The Burgess Shale and the Rise of Animals*. New York: Oxford University Press, 1998.

Simon Conway Morris, ed. *The deep structure of biology, Is convergence sufficiently ubiquitous to give a directional signal?* West Conshohocken, Pennsylvanie: Templeton Foundation Press, 2008.

Durand, Gilbert. *L'imagination symbolique*. Paris: Presses Universitaires de France, coll. « Quadrige Grands Textes », [1964], 2008.

—. *Les structures anthropologiques de l'imaginaire*. Paris: Dunod, [1969], 1992.

Gould, Stephen Jay. *Wonderful Life, The Burgess shale and the nature of history*. New York: W. W. Norton & Company, 1989.

—. La vie est belle, Les surprises de l'évolution, traduction par Marcel Blanc. Paris: Seuil, coll. Points Science, 1991.

—. *Rocks of Ages: Science and Religion in the Fullness of Life*. New York, Ballantine Publishing Group: 1999.

—. *The structure of evolutionary theory*. Cambridge (Massachussetts) and London: The Belknap Press of Harvard University Press, 2002.

[30] Stephen Jay Gould, *La vie est belle*, *op. cit*, 14.

Heams, Thomas *et al. Les mondes darwiniens. L'évolution de l'évolution*. Paris: Editions matériologiques, 2011.

Jurdant, Baudouin. "Communication scientifique et réflexivité", *Espaces réflexifs* [carnet de recherche], http://reflexivites.hypotheses.org/695 (consulté le 17 octobre 2012).

Keller, Olivier. *Une archéologie de la géométrie, Peuples paysans sans écriture et premières civilisations*. Paris: Vuibert, 2006.

Langlois, Cyril. « De Burgess au Gabon : les plus anciennes traces fossiles de pluricellulaires », http://planet-terre.ens-lyon.fr/planetterre/XML/db/planetterre/metadata/LOM-fossiles-pluricellulaires.xml (consulté le 29 janvier 2013).

Lecointre, Guillaume, ed. *Guide critique de l'évolution*. Paris: Belin, 2009.

Levy-Leblond, Jean-Marc. *Le grand écart, La science entre technique et culture,* Paris: Editions Manucius, 2013.

Pestre, Dominique et Atten Michel. *Heinrich Hertz, l'administration de la preuve*. Paris: Presses universitaires de France, 2002.

Shapin, Steven, and Simon Schaffer. *Leviathan et la pompe à air : Hobbes et Boyle entre science et politique*. Traduit de l'anglais par Thierry Plélat avec la collaboration de Sylvie Barjansky, Paris: La Découverte, 1993.

Steemans, Philippe *et al.* "Origin and Radiation of the Earliest Vascular Land Plants," *Science* April 17, 2009, Vol. 324, no. 5925, 353. DOI: 10.1126/science.1169659.

Thomas, Frédéric, Thierry Lefevre et Michel Raymond, eds. *Biologie évolutive*. Bruxelles: De Boeck, 2010.

Walcott, Charles Doolittle. "Evidence of primitive life: Annual Report of the Smithsonian Institution for 1915" Washington, 1916, 235-255.

Whittington, Harry Blackmore and Simon Conway Morris. *Extraordinary Fossil Biotas: Their Ecological and Evolutionary Significance*. London: Royal Society, 1985.

Yochelson, Ellis Leon. *Smithsonian Institution Secretary, Charles Doolittle Walcott*, Kent: Kent State University Press, 2001.

—. *Charles Doolittle Walcott, Paleontologist*. Kent: Kent State University Press, 1998.

York, Richard, and Brett Clark. *The science and humanism of Stephen Jay Gould*. New York: Monthly Review Press, 2011.

TRANSFORMATIONS: THE VISUAL INFLUENCE OF D'ARCY THOMPSON

MATTHEW JARRON

> For the harmony of the world is made manifest in Form and Number, and the heart and soul and all the poetry of Natural Philosophy are embodied in the concept of mathematical beauty.[1]

D'Arcy Thompson's *On Growth and Form* is one of the key works at the intersection of science and the imagination. Hailed as "the greatest work of prose in twentieth century science"[2], it is a book that has inspired scientists, artists and thinkers as diverse as Alan Turing, Richard Dawkins, Claude Lévi Strauss, Jackson Pollock and Norman Foster. It pioneered the science of biomathematics and its influence in architecture, anthropology, geography, cybernetics and many other fields continues to this day. Its author was a biologist and Classical scholar who strongly believed in interdisciplinary thinking and who saw life not as static and defined like the specimens in his museum but as ever-changing, constantly reacting to its environment.

D'Arcy Wentworth Thompson was born in Edinburgh, Scotland in 1860. His father (also named D'Arcy Thompson) was a teacher and the author of *Day-Dreams of a Schoolmaster*, a book in which he expressed his liberal and forward-looking views on education and his dislike of corporal punishment. He advocated equal education for both sexes and also wrote books of verse including *Nursery Nonsense* (1864) which became popular family reading. One of the poems in it, 'A Very Odd Fish', recounts D'Arcy senior's walks to the sea shore with his baby son.

[1] D'Arcy Thompson, *On Growth and Form*, 2nd edition (Cambridge: Cambridge University Press, 1942), 1096-7.
[2] Stephen Jay Gould, Introduction to *On Growth and Form*, Canto edition (Cambridge: Cambridge University Press, 1992), ix.

He was also a great Classicist and became Professor of Greek at Queen's College, Galway.

The young D'Arcy inherited his love of Classics from his father while his passion for natural history was inspired by his grandfather and one of his early teachers. He was encouraged to join the Edinburgh Naturalists' field club, and it was to them that he gave his first public lecture, on micro-organisms called Foraminifera, a topic that would continue to fascinate him later in life.

D'Arcy entered the University of Edinburgh in 1878, intending to pursue a medical degree. One of his tutors was the marine zoologist Charles Wyville Thomson and D'Arcy was given the opportunity to research specimens collected by him on the celebrated Challenger Expeditions. As his interest in biology developed, D'Arcy decided to give up on medicine and pursue a degree in Natural Science at Cambridge. Here he studied zoology under Francis Maitland Balfour and Adam Sedgwick, graduating with First Class Honours in 1883.

Just over a year later, D'Arcy was given the extraordinary opportunity to become a university professor at the age of just twenty-four. University College, Dundee had opened to students in 1883 with Chairs established in Chemistry, Mathematics & Natural Philosophy, Engineering & Drawing, English & Modern History and Classics & Ancient History. The College was a small one initially – just four converted houses and a new chemistry laboratory behind them – but it was eager to expand. In 1884, a Chair in Biology was founded and D'Arcy got the job.

Student numbers started small (D'Arcy had just two students in his first class), but quickly started to grow, particularly after the founding of the Medical School in Dundee, which D'Arcy also did much to bring about. As well as teaching biology students and medical students, he was soon teaching arts students from the Dundee Training College, giving him a taste for interdisciplinary teaching which he evidently relished.

D'Arcy quickly acquired a reputation as a wide-ranging and increasingly eccentric teacher. His successor in Dundee, Alexander Peacock recalled: "Fortunate were all who saw him use sketches, bits of paper and string, and soap bubbles to explain the mathematics of the honeycomb, the nautilus shell and such like recondite things."[3] One of his later students in St Andrews recalled:

[3] A. D. Peacock, MS of BBC lecture broadcast 1/5/1960 (University of Dundee Archive Services, URSF 2/12/3(14)).

> You didn't learn Zoology, but you were educated. You learned a smattering of Latin tags, and you appreciated the immeasurable beauty of the tiny Foraminifera, whose minute chalky shells make so much of our ocean floors. You learned of Aristotle's views on the sea-urchin: and the story of Palolo worms that congregate regularly in the warm seas off Samoa was a fairy story, not merely a useful reference to the migratory habits of the lower animals. Even anatomical differences had their beauty; the fourth tooth of a crocodile fitting into a notch in the upper jaw, and not into a pit as in the alligator, was a fact to be gloried in, not merely to be learned. One felt that God was the Creator, and the wonder was never lost. One knew the beauty of the whole animal kingdom, and one was not limited to learning minute details about single animals.[4]

D'Arcy admired the German universities for pursuing "the idea of Universality of Knowledge" and he deplored the fact that in Scotland "our Universities are limited to the teaching of a somewhat meagre array of subjects, of what is obviously necessary, of what will attract what we choose to all a respectable number of students, in a word, of what may be said to pay."[5] In an address given in 1903, he claimed:

> I grow more and more impatient of the customary syllabus and the conventional examination. The great principles of Biology are pleasant to teach and good to learn, but petty details are so much easier to examine upon. And I am irritated, if not bewildered, when during so many paralysed weeks or days before a University Examination my best pupils can only show a feverish interest in the sequence of a lobster's legs, or a restless anxiety as to the number of teeth proper to a cat... But just as I have gone thousands of times up and down the stairs of my laboratory, and do not yet know how many steps there be, so I have been going these many years up and down the fields of Natural History, and have acquired little memory for facts like these. I can ascertain them when I want, and I am content to do so.[6]

D'Arcy passionately believed in giving students as great a breadth of knowledge as possible, telling them:

---

[4] A. H. Harris, MS of undated talk "An Aristocrat of Learning" (University of St Andrews Library Special Collections, MS 50166/1).

[5] D'Arcy Thompson, "Address delivered at the Opening of Session 1903-4", The College (December 1903): 9.

[6] Thompson, "Address", 7-8.

> if you dream, as some of you, I doubt not, have a right to dream, of future discoveries and inventions, let me tell you that the fertile field of discovery lies for the most part on those borderlands where one science meets another. There is a cry in the land for specialisation… but depend upon it, that the specialist who is not reinforced by a breadth of knowledge beyond his own specialty is apt very soon to find himself only the highly trained assistant to some other man... Try also to understand that though the sciences are defined from one another in books, there runs through them all what philosophers used to call the commune vinculum, a golden interweaving link, to their mutual support and interpretation.[7]

It would be precisely this interweaving link that D'Arcy would demonstrate so powerfully in his masterpiece *On Growth and Form*. But with so much work involved in the running of his department, building up an extensive Natural History Museum in the College and teaching both day and evening classes, D'Arcy struggled to find time for his own research. The specimens he was acquiring for the Museum provided the inspiration for a series of research papers published 1888-90 under the title *Studies from the Museum of Zoology*, but only five of the twelve papers were D'Arcy's work and he was forced to spend far more time helping his students and assistants with their research rather than doing his own.

It was only when D'Arcy's teaching and administrative load lightened that he was able to commence work on the writing of *On Growth and Form*. Although it was not published until 1917, his final year in Dundee, D'Arcy had been thinking about it for many years. In 1889 he wrote to one of his students: "I have taken to Mathematics, and believe I have discovered some unsuspected wonders in regard to the Spirals of the Foraminifera!"[8]

D'Arcy became increasingly convinced that the laws of mathematics could be used to explain the growth and form of living organisms, but when he first shared these ideas with others the result was discouraging. "I confess I am not very much attracted by the line of work, and doubt if it's likely to be very fruitful," his old Cambridge tutor Michael Foster wrote in 1894. "I suppose everyone must admit that there are 'laws of growth'… but after all one does not feel sure how far this is really admitted".[9] One of D'Arcy's assistants, Doris Mackinnon, later recalled that "he had no thought of writing what was in his mind… he would walk up and down

[7] Thompson, "Address", 9.

[8] D'Arcy Thompson, Letter to Mary Lily Walker 18/10/1889 (University of St Andrews Library Special Collections MS 44464).

[9] Ruth D'Arcy Thompson, *D'Arcy Wentworth Thompson: The Scholar-Naturalist* (Oxford: Oxford University Press, 1958), 89-90.

the Laboratory thinking his thoughts aloud and discussing his 'heresies' with her."[10] It would be 1908 before he published in detail on the topic – a paper in *Nature* on 'The Shape of Eggs and the Causes which determine them'. In 1911 he raised the subject at the British Association meeting in Portsmouth, claiming that "the form of an object is a 'diagram of Forces', – in this sense, at least, that from it we can judge or deduce the forces that are acting or have acted upon it".[11] This powerful visual metaphor, restated in *On Growth and Form*, would become one of his most influential ideas.

Finally in 1915 D'Arcy's various ideas were assembled into book form and he sent it to Cambridge University Press, who noted that "the little book on 'Growth and Form' …has now turned out to be a work on a much larger scale".[12] D'Arcy drew extensively on the resources available to him in Dundee, including the specimens in his Museum and the expertise of his colleagues – particularly the Physics professor William Peddie and the Engineering professor Thomas Claxton Fidler. D'Arcy's assistant Doris Mackinnon and one of his former students Helen Ogilvie drew many of the original diagrams for the book.

Partly due to wartime paper shortages and partly due to D'Arcy's insistence on numerous last-minute changes, *On Growth and Form* was finally published in 1917 and attracted immediate attention from a wide range of sources. Although only five hundred copies were printed, reviews quickly appeared in publications as diverse as *Country Life*, the *Times Literary Supplement*, *Veterinary Review* and *Engineering* – D'Arcy thought that the latter was the best one he'd received. CUP could only afford a small number of review copies, and in November they wrote to D'Arcy to say that the *Journal of Decorative Art* had asked for a copy. D'Arcy replied to say: "It does not seem to me that it is a publication which is of much interest or importance to us", clearly suggesting that he had no idea of the kind of artistic significance his work would have.[13]

The review in *Nature* stated: "This book, at once substantial and stately, is to the credit of British Science and an achievement for its distinguished author to be proud of. It is like one of Darwin's books, well-considered, patiently wrought-out, learned and cautious – a disclosure of

---

[10] Ruth Thompson, 161.

[11] D'Arcy Thompson, "Magnalia Naturae; or The Greater Problems of Biology", *Science* volume 34, 875 (1911): 423.

[12] A. R. Waller, Letter to D'Arcy Thompson 25/5/1915 (University of St Andrews Library Special Collections MS 42523).

[13] D'Arcy Thompson, Letter to Cambridge University Press 15/11/1917 (University of St Andrews Library Special Collections MS 42633).

the scientific spirit."[14] The comparison to Darwin is interesting, given that many saw the book as an opposition to Darwinian evolution. D'Arcy certainly admired Darwin greatly, and managed to persuade him to write the preface to D'Arcy's first publication, his translation of Hermann Muller's *The Fertilisation of Flowers*, published in 1883 shortly after Darwin's death. But at the British Association meeting in 1894, D'Arcy gave a paper entitled 'Some Difficulties in Darwinism'. It was never published, but *Nature* summarised its content:

> He doubts the efficacy of the struggle for existence in the case of humming-birds, &c., and in these cases he regards the profusion of forms, colours, and other modifications as due merely to laws of growth, and thinks that growth may be more exuberant in the absence of struggle and hardship.[15]

Writing about *On Growth and Form*, D'Arcy claimed: "I have tried to make it as little contentious as possible. That is to say where it undoubtedly runs counter to conventional Darwinism, I do not rub this in, but leave the reader to draw the obvious moral for himself."[16] The "obvious moral" was that Darwin was wrong in seeing evolution purely as a slow, gradual process dictated by natural selection. D'Arcy's Theory of Transformations, the most famous and radical chapter in the book, proposed that physical forces could cause a transformation from one species into another based on mathematical principles. It sought to demonstrate "that discontinuous variations are a natural thing, that 'mutations' – or sudden changes, greater or less – are bound to have taken place".[17]

To demonstrate this D'Arcy drew on his knowledge of art, taking as his starting point Albrecht Dürer's work on geometry and proportion (Figure 5-1):

[14] J. Arthur Thomson, Review of On Growth and Form, by D'Arcy Thompson, *Nature* volume 100, 2498 (1917): 21.

[15] Unknown author, "Biology at the British Association", *Nature* volume 50, 1296 (1894): 435.

[16] Ruth Thompson, Scholar-Naturalist, 161-2.

[17] D'Arcy Thompson, *OG&F*, 1095.

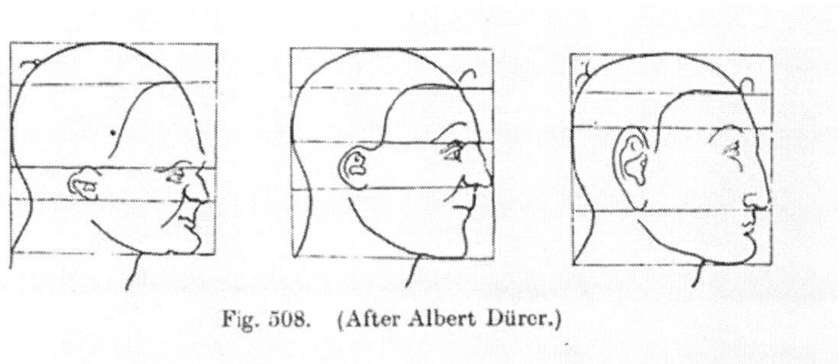

Figure 5-1 : Diagram after Albrecht Dürer from *On Growth and Form* (Cambridge University Press, 1917)

In his *Four Books on Human Proportion* (1512-28), according to D'Arcy, "the manner in which the human figure, features, and facial expression are all transformed and modified by slight variations in the relative magnitude of the parts is admirably and copiously illustrated."[18] As well as dealing with transformations of the human form, Dürer's work on proportion also encompassed his theories of ideal beauty, claiming that while beauty was ordered by as-yet undefined laws, it was not an objective concept but was based on the infinite variety of nature. This must have struck a chord with D'Arcy, whose work aimed to reveal the fundamental laws that governed the extraordinary variety of nature's patterns. In contrast to most scientific works, *On Growth and Form* frequently describes nature in terms of its aesthetic qualities.

D'Arcy combined Dürer's techniques with those of René Descartes, using his method of co-ordinates to turn Dürer's proportional drawings into scientific diagrams. By applying a Cartesian grid to the form of an animal or a part of an animal, and subjecting it to increasingly complex mathematical transformations, he was able to demonstrate that laws of growth rather than evolution could be used to explain the different forms of related species (Figure 5-2).

[18] D'Arcy Thompson, *OG&F,* 1053.

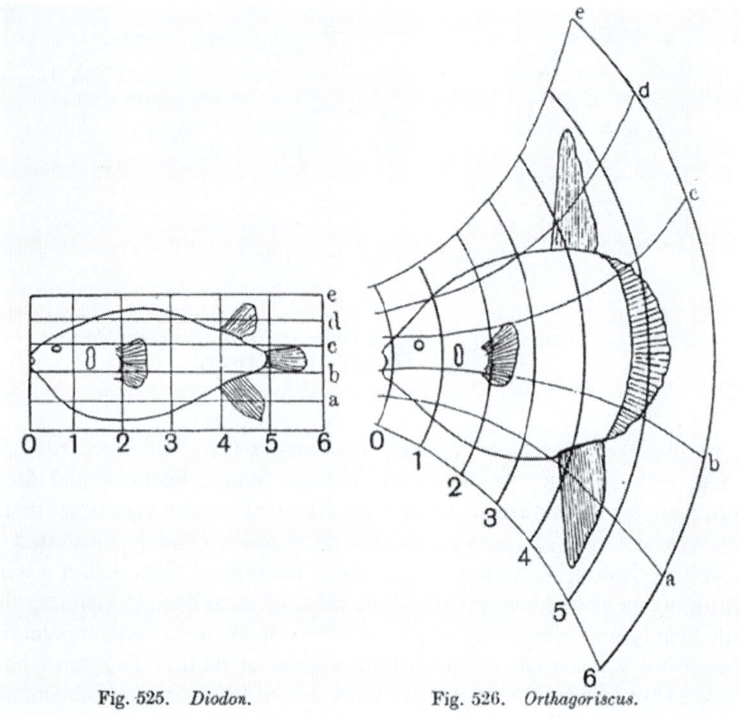

Figure 5-2 : Diagram from *On Growth and Form* illustrating the mathematical relationship between a porcupine fish and a sunfish (Cambridge University Press, 1917).

For much of the twentieth century, D'Arcy's ideas seemed to run counter to biology's increasing focus on genetics. Peacock claimed that "D'Arcy's originality and versatility made him unclassifiable by formal standards and his individualism inhibited his founding of a school for the development of his ideas."[19] But the ideas lingered, and found new followers, who styled themselves developmental biologists rather than evolutionary ones. They included Julian Huxley, C H Waddington and the father of modern computing, Alan Turing.

On Growth and Form shows how complex biological systems can be understood by fundamental mathematical principles, and as such D'Arcy's ideas provided the basis for the development of modern systems theory.

[19] Peacock, BBC lecture.

Turing's experimental thinking machines rely on similar ideas, and in the early 1950s Turing turned his attention to biology with his landmark paper 'The Chemical Basis of Morphogenesis'. This began an investigation into animal patterning that D'Arcy had raised the possibility of but had not himself attempted. Turing's work demonstrated how mathematical equations applied to different chemicals in the body that he called morphogens could create visual patterns such as this. In 1988 his ideas were taken further by James Murray in his celebrated paper 'How the Leopard Gets its Spots', which proposed that a single mathematical model called reaction-diffusion could possibly explain most if not all of the wide variety of animal coat markings found in nature.

By this time the two opposing biological camps were finally being brought back together and the growth of evolutionary-developmental biology (popularly known as evo-devo) caused D'Arcy's work to be looked at again by the evolutionists and geneticists that had hitherto dismissed it. D'Arcy had been all too aware of the limits of the mathematics of the time to undertake the task he had begun, admitting that:

> The organic forms which we can define, more or less precisely, in mathematical terms, and afterwards proceed to explain and to account for in terms of force, are of many kinds, as we have seen; but nevertheless they are few in number compared with Nature's all but infinite variety.[20]

But he believed that this variety was the result of "numerous, though perhaps still very simple, causes" and the work of Turing, Murray and many others have helped to define these. Through the development of ideas such as chaos theory, fractals and self-organising systems, D'Arcy's view of nature is becoming ever more understood. Developments in mathematical modelling techniques have allowed the theories of *On Growth and Form* to be tested scientifically for the first time, and even arch-geneticists like Richard Dawkins now freely acknowledge D'Arcy's significance. Prof Pedro Miramontes of the National Autonomous University of Mexico, one of the many institutions around the world now specialising in biomathematics, recently stated that D'Arcy "simply will be the most influential character in the future of biology."[21]

D'Arcy's influence has gone far beyond biology. His importance to systems theory was also recognised by Norbert Wiener, the pioneer of cybernetics, acknowledged in his landmark book *Cybernetics or Control*

---

[20] D'Arcy Thompson, *OG&F*, 1030.

[21] Personal communication with author, 2011.

and Communication in the Animal and the Machine (1948). It also proved an inspiration to the anthropologist Claude Levi-Strauss, who refers to D'Arcy in his important book on *Structural Anthropology* (1963). Many modern anthropologists view *On Growth and Form* as an important source text, and today's forensic anthropology also uses techniques derived from the Theory of Transformations. Within geography, the book has also proved a significant influence, for example in the algorithms used to create the Worldmapper series of cartograms in 2007, a way of visualising data on world poverty and health using D'Arcy's transformation diagrams as their starting point.

In architecture and engineering, *On Growth and Form* has inspired creators and practitioners from Le Corbusier and Mies van der Rohe to Norman Foster and Cecil Balmond. The famous diagrams comparing the Forth Bridge to animal and dinosaur skeletons made many designers think about structures in a new way, and D'Arcy's work on the mechanical efficiency of soap bubbles and the structural tension of dragonfly wings directly inspired the development of lightweight structures such as Buckminster Fuller's geodesic domes and Frei Otto's Olympic stadium in Munich. In particular, the idea of organisms constantly subject to transformation through external pressures, lies at the root of the theories of emergence, organic architecture and natural design that form a fundamental part of current architectural theory.

Like architects, artists were much quicker than scientists in recognising the potential of D'Arcy's ideas. The sculptor Henry Moore discovered *On Growth and Form* as a student and its influence can undoubtedly be seen in the series of 'Transformation Drawings' that he created in the early 1930s, which use overlapping pencil lines to depict organic forms apparently in the act of morphing from one state to another. These drawings then informed the sketches Moore created while working out ideas for his sculptures, and similar transitory or chimera-like biological forms can also be seen in his early print work.

It was also in D'Arcy's book that Moore first learnt of the great German poet Goethe's work in biology. It was Goethe that had introduced the term 'morphology' to define the study of form in living organisms, though (as D'Arcy was quick to note) he "ruled mathematics out of place in natural history."[22] Moore was less interested in mathematics but more in the ever-changing possibilities of natural forms, and indeed the term 'biomorphism' (used to describe the "fluid, organic, psychologically

[22] D'Arcy Thompson, *OG&F*, 2.

potent shapes" of artists such as Jean Arp and Joan Miró) was originally used to describe Moore's work.[23]

Moore was soon discussing *On Growth and Form* with the influential art critic Herbert Read, who introduced it to the St Ives circle of artists including Paul Nash, Barbara Hepworth and Naum Gabo. D'Arcy's work soon found its way around other avant-garde artists, including Laszlo Moholy-Nagy who was based in Britain in 1935-7 and referred to D'Arcy in his posthumously published book *Vision in Motion* (1947).

The second edition of *On Growth and Form* was also enthusiastically taken up by a group of students at the Slade School of Art in London in the 1940s, including Nigel Henderson, Richard Hamilton, Eduardo Paolozzi and the Dundee-born William Turnbull. In 1951, Hamilton staged an influential exhibition called *Growth and Form* at Herbert Read's new Institute of Contemporary Art (ICA). In 2003, Hamilton described the impact of D'Arcy's book:

> It opened my eyes to the idea that the world is as it is because it must follow certain mathematical principles. He describes phenomena, like the spirals on a cauliflower, so that you see it has to be this way, because time and the activity of growth are related. It is a beautiful notion; it hangs in the mind as an explanation of life itself. Organisms start as a single cell, and then this cell splits into two, then two into four and then 8 and 16 and 32 and so on… That this procedure should create something as complex as a human being is magical. That it should produce this extraordinary object at the end is one of the mysteries of existence. Thompson describes it in that way.[24]

Along with another D'Arcy enthusiast, Victor Pasmore, Hamilton would go on to become an important teacher at the Department of Fine Art in King's College, Newcastle. The new Basic Design Course they introduced proved to be hugely influential on art schools around the country, and it included *On Growth and Form*-based exercises. Through teaching such as Hamilton's and exhibitions at the ICA, the work that D'Arcy wrote in Dundee in the 1910s became fully integrated into British contemporary art years after his death. Martin Kemp has recently claimed:

[23] David Cohen, "Who's Afraid of Henry Moore?" Henry Moore: Sculpting the 20th Century, ed. Dorothy Kosinski (New Haven: Yale University Press, 2001), 264.

[24] Richard Hamilton, interview by Hans-Ulrich Obrist in Tate Magazine 4 (2003), accessed 18 December 2009, http://www.tate.org.uk/magazine/issue4/popdaddy.htm.

"In the worlds of art and architecture, D'Arcy Thompson probably exercised a greater impact than any scientist of the 20th century".[25]

D'Arcy himself probably had only a slight realisation of the creative impact his work had had, though Herbert Read had written to him towards the end of his life, telling him: "you have built the bridge between science and art. The danger lies in one's impulse to cross it too impetuously, without a proper understanding of the country on the other side."[26] D'Arcy was one of the few people who could cross this bridge fearlessly, having a strong interest in art as well as biology and Classics.

There are two important influences to look at in trying to understand D'Arcy's own artistic interests. Firstly, D'Arcy's childhood in Edinburgh was an unusually cultured one, with several of Scotland's leading artists as close family friends. These included the landscape painter Sam Bough, the portrait painter Robert Herdman, the painter and photographic pioneer David Octavius Hill, the historical and mythological painter Joseph Noel Paton (whose son Diarmid was at school with D'Arcy and would later become a notable physiologist) and the painter and sculptor George Clark Stanton, who was married to D'Arcy's aunt.

Secondly it's important to note the influence that D'Arcy's surroundings in Dundee must have had. The city was particularly vibrant in the visual arts at that time, and was described by one leading Scottish painter as "perhaps the most vital centre of art appreciation in Scotland."[27] Since opening in 1873 the Albert Institute (now The McManus: Dundee's Art Gallery & Museum) had been home to major blockbuster art exhibitions that were the biggest of their kind outside London and attracted artists across the country. The extraordinary success of the jute industry meant that Dundee was home to numerous wealthy businessmen keen to buy contemporary art, often commissioning works direct from leading painters such as William McTaggart. At the same time, Dundee was home to one of the country's largest newspaper industries, and in 1880 the *Dundee Advertiser* became the first daily paper in Britain to employ a regular staff artist. Art education was also expanding at that time thanks to the Dundee Technical Institute, out of which would grow today's Duncan of Jordanstone College of Art & Design.

All this gave a huge impetus to the city's own artists, and in 1890 the Dundee Graphic Arts Association was formed to promote their work,

---

[25] Personal communication with author, 2011.

[26] Herbert Read, Letter to D'Arcy Thompson, undated but probably 1944 (University of St Andrews Library Special Collections MS 25480).

[27] Dundee Graphic Arts Association Annual Report 1890 (Dundee Central Library Local History Centre).

holding annual exhibitions from 1893. D'Arcy would certainly have known of their work through the active engagement of his close friend and colleague Patrick Geddes, who joined D'Arcy's department as Professor of Botany in 1888 and who shared D'Arcy's wide-ranging interests. In 1891 he commissioned members of the Graphic Arts Association to create a series of paintings of plants in their natural settings to use as teaching aids for his students. It may have been through this scheme that Geddes first met one of the leading Dundee painters John Duncan, who became an important figure in Geddes's Celtic Revival movement in art and culture, painting richly decorative works based on traditional Celtic folktales, and acting as lead illustrator on Geddes's seasonal publication *The Evergreen*.

One of the most promising of the so-called Celtic School of artists in Dundee was George Dutch Davidson, who shared a studio with John Duncan and created surreal but beautiful decorative work such as *The Hills of Dream* (1899, Dundee Art Galleries & Museums) and *The Tomb* (1900, Dundee Art Galleries & Museums). In 1900, D'Arcy commissioned Davidson to design a series of three decorative panels for his study, the subjects to be Orpheus, Neptune and Juno, each surrounded by animals. Tragically, Davidson died (aged just twenty-one) before these could be carried out, leaving only a pencil study for the Orpheus panel (now with Dundee Art Galleries & Museums). In its use of the four elements, the twelve animals of the zodiac and possible references to the Golden Ratio, it shows that Davidson clearly understood D'Arcy's interests in natural history, mathematics and Classics, and is arguably the first example of an artist drawing inspiration from D'Arcy's work, albeit seventeen years before the publication of *On Growth and Form*.

D'Arcy clearly had an interest in supporting local artists. Soon after taking up his post he began employing James Eadie Reid, one of the early illustrators for the *Dundee Advertiser*, to create drawings and diagrams for him. He soon got to know the other members of the Graphic Arts Association, which in 1904 was renamed the Dundee Art Society. At the opening of one of their annual exhibitions he gave the introductory speech, referring to "my artist friends".[28] Throughout his life D'Arcy continued to befriend and correspond with artists. As well as his Dundee friends, he also knew Dame Laura Knight well, and friendly correspondence survives with, among others, the English painter William Rothenstein, the Irish-based painter Elizabeth Rivers and the Scottish painters John Bowie and Keith Henderson. Other notable figures from the art world that D'Arcy corresponded with included Stanley Cursiter of the National Gallery of

[28] D'Arcy Thompson, Typed MS of notes for address to the Dundee Art Society (University of St Andrews Library Special Collections MS 45704).

Scotland; Daniel V Thompson of the Courtauld Institute; Paul Lambotte of the Musée des Beaux-Arts in Paris; and Beaumont Newhall of the Museum of Modern Art in New York.

In his address to the Dundee Art Society, D'Arcy claimed: "I have neither technical nor historic knowledge of art to impart to you, and nothing in the world but my share of that love of art and of natural beauty which brings us all here together."[29] In one area, however, D'Arcy did have a detailed knowledge of art, one which combined his knowledge of zoology and Classics. In 1894 he presented a paper to the Royal Society of Edinburgh on 'Bird and Beast in Ancient Symbolism' and another two years later on 'The Emblems of the Crab in Relation to the sign Cancer'. In the latter he described "the representations of animal form in Ancient Art" as "a leisure-time study of mine".[30] In 1896 Patrick Geddes had written to him saying "I am glad to see that you are holding forth on Symbolism, and write to remind you of my suggestion that you should contribute a short paper to the *Summer* [issue of the] *Evergreen* on this subject."[31] Sadly D'Arcy never took Geddes up on this invitation.

D'Arcy's fascination with Classical literature and art was not just a spare-time hobby, but produced several important publications including *A Glossary of Greek Birds* (1895), *A Glossary of Greek Fishes* (1947) and a translation of Aristotle's *Historia Animailum* (1910) with detailed annotations. Given his obsession with placing ancient literature and art into a scientific context, we can imagine that texts such as Ovid's *Metamorphoses* (referred to frequently in his *Glossaries*) and the many chimeric forms of early animal art would have had a significant influence on D'Arcy's interest in morphogenesis and his Theory of Transformations.

The imaginative possibilities of biomorphism are at the root of many artists' fascination with *On Growth and Form*. D'Arcy's transformation diagrams are referred to in Jean Clair's recent book *Hubris – La fabrique du monstre dans l'art moderne* (2012), in which he links them with the neurosurgeon Wilder Penfield's cortical homunculus drawings, describing « ces jeux de transformations cartesiennes qui unissent les fantasmagories d'une evolution des especes a la rigueur des mathématiques et dans lesquelles la vie, le desir et la mort sont absents. »[32]

[29] D'Arcy Thompson, Dundee Art Society.

[30] D'Arcy Thompson, "The Emblem of the Crab in Relation to the sign Cancer", *Transactions of the Royal Society of Edinburgh* volume 39, 3 (1896): 603.

[31] Patrick Geddes, Letter to D'Arcy Thompson 25/1/1896 (University of St Andrews Library Special Collections MS 16359).

[32] Jean Clair, *Hubris : la fabrique du monstre dans l'art moderne* (Paris: Editions Gallimard, 2012), 70. The text can be translated as "these Cartesian transformation

This unity of phantasmagoria and rigour can be seen in a recent research project, *Coded Chimera*, by sculptor Bruce Gernand. Thanks to funding from the UK Arts & Humanities Research Council, Gernand (a Senior Research Fellow at Central St Martins College of Art & Design, London) sought to explore new ways of sculptural form-making using D'Arcy's biomathematical legacy as a starting point:

> I wanted to locate myself at a nexus of the biological, the computational and the aesthetic, and through the concept of morphogenesis – the study of biological dynamics – make work which embodied the intertwining of these vectors.[33]

Like other sculptors drawn to *On Growth and Form*, Gernand's interest was in the process of form-making as much as its final outcome:

> I am intrigued by the relationship between an activity and its outcome, between means and ends, and the need to resolve an activity into a finished work. With sculpture we deal with the facts and properties of materials; these require an on-going investment in time and effort, which reinforces the propensity to speculate on process. To be sure, we stalk and are haunted by poetic images. Sculptors break their fingernails on them.[34]

This led Gernand to morphogenesis as a model for exploring sculptural form-making, which in turn opened up an anthropological dimension to the project – "our relationship to animals and our relation to the mediated images of animal form."[35] As D'Arcy's own research showed, mankind has been making visual representations of animals for millennia, and many of them had complex symbolic meanings for their image-makers. Gernand was particularly drawn to the profusion of chimeric forms in sculpture:

> The chimera has cultural topicality – a hybridization that confounds the conventions of taxonomy which are being contested from a variety of positions and agendas. As a composite of difference the chimera generates new forms, provoking new thoughts and experience.[36]

---

games that unite the phantasmagoria of species evolution with the rigour of mathematics, and in which life, desire and death are absent."

[33] Bruce Gernand, Alan Blackwell and Norman MacLeod, *Coded Chimera* (Cambridge: Crucible Network, 2011), 3.

[34] Gernand, *Coded Chimera*, 5.

[35] Gernand, *Coded Chimera*, 5.

[36] Gernand, *Coded Chimera*, 6.

D'Arcy's transformation diagrams thus became the starting point for a project to use computer modelling techniques to morph different animal forms as a way of exploring sculptural form-making. Rather than revealing the mathematical relationship between two related forms, Gernand sought to use animals that were culturally rather than taxonomically connected, such as tortoise and hare, or cat and crocodile. What attracted him to D'Arcy's diagrams "was their apparent flexibility and the metaphoric possibilities for the reciprocities between object and environment."[37] Indeed, in *On Growth and Form*, D'Arcy drew sculptural analogies, acknowledging: "We might suppose that by the combined action of appropriate forces any material form could be transformed into any other: just as out of a 'shapeless' mass of clay the potter or the sculptor moulds his artistic product."[38]

The project involved both the physical and the virtual – taxidermied museum specimens were scanned in 3D and the digital forms subjected to various computer morphing techniques, which in turn generated rapid prototype models of particular stages in the transformation process. Gernand worked with Prof Norman MacLeod of the Natural History Museum in London and Alan F Blackwell of the University of Cambridge Computer Laboratory to realise the project, each approaching it from a different viewpoint. As Blackwell described it:

> Bruce's interest in the chimera could be considered… as a taxonomic challenge, in the sense that it described a solution reached by no evolutionary path. But conventional artistic descriptions of chimerae, mixing the forms and features of familiar beasts, cannot be recovered simply by simulating alternative evolutionary paths. When described computationally, our challenge was to take detailed scans of real animals, and from these deduce the detailed shape of a new chimeric form.[39]

This involved creating specific morphing tools that recognised those aspects of each animal that held particular human meanings, to prevent the results becoming simply amorphous blobs. This act of subjectively defining meaning linked the computer programming to the experience of both taxonomist and artist. Again, D'Arcy's "diagram of forces" can be seen in the physical forms that resulted. "Rather than looking at the chimera in a literal, prosaic way," Gernand recalled, "I responded more to

[37] Gernand, *Coded Chimera*, 22.

[38] D'Arcy Thompson, *OG&F*, 1034.

[39] Blackwell in Gernand, *Coded Chimera*, 13.

the renderings which expressed a fluidity, a transforming from one condition to another, a sense of becoming another, an other."[40]

The resulting sculptures, together with documentation of the processes involved, were exhibited in the Museum of Zoology, University of Cambridge in 2011 and the Tower Foyer Gallery, University of Dundee in 2012 (Fig. 4). The latter exhibition was part of an on-going project exploring the artistic influence of D'Arcy Thompson, with funding from the Art Fund's RENEW scheme. The project seeks to build a collection of relevant historic and contemporary artworks for the University of Dundee Museum Collections, to introduce new artists to the ideas of D'Arcy Thompson and to encourage greater use of the D'Arcy Thompson Zoology Museum at the University by artists. Further exhibitions and events will be held over the next two years, helping to ensure that D'Arcy's work continues to act as an artistic inspiration as well as a scientific one.

Works cited

Clair, Jean. *Hubris : la fabrique du monstre dans l'art moderne*. Paris: Editions Gallimard, 2012.

Gernand, Bruce, Blackwell, Alan and Noram MacLeod. *Coded Chimera*. Cambridge: Crucible Network, 2011.

Hamilton, Richard, interview by Hans-Ulrich Obrist in *Tate Magazine* 4 (2003), accessed 18 December 2009, http://www.tate.org.uk/magazine/issue4/popdaddy.htm.

Kosinski, Dorothy, ed. *Henry Moore: Sculpting the 20th Century*. New Haven: Yale University Press, 2001.

Thompson, D'Arcy. "The Emblem of the Crab in Relation to the sign Cancer", *Transactions of the Royal Society of Edinburgh* volume 39, 3 (1896): 603-11.

—. "Address delivered at the Opening of Session 1903-4", *The College* (December 1903): 4-10.

—. "Magnalia Naturae; or The Greater Problems of Biology". *Science* volume 34, 875 (1911): 417-428.

—. *On Growth and Form,* second edition. Cambridge: Cambridge University Press, 1942.

—. *On Growth and Form,* Canto edition. Cambridge: Cambridge University Press, 1992.

Thompson, Ruth D'Arcy. *D'Arcy Wentworth Thompson: The Scholar-Naturalist*. Oxford: Oxford University Press, 1958.

---

[40] Gernand, Coded Chimera, 7.

Thomson, J. Arthur. Review of On Growth and Form, by D'Arcy Thompson, *Nature* volume 100, 2498 (1917): 21-2.
Unknown author, "Biology at the British Association", *Nature* volume 50, 1296 (1894): 433-6.

L'INFLUENCE DE LA PSEUDO-SCIENCE DANS LA LITTERATURE FANTASTIQUE : UNE CHIMERE SCIENTIFIQUE DU XIXEME SIECLE

ISABELLE PERCEBOIS

> Nos chimères sont ce qui nous ressemble le mieux
> —Victor Hugo, *Les Misérables*.

Si l'on parle volontiers de « la » Science au XIXème siècle, mère du progrès technique, l'emploi du singulier se révèle trompeur : une fracture apparaît entre la science positive défendue par Auguste Comte, et celle que l'on désigne dès cette époque par les termes de « pseudo-science », « science molle » ou encore « fausse science ». Alors que le positivisme triomphant s'érige en gardien de la vérité, les disciplines telles que le magnéto-spiritisme, la phrénologie ou la physiognomonie sont taxées de charlatanisme et d'hérésie. Elles sont soudain mises « à *l'index* de la science moderne »[1], comme l'explique dès 1859 Louis Figuier, dans son *Histoire du merveilleux*. Pourtant, le peuple se passionne pour ces théories extraordinaires qui promettent de révéler la clef des mystères de l'univers ; tous veulent croire en ces « chimères » et au progrès humain, qui semble désormais illimité. En devenant un phénomène de société, la pseudo-science devient également un phénomène littéraire, inspirant les écrivains du XIXème siècle par son caractère magique et fascinant. Et c'est dans le fantastique qu'elle trouve sa pleine expression, en offrant aux auteurs le moyen d'expliquer l'inexplicable et d'unir enfin la science et la fiction.

---

[1] Louis Figuier, *Histoire du merveilleux*, vol.III (Paris : Hachette et Cie, 1860), 343.

1- Les pseudo-sciences, des monstres enfantés par la science et la fiction :

Définir ce qu'est la pseudo-science n'est pas chose facile car ce titre englobe une multitude de spécialités insolites, qui prêtent à rire de nos jours : en effet, quel médecin songerait désormais à établir un diagnostic en fonction de la forme du crâne de son patient, ou d'un quelconque déséquilibre de son « fluide vital » ? L'essai d'Henri-Géry Hers intitulé *Science, non-science et fausse science*, propose de définir les pseudo-sciences comme des « doctrines créées de toutes pièces dans l'imaginaire de l'un ou l'autre visionnaire et [...] considérées par leurs adeptes comme des lois de la nature »[2]. Cette formulation des plus précises met en évidence le caractère profondément fictif de ces théories, proches de l'univers de la magie et du rêve. Dès la fin du XVIIIème siècle, nombreux sont les savants qui voient en elles des « chimères », de pures créations de l'esprit, sans fondement tangible, incapables de passer l'épreuve de l'expérience. En outre, le mot de « chimère » ressurgit alors sous la plume d'hommes de science tels qu'Antoine Lavoisier. Le célèbre chimiste s'emporte notamment contre les adeptes du magnétisme animal et déclare la doctrine de Franz Anton Mesmer « parfaitement chimérique »[3]. La physiognomonie, qui entend déterminer « l'intérieur » de l'homme d'après son « extérieur » (autrement dit son apparence), n'échappe pas non plus à cette accusation. Les critiques sont tellement virulentes à l'encontre de cette doctrine, qu'elles obligent ses partisans à répliquer à l'écrit, comme le fait Johann Kaspar Lavater dans *La Physiognomonie ou l'art de connaître les hommes d'après les traits de leur physionomie* : « Mais pourquoi, me demandera-t-on, la plupart des méchants se déclarent-ils publiquement contre la physiognomonie ? – Pourquoi ? Parce qu'ils y croient secrètement. [...] Il est de leur plus grand intérêt de rejeter cette science comme une chimère et d'en faire un objet de plaisanterie »[4]. De fait, ce serait une erreur de croire que les pseudo-sciences ne regroupent qu'une poignée d'illuminés plus magiciens que savants, car au XIXème siècle, les esprits les plus sérieux se penchent sur les questions mystérieuses des fantômes ou des tables tournantes, qui connaissent une vogue sans précédent en Europe. Les expressions de « sciences occultes »

---

[2] Henri-Géry Hers, *Science, non-science et fausse science, réflexion sur les chemins de la connaissance* (Paris : l'Harmattan, essai,1998), 113.

[3] Mots d'Antoine Lavoisier cités dans *L'Antimagnétisme* de J.J. Paulet (Paris : Desennes; 1784), 249.

[4] Johann Kaspar Lavater, *La Physiognomonie ou l'art de connaître les hommes d'après les traits de leur physionomie* (Delphica : l'Age d'homme, 1998), 11.

et « fausses sciences », couramment utilisées comme synonymes de pseudo-sciences, soulignent par leur caractère oxymorique cette tension entre le monde réel et le monde imaginaire. Et si la confusion règne à l'époque, c'est que l'on peine à définir les critères mêmes de la science, en mélangeant encore vérité et vraisemblance : « pseudo-science cannot be determined […] without at the same time determining what is properly scientific […] ; it is what either appears *as* science, or represents itself *as* science »[5].

Le XIXème siècle offre un contexte favorable à la fois aux pseudo-sciences et à la littérature fantastique qui s'en nourrit. En effet, le « merveilleux scientifique » envahit les foyers et le commun des mortels découvre les pouvoirs de la fée électricité. La révolution électrique contribue à changer les mentalités et prépare le terrain des pseudo-sciences en demandant à chacun de croire en une puissance invisible, en une force que le savant serait le seul à dompter. La science n'est donc plus seulement un objet de savoir, mais de croyance, et ouvre la voie à l'occultisme. Grâce au Progrès, les scientifiques accomplissent des miracles quotidiens, sous l'œil médusé du profane, et accèdent au rang de dieux vivants. La fin du siècle des Lumières annonce ainsi l'ère du pouvoir invisible, comme l'explique Robert Darnton dans son ouvrage consacré au mesmérisme :

> A une époque où Voltaire rend intelligible la théorie de la gravitation universelle de Newton, où Franklin applique les propriétés de l'énergie électrique à l'invention du paratonnerre, où Montgolfier stupéfie l'Europe en soulevant l'homme dans les airs, le fluide invisible de Mesmer ne semble pas tellement miraculeux[6].

Ce contexte inédit explique le développement des pseudo-sciences comme le magnéto-spiritisme qui repose sur deux principes essentiels : d'une part l'existence d'un fluide unique parcourant l'univers, d'autre part la nature semi-matérielle de ce fluide, véritable incarnation de l'âme humaine. Puisque la science a repoussé les limites du savoir, elle doit pouvoir définitivement percer les secrets de la Nature, défier l'ordre divin et se jouer de la Mort ! C'est ce que croient les adeptes des sciences

[5] *Science, Pseudo-science and Society* (Waterloo, Canada : University Press, 1980), ed. Marsha P. Hanen, *et al.*, 4: « on ne peut déterminer ce qu'est la pseudo-science sans définir en même temps ce qui relève réellement du scientifique […] ; c'est aussi bien ce qui a l'apparence de la science que ce qui prétend être une science ».

[6] Robert Darnton, *La fin des Lumières, le mesmérisme et la Révolution* (Paris: O. Jacob, 1995), 21.

occultes, persuadés que leurs théories, un temps dénigrées, seront un jour prochain validées par la science, quand le progrès aura trouvé les réponses aux questions existentielles. Jules Grasset définit d'ailleurs l'occultisme comme un « merveilleux préscientifique », autrement dit en avance sur son temps : « Je limite l'occultisme à l'étude des phénomènes qui : 1/ n'appartiennent pas encore à la science ; 2/ peuvent sans contradiction logique en faire partie plus tard »[7]. Afin d'asseoir leur crédibilité, les pseudo-sciences n'hésitent pas à mettre le langage et les instruments de la science au service de leur doctrine. Elles comptent sur le progrès technique, qui permet désormais d'explorer l'infiniment petit au travers du microscope, pour dépasser les limites des sens humains et sonder l'invisible, « expliquer l'inexplicable ».

Cette tendance est particulièrement visible dans le spiritisme où le prétendu *savant* utilise les inventions révolutionnaires de son temps pour explorer le monde des esprits. Dans son article intitulé « La Langue de l'Au-delà », Michel Pierssens souligne ainsi que « le monde matériel, technicisé, occupe une place de plus en plus grande dans le processus de communication avec l'au-delà : il s'instrumentalise, depuis les tables tournantes jusqu'à la photographie spirite ou aux machines à écrire »[8]. Quelle meilleure illustration de la pseudo-science que la photographie spirite au XIXème siècle ? Cette mode qui entend capturer l'énergie des esprits sur le papier photographique, en laissant à l'appareil le soin de révéler ce que l'œil ne pourrait voir ? Cette incroyable supercherie, qui donne lieu à l'époque à des procès pour escroquerie, en France comme en Angleterre, illustre bien le sentiment de toute-puissance qu'inspire la science. Car si les récits de fantômes sont d'ordinaire classés dans la littérature fantastique et dans la fiction pure, l'alibi scientifique les ramène dans la sphère du « vrai » et efface la frontière entre réalité et fiction. On peut dire, en reprenant les mots de Daniel Grojnowski, que « pour les esprits spéculatifs comme pour les écrivains, la photographie dans la deuxième moitié du XIXème siècle est le lieu où s'accomplissent d'étranges sortilèges »[9]. En France, c'est le docteur Hippolyte Baraduc, médecin respecté et passionné par l'âme humaine, qui devient l'une des plus importantes cautions scientifiques de la photographie de l'invisible.

---

[7] Jules Grasset, *L'Occultisme hier et aujourd'hui. Le merveilleux préscientifique* (Paris: Coulet et Fils, 1908), 49.

[8] Michel Pierssens, « La Langue de l'Au-delà » dans *Altérations, créations dans la langue : les langages dépravés*, ed. Anne Tomiche (Clermont-Ferrand : Presses Universitaires Blaise Pascal, CRLMC, 2001), 30.

[9] Daniel Grojnowski, « Sortilèges photographiques, de Villiers à Strindberg », *Romantisme* n°105, 1999 : 72.

En cherchant à saisir dans ses clichés le fluide vital émanant des vivants, il inspire une génération d'écrivains comme Villiers de l'Isle-Adam, qui exploite la dimension fantastique de ce procédé dans *Claire Lenoir*, ainsi que nous l'étudierons par la suite. L'invention de la photographie au XIXème siècle, récupérée par les spirites, illustre parfaitement cette nouvelle science prodigieuse à laquelle rien ne saurait résister : « l'objectif et le procédé photographique [...] se trouvent eux-mêmes érigés en révélateurs de l'âme humaine, laquelle impressionne les plaques par sa puissance et ses émanations fluidiques »[10]. Si l'imaginaire littéraire influence la science, les pseudo-sciences vont donc entraîner, à leur tour, une véritable « scientifisation » du fantastique.

En dépit de leurs théories irrationnelles et souvent fantasques, les pseudo-sciences refusent d'être cantonnées aux rôles de chimères ou « d'Objets Scientifiques Non-Identifiés ». Toutes réclament le statut de science véritable, même la doctrine spirite emmenée par Allan Kardec –ce qui peut sembler étrange quand on sait que Kardec, de son vrai nom Hippolyte Rivail, était persuadé d'avoir été druide dans une vie antérieure ! Dans « Honorer l'incertain : la science positive du XIXème siècle enfante le spiritisme », Françoise Parot explique que « Le spiritisme de Kardec est une science « naturelle », et non une science du « surnaturel » : l'ambition de son créateur est précisément de démontrer que les « phénomènes » livreront leur secret à la science, à laquelle rien ne résiste ; que ce que l'on considérait auparavant comme miracle doit pouvoir être scientifiquement expliqué »[11]. Alors que sa doctrine prétend entrer en communication avec les esprits, Kardec se montre même sceptique envers la photographie spirite et refuse de croire aveuglément en ce nouveau prodige. Il cherche à préserver la pureté et la « vérité » du spiritisme, en se méfiant des contes qui pourraient le discréditer. En outre, en mars 1863, il rédige un article pour *La Revue Spirite*, en enjoignant ses contemporains à la prudence : « Si la chose est vraie, elle se vulgarisera ; en attendant, il faut se garder de donner créance à tous les récits merveilleux que les ennemis même du Spiritisme se plaisent à répandre pour le rendre ridicule ainsi que ceux qui les acceptent trop facilement »[12]. Cette déclaration est d'autant plus paradoxale, pour les lecteurs que nous sommes, que le spiritisme est bel et bien une science « merveilleuse »

[10] Cf. « Photographie de l'âme ou âme de la photographie » par Jean-David Jumeau-Lafond, MUCRI (Musée Critique de la Photographie de la Sorbonne), http://mucri-photographie.univparis1.fr/article.php?id=10.

[11] Françoise Parot, « Honorer l'incertain : la science positive du XIX° siècle enfante le spiritisme », *Revue d'histoire des sciences* n°57, 2004 : 51.

[12] Cf. *Revue Spirite* n°3, mars 1863, 93.

cherchant à percer le mystère métaphysique de l'âme. Ainsi, malgré ces incohérences, la démarche d'Allan Kardec est réellement scientifique et pose la question de la légitimité des pseudo-sciences. Bernadette Bensaude-Vincent tente de répondre à cette interrogation dans l'ouvrage intitulé *Des savants face à l'occulte* :

> Ni les magnétiseurs, ni les spirites, ni les métapsychistes du XIXème siècle ne sont jamais parvenus à obtenir la reconnaissance scientifique qu'ils n'avaient cessé de solliciter avec ardeur et persévérance. Pourquoi ? La raison principale n'est pas l'extravagance, réelle ou supposée de leurs hypothèses [...]. Ce n'est pas non plus le « défaut de méthode » [...]. De fait, il faut bien convenir que tous les critères de scientificité –d'ailleurs variables et controversés dans le champ du savoir légitime- ont bien été respectés[13].

Il nous faut donc distinguer l'aspect scientifique et la scientificité réelle, car si les pseudo-sciences emploient le langage voire le protocole des sciences exactes, elles sont fondées sur des conceptions erronées, bien qu'intellectuellement séduisantes. Dans l'histoire humaine, elles ne sont certes pas les seules constructions mentales irréalistes et nombreuses sont les « erreurs à caractère *collectif* [...], les erreurs 'faisant école' », selon les mots de Jean Rostand[14], à s'être imposé au fil des siècles. Depuis la théorie de la terre plate jusqu'aux prétendus rayons N, l'histoire a été jalonnée par ces fausses hypothèses qui ont, elles aussi, influencé la littérature et les arts. Mais si le phénomène prend de l'ampleur au XIXème siècle, c'est qu'il est contemporain du scientisme et offre l'image d'une époque aux deux visages : tandis que le positivisme inspire les écrivains réalistes et naturalistes, le courant fantastique puise, quant à lui, dans ces pseudo-sciences qui repoussent les limites du réel.

Qu'elles le veuillent ou non, les pseudo-sciences sont bien des « chimères » au sens mythologique du terme, des monstres réunissant en un même corps le langage des sciences et l'imaginaire littéraire. Elles créent un univers onirique et entretiennent une atmosphère quasi magique, pour mieux impressionner les esprits faibles. Les magnétiseurs et les spirites du XIXème siècle jouent notamment sur le décorum, en accentuant l'aspect funèbre ou oriental de leurs cabinets. Dans son étude consacrée au mesmérisme, Franklin Rausky décrit cette « mise en scène [...] incontestablement théâtrale, avec les habits en soie et les pantoufles dorées du magnétiseur, la pénombre des séances, la musique émouvante et

[13] Bernadette Bensaude-Vincent, *Des savants face à l'occulte, 1870-1940* (Pari : Ed. La Découverte, 2002), 41.

[14] Jean Rostand, *Science fausse et fausses sciences* (Paris: Gallimard, 1958), 13.

passionnée, les grands miroirs, reflets de la lumière magnétique, les signes astrologiques du décor »[15]. Le critique recrée, sous les yeux du lecteur, l'atmosphère mystérieuse des salons qui fleurissent dans les grandes villes européennes comme Londres ou Paris, et où se presse la bonne société avide de frisson. Mais sa description trouve également un écho dans la littérature fantastique, par exemple dans la nouvelle de Théophile Gautier intitulée *Avatar,* qui offre un semblable décor. En effet, dans le cabinet de l'étrange Balthazar Charbonneau, docteur des âmes, l'on retrouve de riches étoffes « historiées d'éléphants chimériques et d'oiseaux fabuleux », ainsi que d'énigmatiques « livres sanscrits tracés au poinçon ». L'auteur tombe volontiers dans le leitmotiv et semble éprouver le besoin de justifier son héros, comme s'il s'agissait d'un véritable magnétiseur :

> M. Cherbonneau n'était rien moins que charlatan et ne cherchait pas la mise en scène, mais cependant il était difficile de pénétrer dans cette retraite bizarre sans éprouver un peu de l'impression que devaient causer autrefois les laboratoires d'alchimie[16].

Ainsi, une interaction féconde s'instaure entre la science et la fiction qui deviennent de véritables « vases communicants », comme l'explique Bernadette Bensaude-Vincent en précisant que « le savoir de l'occulte a dû subir [...] la contamination des formes contemporaines de la littérature, en pleine transformation au XIXème siècle »[17]. Des tables tournantes aux récits de revenants, des théories de Mesmer ou Kardec à la littérature fantastique, il n'y a qu'un pas, aisément franchi par les auteurs comme Gautier, Hoffmann ou Villiers de l'Isle-Adam, eux-mêmes adeptes des pseudo-sciences.

2- Des pseudo-sciences au fantastique :

L'œuvre de Villiers de l'Isle-Adam illustre bien cette interaction car elle laisse une grande place au langage savant, qui apparaît à la base des intrigues fantastiques. Dans un texte comme l'*Eve future*, la science devient même un personnage puisque l'écrivain transforme l'illustre Thomas Edison en héros de papier, s'abandonnant à d'infinies explications sur ses découvertes. Mais si la science inspire l'*Eve future*, c'est la

---

[15] Franklin Rausky, *Mesmer ou la révolution thérapeutique* (Paris : Payot, 1977), 70.
[16] Théophile Gautier, *Avatar, dans Contes et récits fantastiques* (Paris : Librairie Générale Française, 1990), 359.
[17] Bensaude-Vincent, 42.

pseudo-science que l'on retrouve dans le texte de *Claire Lenoir*, où le docteur Bonhomet se voit confronté à d'étranges phénomènes. Dans cette nouvelle, Tribulat Bonhomet incarne un médecin cartésien, hostile aux croyances de tout ordre, hébergé à Saint-Malo par l'un de ses confrères, le docteur Lenoir. Ce dernier, versé dans les pseudo-sciences, est passionné par la question de l'âme et de sa possible survie après la mort. Ses maîtres à penser son « Mesmer », « Eliphas Lévi » (auteur du *Dogme et rituel de la haute magie* en 1856), ou les occultistes « Raymond Lulle [...] et Guillaume Postel »[18] qui ont marqué l'histoire du XIIIème et du XVIème siècle, mais dont le commun des mortels ignore l'existence. En auteur féru d'occultisme, Villiers est familier de ces savants et de leurs théories ; il consacre même un chapitre entier aux pseudo-sciences, le chapitre VIII de la nouvelle, sobrement intitulé « Spiritisme ». C'est dans cette section que l'on découvre les lectures de Césaire Lenoir, dont le docteur Bonhomet se plaît à dresser la liste, en se moquant du caractère infantile et crédule de son ami :

> Il ne jurait que par Auréole Théophraste Bombaste, dit le « divin Paracelse ». Gaffarel et le populaire Swedenborg le ravissaient jusqu'au délire. [...] Les modernes, Mirville, Crookes, Kardek, le plongeaient dans de profondes rêveries. Il croyait aux *Ressuscités* d'Irlande, aux vampires valaques, au mauvais œil ; il me citait des passages tirés du cinquième volume de la *Mystique* de Görres, à l'appui de ses propositions[19].

Dans l'esprit du docteur Lenoir, la philosophie côtoie l'alchimie, et le spiritisme se mêle aux croyances fantastiques et aux superstitions. Les références qui s'enchaînent créent un effet de désordre et discréditent aussi bien le personnage que les prétendues sciences qu'il étudie. En effet, comment prêter foi aux théories de Kardec si elles valent les histoires de vampire ? Villiers s'abandonne à ce que Umberto Eco nomme « le vertige de la liste » et cherche à « provoquer une impression de scientificité » en multipliant les noms et les citations, dont la plupart échappent au commun des lecteurs. L'on retrouve ainsi dans *Claire Lenoir* l'utilisation « hyperbolique » du langage scientifique qu'Anne Lefeuvre a analysé dans l'*Eve future* : « une exploitation, non dénotative, mais purement connotative (et partant poétique) de la science, puisque seul compte l'effet à produire... »[20]. Dans la liste dressée par le docteur Bonhomet, on

[18] Villiers de l'Isle-Adam, *Claire Lenoir*, dans les *Œuvres Complètes* vol.2 (Paris: Gallimard, Editions de la Pléiade, 1986), 173.
[19] Ibid.
[20] Anne Lefeuvre, « Le Discours scientifique dans l'Eve future », extrait du colloque Frontières de la fiction : http://www.fabula.org/colloque99.pdf .

reconnaît les noms de célèbres spirites du XIXème siècle comme Mirville ou Sir William Crookes, connu en Europe pour ses séances de tables tournantes et ses expériences de photographie spirite. Mais c'est l'œuvre d'Allan Kardec, négligemment cité au détour d'une phrase, qui inspire le plus Villiers dans cette nouvelle.

Le texte de *Claire Lenoir* emprunte au *Livre des Esprits* de Kardec la théorie du « fluide nerveux » ou « périsprit » sur laquelle repose l'intrigue fantastique. Dans la doctrine spirite, ce fluide est conçu comme un principe intermédiaire, une substance semi-matérielle qui permettrait d'unir l'âme au corps : « Le périsprit est le lien qui unit l'Esprit à la matière […], il tient à la fois de l'électricité, du fluide magnétique et, jusqu'à un certain point, de la matière inerte »[21]. Si Kardec est bien l'inventeur de ce terme de « périsprit », l'idée d'une enveloppe entourant l'âme humaine ne date pas du XIXème siècle : dès l'Antiquité, les philosophes grecs évoquent l'existence d'un « véhicule léger » ou d'un « corps lumineux », auxquels les Evangiles donneront le nom d'« aura » ou de « pneuma ». Le spiritisme confère à cet élément fluidique un pouvoir extraordinaire : celui de survivre à l'enveloppe corporelle, et de repousser les limites de la mort. Dans le *Livre des Esprits*, l'on peut lire ainsi que « la mort est la destruction du corps seul et non de cette seconde enveloppe qui se sépare du corps quand cesse en celui-ci la vie organique »[22].

Dès lors, la pseudo-science vient légitimer les récits de fantômes et apporte une caution scientifique au fantastique ; elle rend tout à coup vraisemblables les histoires de revenants, errant sur terre pour hanter les vivants, et ouvre une brèche scientifique dans laquelle vont s'engouffrer les écrivains. Lorsque le docteur Lenoir agonise dans les bras de Tribulat Bonhomet, il ne meurt pas tout à fait et demeure sous cette forme de périsprit, pour mieux accomplir sa vengeance. La mort n'est pas une fin pour le personnage, mais une sorte de désincarnation ou de métamorphose puisqu'il réapparaît sous les traits d'un ottysor-vampire. Villiers emprunte au folklore polynésien l'image de cette créature assoiffée de sang, à la stature monstrueuse, qui semble tout droit sortie d'une légende : « J'avais entendu parler de ces très rares Ottysors couleur de jais, ou *guetteurs de naufrage*. […] Ces féroces cannibales sont enveloppés d'un mystère non pénétré encore »[23]. L'ottysor apparaît comme un être cruel et n'a rien de commun avec l'homme qu'était Césaire Lenoir, si ce n'est le désir de

[21] Allan Kardec, *Le Livre des Esprits, concernant les principes de la doctrine spirite*, livre II, chapitre VI, item 257 (Paris : Didier et Cie, 1866), 119.
[22] Ibid., 69.
[23] Villiers de l'Isle-Adam, 207-208.

venger son honneur. Une fois libéré de son enveloppe corporelle, il traque l'infortuné Henry Clifton, l'amant de son épouse, jusqu'aux confins du monde, jusqu'aux rivages de l'Océanie où il le décapite sauvagement. Pour accomplir ce meurtre, l'ottysor a besoin de prendre une forme matérielle afin de pouvoir agripper sa victime et se saisir de son arme ; il devient visible aux yeux des compagnons d'Henry Clifton, restés sur le bateau, qui regardent, impuissants, « le *fantastique* indigène [...] agitant par les cheveux, en son poing levé tout droit, la tête sanglante »[24] de leur ami. On pourrait croire, à première vue, que le texte de *Claire Lenoir* présente une contradiction puisque l'ottysor n'apparaît pas au moment du décès du docteur, et reste invisible pour Tribulat Bonhomet. En réalité, Villiers applique à la lettre l'enseignement d'Allan Kardec et s'appuie sur la nature « intermédiaire » du périsprit pour rendre son récit vraisemblable. Dans le *Livre des Médiums*, le savant explique en effet que « l'Esprit, que l'Homme soit mort ou vivant, a toujours une enveloppe semi-matérielle qui [...] peut acquérir la visibilité et la tangibilité »[25]. L'ottysor inventé par l'écrivain n'est donc pas un fantôme traditionnel mais un monstre possédant le pouvoir extraordinaire de se rendre visible ou invisible à loisir. Sous cette forme démoniaque, il incarne les pulsions refoulées par le docteur Lenoir et soudain libérées par sa mort brutale.

Le seul aspect vraiment fantastique du récit reste cette transformation du périsprit, qui se distingue très nettement du corps qui était son hôte. Sous les traits de ce pirate à la peau de « jais », impossible de reconnaître le personnage de Césaire Lenoir, qui apparaissait comme un homme calme et aimant au début de la nouvelle. Cette fois, Villiers semble avoir pris des libertés par rapport à la pseudo-science car, si l'on en croit le *Livre des Esprits*, le périsprit devrait « représente[r] l'apparence de sa dernière incarnation »[26]. La dimension fantastique du texte s'exprime donc dans l'image du monstre qui reste un leitmotiv du genre; en décrivant un « ottysor-vampire », l'auteur cède assurément à la mode du vampirisme qui traverse le XIXème siècle et les œuvres de John William Polidori, Théophile Gautier ou Bram Stoker. Toutefois, en parcourant les textes fondateurs du spiritisme, l'on se rend compte que Villiers ne fait que prolonger la pensée d'Allan Kardec et que le fantastique est déjà en germe dans la pseudo-science. En effet, en 1868, *La Genèse, les miracles et les prédictions selon le spiritisme* avance la théorie d'une possible métamorphose du fluide nerveux, dont la forme dépendrait de l'esprit

[24] Ibid.
[25] Allan Kardec, *Le Livre des Médiums, ou guide des médiums et des évocateurs* (Paris: Librairie de la Revue Spirite, 1869), 142.
[26] Allan Kardec, *Le Livre des Esprits,* 66.

auquel il est attaché : « La nature de l'enveloppe fluidique est toujours en rapport avec le degré d'avancement moral de l'Esprit. Les Esprits inférieurs ne peuvent changer à leur gré »[27]. D'après la classification établie par Kardec, l'esprit de Césaire Lenoir serait donc assez puissant pour changer d'aspect et se doter d'une forme nécessaire à la vengeance. Une fois de plus, le fantastique est disséqué, expliqué et comme récupéré par la pseudo-science qui crée un univers onirique où le mystère peut enfin être éclairci. Si la science et la littérature ont pu être qualifiées de « sœurs ennemies » dans l'histoire, le spiritisme et le fantastique sont bien frères au XIXème siècle.

En outre, c'est bien un instrument scientifique, l'ophtalmoscope, qui permet au docteur Bonhomet d'apporter la preuve de l'existence des esprits. A la fin de la nouvelle, lorsque le héros se tient au chevet d'une Claire Lenoir mourante, cet objet parvient à dissiper le brouillard du doute en révélant l'invisible. Villiers exploite, dans le texte fantastique, une découverte récente puisque l'ophtalmoscope n'est mis au point par Hermann Von Helmholtz qu'en 1851, quelques années à peine avant la parution de son œuvre. L'auteur confère même à cet instrument d'optique des pouvoirs magiques car l'appareil d'observation se transforme en révélateur de forces psychiques. C'est dans l'œil de Claire Lenoir, ou plutôt de son cadavre, que le docteur Bonhomet découvre l'image terrifiante de l'ottysor auquel il ne voulait croire. En bafouant le respect dû aux morts au nom de la science et en faisant de sa dépouille un sujet d'expérience, il se trouve pris à son propre piège. Alors qu'il attribuait les craintes de Claire Lenoir aux divagations d'une mourante, il découvre grâce à l'ophtalmoscope la silhouette du monstre vengeur, imprimée sur son œil comme sur un cliché photographique :

> En examinant les yeux de la morte, je vis, distinctement, d'abord se découper, comme un cadre, le liseré de papier violet qui bordait le haut de la muraille. Et, dans ce cadre, réverbéré de la sorte, j'aperçus un tableau que toute langue, morte ou vivante […] est, sous le soleil et la lune, hors d'état d'exprimer[28].

L'image de l'ottysor au regard haineux, tenant encore dans sa main la tête sanglante d'Henry Clifton, entraîne Tribulat Bonhomet aux portes de la folie. Sa conception de la réalité et de la science s'effondre soudain, lorsqu'apparaît l'empreinte reconnaissable du périsprit. Villiers s'inspire à

---

[27] Allan Kardec, *La Genèse, les miracles et les prédictions selon le spiritisme* (Paris: Librairie Internationale, 1868), 297.
[28] Villiers de l'Isle-Adam, 219.

l'évidence des photographies spirites pour rédiger cet épisode qui mêle étroitement le scientifique et le fantastique. Il choisit à dessein de mettre en scène un ophtalmoscope, peu connu du grand public, et auquel on prête des pouvoirs extraordinaires. Alors que l'œil ne possède qu'une vision limitée de la réalité, cette lunette semble capable, tout comme le microscope, de dévoiler les merveilles de l'univers et de repousser les frontières du réel. Dans la nouvelle, le lecteur ne sait plus vraiment où s'arrête la science et où commence la fiction car Villiers exploite également de réelles découvertes, comme l'explique Daniel Grojnowski : « [L'auteur] s'est très probablement inspiré d'une information parue dans *Le Publicateur des Côtes-du-Nord* (26 septembre 1863). Selon ce journal, un Anglais a fixé sur le collodion l'image impressionnée sur l'œil d'un bœuf »[29]. L'écrivain ne se contente donc pas d'un vague cadre réaliste, et utilise la dimension prodigieuse de la science comme de la pseudo-science. Sous sa plume, les esprits prennent forme humaine et les savants les plus cartésiens sont contraints de céder à l'appel du fantastique.

La question du fluide nerveux ou périsprit, évoquée dans le texte de *Claire Lenoir*, traverse les XVIIIème et XIXème siècles en suscitant le débat dans la communauté scientifique. Nombreuses sont les voix qui s'élèvent, dans le corps médical, pour condamner cette hypothèse sans fondement, que les Encyclopédies anatomiques qualifient d'hérésie. Le doute demeure néanmoins car de grands noms défendent cette idée, et pas seulement parmi les adeptes du magnéto-spiritisme. Benjamin Franklin développe, notamment, la théorie d'un « fluide unique » de nature électrique, que les auteurs assimilent abusivement au fluide de Mesmer, en brouillant une fois encore la frontière entre science et pseudo-science. Bien avant le texte de Villiers, on en retrouve la trace dans *Frankenstein*, le plus célèbre des romans de Mary Shelley ; lorsque Victor Frankenstein donne vie à sa créature monstrueuse, il lui transmet, en effet, une force ou une énergie proche de ce fluide vital. Certes, le mot de « fluide » n'apparaît pas tel quel dans le roman, mais les métaphores employées par l'auteur rappellent les travaux de Franklin ou Mesmer. Le principe de vie découvert par le héros s'apparente ainsi à une force lumineuse, décrite comme une « étincelle » par Mary Shelley : « With an anxiety that almost amounted to agony, I collected the instruments of life around me, that I might infuse a spark of being into the lifeless thing that lay at my feet »[30].

[29] Daniel Grojnowski, « Sortilèges photographiques, de Villiers à Strindberg », *Romantisme* n°105 (1999) : 74.

[30] *Mary Shelley, Frankenstein or the Modern Prometheus* (Londres : Penguin Books, 1994), 63 : « Avec une anxiété qui me mettait à l'agonie, je disposai à

Afin de préserver la vraisemblance du texte, l'écrivain ne s'attarde pas sur la technique révolutionnaire du docteur Frankenstein, bien décidé à emporter son fatal secret dans la tombe. Toutefois, la parenté scientifique du récit apparaît clairement dans ce principe vital qui semble de nature électrique, comme celui de Franklin, et se transmet comme le fluide des magnétiseurs. Si la référence à la science et à la pseudo-science est moins évidente dans *Frankenstein* que dans *Claire Lenoir*, on y retrouve néanmoins ce mariage entre le langage littéraire et le langage de la science.

On comprend aisément le lien qui unit une discipline comme le magnéto-spiritisme et la littérature fantastique, car toutes deux abordent les mystères de l'univers, que ce soit d'un point de vue scientifique ou fictionnel. Le rôle joué par la physiognomonie dans ce courant est, en revanche, moins évident car cette pseudo-science, héritière de la phrénologie de Gall, cherche à déterminer le caractère d'un individu d'après sa physionomie et les traits de son visage. L'œuvre de Johann Kaspar Lavater ouvre, néanmoins, une piste fantastique en définissant la physiognomonie comme « la connaissance du rapport qui lie l'extérieur et l'intérieur, la surface visible à ce qu'elle couvre d'invisible ». C'est l'invisible qu'explorent les écrivains fantastiques, en donnant forme aux forces tapies dans l'ombre, aux pulsions cachées dans l'âme humaine et que le masque social ne saurait totalement dissimuler. La physiognomonie sert donc à légitimer la théorie de l'*homo duplex*, et permet aux auteurs de renouveler le thème littéraire du double. Cette pseudo-science décrit, en effet, un certain nombre de « types » humains, proches de l'animal, correspondants à des caractères bien définis. L'on retrouve cette classification dès la fin du XVIème siècle dans le *De Humana Physiognomonia* de l'italien Giambattista Della Porta, où apparaissent notamment le type de l'ours, du cerf, du chien ou encore du lion. Chaque animal représente un caractère différent : le lion symbolise, par exemple, l'énergie et la force, tandis que l'ours incarne « la fureur, le pouvoir de déchirer et l'humeur misanthrope »[31], si l'on en croit le *Dictionnaire des sciences occultes* publié par l'Abbé Migne en 1848. Au XIXème siècle, la physiognomonie a encore de nombreux adeptes et influence les écrivains fantastiques comme Villiers de l'Isle-Adam ou Mérimée qui illustrent ses théories. Si le nom de cette pseudo-science n'apparaît pas dans *Claire Lenoir*, la référence reste transparente et prouve, une nouvelle fois,

---

portée de ma main les instruments qui allaient me permettre de transmettre une étincelle de vie à la forme inerte qui gisait à mes pieds ».

[31] Abbé Migne, *Dictionnaire des sciences occultes* vol.49 (Paris: Ateliers catholiques du Petit-Montrouge, 1848), 503.

l'empreinte du discours scientifique dans la fiction. A plusieurs reprises, le docteur Lenoir défend les thèses physiognomonistes et tente de rallier Tribulat Bonhomet à sa cause : « Observez attentivement les mouvements familiers, les instincts, les tendances de l'individu chez lequel prédominera le type de *l'ours*, par exemple, ou du *tigre* ».[32] Cette théorie obsède littéralement le personnage parce qu'il sent déjà en lui la présence de l'ottysor, d'une force animale incontrôlable cherchant à se libérer de ses chaînes. Au chapitre XIV de la nouvelle, la discussion philosophique laisse place à la peur, lorsque le docteur dévoile cette duplicité :

> Je sens en moi des instincts dévorateurs ! J'éprouve des accès de ténèbres, de passions furieuses !... des haines de Sauvage, de farouches soifs de sang inassouvies, *comme si j'étais hanté par un cannibale* !... […] Et, lorsque je quitte le royaume de l'Esprit, je distingue très bien cette nature infernale en moi !... C'est la *vraie* ![33]

Cet extrait souligne bien l'interprétation fantastique du discours savant, qui passe de la métaphore animale à la métamorphose des personnages. Tandis que la physiognomonie se contente d'établir des ressemblances entre l'humain et l'animal, la littérature donne corps à cette théorie. Dans le fantastique, le langage scientifique prend donc un sens littéral et l'image s'anime sous les yeux du lecteur. Le docteur Lenoir, rongé par les soupçons d'adultère et les désirs de vengeance, est véritablement habité par un monstre et livre un combat intérieur pour maintenir sa santé mentale. En brisant sa prison de chair, la mort libère cette force démoniaque et relâche le monstre sur terre.

Le thème du héros possédé par l'animal est récurrent dans la littérature fantastique, notamment dans les récits de loups-garous où un homme se voit contaminé par un esprit sauvage. On le retrouve, par exemple, dans une nouvelle comme *Lokis* de Mérimée, où le narrateur suggère que le personnage du mystérieux comte Szémioth se transforme peu à peu en ours sanguinaire. Ce texte renouvelle, en outre, le thème traditionnel de la lycanthropie en remplaçant le loup par l'ours, considéré comme le roi de la forêt dans les anciennes légendes. Nul doute que Mérimée s'est inspiré des pseudo-sciences pour rédiger cette œuvre fantastique, car le comte semble donner vie au « type de l'ours » évoqué dans les traités de physiognomonie. Comme le docteur Lenoir, ce personnage a l'intuition de sa profonde dualité et interroge les savants qui l'entourent –à savoir le professeur Wittembach et le docteur Froeber-, pour trouver une explication

[32]Villiers de l'Isle-Adam,198.
[33]Villiers de l'Isle-Adam,199-200.

à ses maux : « Comment expliquez-vous, monsieur le professeur, [...] comment expliquez-vous la *dualité* ou la *duplicité* de notre nature ? »[34]. Une fois de plus, la pseudo-science n'est pas citée dans le texte, mais le portrait du personnage ne laisse pas place au doute. En effet, au début de la nouvelle, le narrateur se lance dans une description analytique du visage du comte, qui rappelle les ouvrages de physiognomonie : « Il avait le front haut [...], quoique un peu étroit. Ses traits étaient d'une grande régularité, seulement ses yeux étaient trop rapprochés, et il me sembla que d'une glande lacrymale à l'autre il n'y avait pas la place d'un œil, comme l'exige le canon des sculpteurs grecs »[35]. Les détails de ce portrait renvoient aux développements de Della Porta ou de Lavater car les yeux « trop rapprochés » soulignent un caractère cruel, sauvage, voire criminel, pour les physiognomonistes. Ces indices suggèrent la nature bestiale du comte Szémioth et annoncent la chute du récit, lorsque sa jeune épouse sera retrouvée morte sur son lit de noces, « la figure horriblement lacérée, la gorge ouverte, inondée de sang ». Dans *Lokis*, le « type de l'ours » cache donc l'ours véritable et offre un prolongement imaginaire aux théories scientifiques. Tandis que le magnéto-spiritisme justifie les récits de fantômes, la physiognomonie légitime, quant à elle, les histoires de loups-garous (ou d'ours-garous) et prend clairement une dimension fantastique.

A travers ces quelques œuvres, se dessine une tendance de la littérature du XIXème siècle : un fantastique d'inspiration savante, proche de l'univers onirique des pseudo-sciences et capable d'assimiler totalement le langage scientifique ; un fantastique où le surnaturel rejoint le réel, où l'invisible devient visible et où l'inexplicable est soudain expliqué. Le siècle positiviste voit la naissance de ces *chimères*, créées par l'imagination humaine, et ce n'est pas un hasard si la pseudo-science se mêle alors à la littérature fantastique. Elle trouve en elle une sœur, capable de repousser les limites de la réalité et d'ouvrir une fenêtre de rêve sur ce monde devenu trop *positif*. Dans ce siècle de progrès où l'homme s'est imposé de nouvelles normes, la « pseudo-science » ne pouvait qu'inspirer la « para-littérature ».[36]

[34] Prosper Mérimée, *Lokis*, dans *Théâtre de Clara Gazul, Romans et Nouvelles* (Paris: Gallimard, Bibliothèque de la Pléiade, 1978), 1080.

[35] Ibid., 1058.

[36] Bien que controversée, la division entre littérature et para-littérature alimente la critique, comme en témoigne la publication en 2005, chez Vuibert, de l'ouvrage de Daniel Fondanèche intitulé *Paralittératures*.

Œuvres citées

Bensaude-Vincent, Bernadette. *Des savants face à l'occulte, 1870-1940*. Paris: Ed. La Découverte, 2002.

Darnton, Robert. *La fin des Lumières, le mesmérisme et la Révolution*. Paris: O. Jacob, 1995.

Figuier, Louis. *Histoire du merveilleux*, vol.III. Paris: Hachette et Cie, 1860.

Fondanèche, Daniel. *Paralittératures*. Paris: Vuibert, 2005.

Gautier, Théophile. *Avatar,* dans *Contes et récits fantastiques*. Paris: Librairie Générale Française, 1990.

Grasset, Jules. *L'Occultisme hier et aujourd'hui. Le merveilleux préscientifique*. Paris: Coulet et Fils, 1908.

Grojnowski, Daniel. « Sortilèges photographiques, de Villiers à Strindberg. » *Romantisme* n°105, 1999.

Hanen, Marsha P., Margaret J. Osler and Robert G. Weyant, eds. *Science, Pseudo-science and Society*, Waterloo, Canada: Wilfrid Laurier University Press, 1980.

Hers, Henri-Géry. *Science, non-science et fausse science, réflexion sur les chemins de la connaissance*. Paris: l'Harmattan, 1998.

Jumeau-Lafond, Jean-David. « Photographie de l'âme ou âme de la photographie », MUCRI (Musée Critique de la Photographie de la Sorbonne), http://mucri-photographie.univparis1.fr/article.php?id=10

Kardec, Allan. *La Genèse, les miracles et les prédictions selon le spiritisme*. Paris: Librairie Internationale, 1868.

—. *Le Livre des Esprits, concernant les principes de la doctrine spirite*, livre II, chapitre VI, item 257. Paris: Didier et Cie, 1866.

—. *Le Livre des Médiums, ou guide des médiums et des évocateurs*. Paris: Librairie de la Revue Spirite, 1869.

Lavater, Johann Kaspar. *La Physiognomonie ou l'art de connaître les hommes d'après les traits de leur physionomie*. Delphica: l'Age d'homme, 1998.

Lefeuvre, Anne. « Le Discours scientifique dans l'*Eve future* », extrait du colloque *Frontières de la fiction* : http://www.fabula.org/colloque99.pdf.

Mérimée, Prosper. *Lokis*, dans *Théâtre de Clara Gazul, Romans et Nouvelles*. Paris: Gallimard, Bibliothèque de la Pléiade, 1978.

Migne, Abbé. *Dictionnaire des sciences occultes*, vol.49. Paris: Ateliers catholiques du Petit-Montrouge.

Parot, Françoise. « Honorer l'incertain : la science positive du XIXème siècle enfante le spiritisme. » *Revue d'histoire des sciences* n°57-1, 2004.

Paulet, Jean-Jacques. *L'Antimagnétisme*, Paris: Desennes, 1784.

Pierssens, Michel. « La Langue de l'Au-delà » dans *Altérations, créations dans la langue : les langages dépravés*, ed. Anne Tomiche, pp.27-37, Clermont-Ferrand: Presses Universitaires Blaise Pascal, CRLMC 2001.

Rausky, Franklin. *Mesmer ou la révolution thérapeutique*. Paris: Payot, 1977.

Revue Spirite n°3, mars 1863.

Rostand, Jean. *Science fausse et fausses sciences*. Paris: Gallimard, 1958.

Shelley, Mary. *Frankenstein or the Modern Prometheus*. London: Penguin Books, 1994.

Villiers de l'Isle-Adam, *Claire Lenoir*, dans les *Œuvres Complètes* vol.2. Paris: Gallimard, Editions de la Pléiade, 1986.

PART B:

HOLISTIC VISIONS

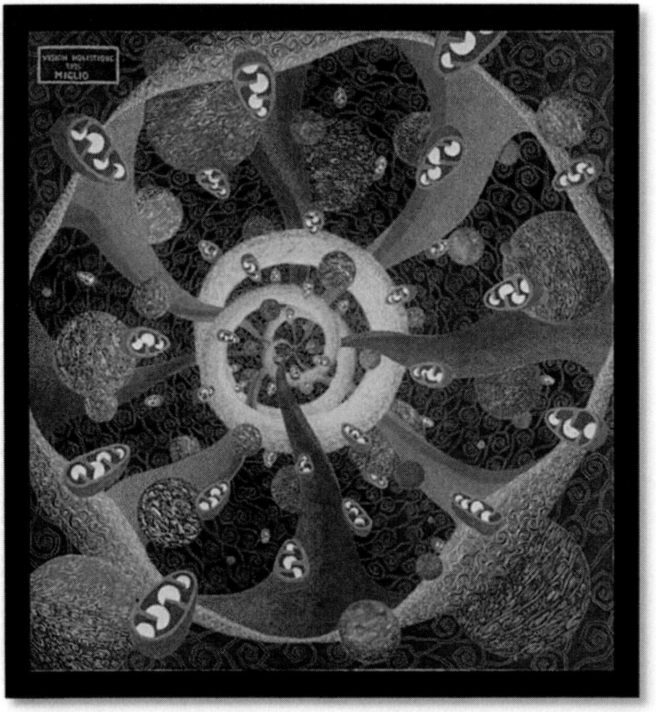

ILL. B-1 : François Miglio, *Holistic Vision*, detail from Triptyque EXPLORATION I (H 85cm x L 80cm)

SECTION I

L'ART CONTEMPORAIN ET LA MODE DE L'OCCULTE

MARCO PASI

En 1948, Terence Harold Robsjohn-Gibbings (1905-1976), un architecte d'intérieur d'origine anglaise très apprécié et recherché par les stars d'Hollywood publia un livre sur l'art moderne, intitulé *Mona Lisa's Moustache*[1]. Dans ce livre, il présenta une interprétation originale de l'art moderne comme étant fortement influencée par « l'occulte ». C'était en effet une longue tirade contre les tendances les plus significatives de l'art moderne, y compris le post-impressionnisme de Gauguin, le cubisme, le futurisme, l'expressionnisme et le surréalisme. Tous ces courants étaient vus par Robsjohn-Gibbings comme rien d'autre qu'une tentative de retour à la mentalité primitive de la magie, s'opposant donc à ce qu'il considérait comme des valeurs fondamentales de la culture occidentale, à savoir « la logique, l'ordre, la vérité, et la raison. »[2] Pour Robsjohn-Gibbings, « [l]'art *moderne n'est pas moderne du tout.* C'est plutôt la reviviscence de l'un des systèmes les plus anciens pour obtenir le pouvoir. C'est la reviviscence de la magie »[3].

Les affirmations de Robsjohn-Gibbings, et leur connotation politique évidente, devraient être situés dans le contexte de l'atmosphère particulière qui a été dominante en Europe au lendemain de la Deuxième Guerre mondiale. Logiquement, il y avait un besoin urgent de tourner la page, de passer à autre chose et de laisser derrière les fantômes tragiques du passé récent. Mais il y avait aussi le besoin de comprendre les raisons de cette catastrophe culturelle, sociale, politique, et tout simplement humaine. Dans ces mêmes années de l'après-guerre Theodor Adorno, dans ses fameuses *Thèses contre l'occultisme* (1949), fut parmi les premiers à proposer cette équation entre l'occultisme et le fascisme qui est aussi au

---

[1] T.H. Robsjohn-Gibbings, *Mona Lisa's Mustache. A Dissection of Modern Art* (New York: Alfred A. Knopf, 1947). James Webb cite le livre dans son ouvrage classique *The Occult Establishment* (La Salle: Open Court, 1985), 420, là où il aborde le rapport entre l'art moderne et l'occulte.

[2] Ibid., 17.

[3] Ibid., 13.

cœur du livre de Robsjohn-Gibbings[4]. La magie et l'occultisme étaient perçus comme dangereux pour un progrès sain de la culture occidentale, et pour Robsjohn-Gibbings (contrairement à Adorno) l'enjeu était la survie des valeurs bourgeoises, tel l'ordre, la rationalité et la beauté classique (qu'il voyait clairement représentées par l'esthétique des décorations d'intérieur qu'il dessinait). Il y a là bien sûr un paradoxe, qui consiste en ceci : alors que Robsjohn-Gibbings voyait un lien intrinsèque entre l'art moderne et le fascisme, car pour lui les deux étaient fondés sur le retour d'une mentalité magique et primitive, la véritable cible de son livre était en effet l' art même qui avait été recueilli et exposé par les nazis en 1937 sous l'étiquette méprisante de *entartete Kunst* (« art dégénéré »). Et pourtant, malgré des nombreuses exagérations et le ton quelque peu hystérique du livre, Robsjohn-Gibbings avait mis en lumière certains aspects du développement de l'art moderne qui avaient tout de même

[4] Cf. Theodor W. Adorno, "Theses against Occultism," in The Stars down to Earth and Other Essays on the Irrational in Culture, (London - New York: Routledge, 1994). Sur ce point cf. aussi Marco Pasi, " The Modernity of Occultism: Reflections on Some Crucial Aspects," in *Hermes in the Academy: Ten Years' Study of Western Esotericism at the University of Amsterdam*, Wouter J. Hanegraaff and Joyce Pijnenburg, eds., (Amsterdam: Amsterdam University Press, 2009), 59-74. Carlo Ginzburg a suggéré qu'un certain nombre de textes écrits par des intellectuels européens pendant la Seconde Guerre Mondiale, avant la défaite de l'Allemagne, peuvent être définis comme " écrits de l'Année Zéro ", justement parce qu'ils dérivent d'une "expérience partagée : le sentiment d'un écroulement imminent de la civilisation, dû à l'avancée apparemment irrésistible de l'Allemagne et de ses alliés. Comment – chacun de ces écrivains se demandait – tout cela pouvait-il se produire ? " (C. Ginzburg, " Mircea Eliade's Ambivalent Legacy," in *Hermeneutics, Politics, and the History of Religions. The Contested Legacies of Joachim Wach and Mircea Eliade*, C.K. Wedemeyer and W. Doniger eds, (Oxford - New York: Oxford University Press, 2010), 318; toute traduction de l'anglais, ici comme ailleurs, est par mes soins). Il me semble pourtant que ce questionnement ne s'arrêta pas avec la défaite de l'Allemagne. La catastrophe morale et politique d'un pays qui avait tellement contribué au développement culturel de l'Europe, ainsi que les dilemmes moraux posés par le bombardement atomique de Hiroshima et Nagasaki, ne pouvaient pas être résolus si facilement par la victoire finale des forces alliées. "Theses Against Occultism" écrit peu après la fin de la guerre, montre clairement que, même après la défaite militaire de l'Allemagne, les intellectuels européens continuèrent d'être hantés par cette question, à savoir comment l'Occident avait pu arriver à ce point de dégradation. Un autre exemple significatif que l'on pourrait citer est l'impressionnante *Zerstörung der Vernunft* (La destruction de la raison) de György Lukács, publiée en première édition en 1953.

quelques fondements dans la réalité et qui méritaient davantage d'attention.

En fait, même si le livre fut ignoré ou traité avec mépris dans les comptes rendus de la presse spécialisée, et fut donc vite oublié, il anticipait une certaine interprétation de l'art moderne qui serait devenu beaucoup plus à la mode quelques années plus tard. En 1966, le critique et historien d'art finlandais Sixten Ringbom ouvrit la voie, avec un important article publié dans le *Journal of the Warburg and Courtauld Institute,* à une nouvelle approche qui tenait compte de l'importance des idées occultes et spiritualistes dans l'évolution historique de l'art moderne.[5] Cette nouvelle approche reçut son sacre vingt ans plus tard dans l'exposition qui se tint à Los Angeles en 1986, *The Spiritual in Art*, et a été consolidée et corroborée depuis par un grand nombre de monographies et d'expositions, qui ont sensiblement modifié notre perception de l'histoire de l'art moderne. Aujourd'hui, il est beaucoup moins controversé de soutenir que certains parmi les artistes et les courants les plus importants de l'art moderne ont été inspirés par des idées ésotériques, occultes, ou plus généralement spiritualistes.

Mais ce n'est pas seulement dans les avant-gardes du début du XXème siècle que cette fascination peut être perçue. La présence fantomatique de l'occulte et du spirituel est devenue de plus en plus visible dans les dix dernières années dans l'art contemporain, même si son importance, sa fonction, voire sa cohérence, peuvent encore échapper aux critiques et aux observateurs.[6] En témoigne un certain nombre d'expositions récentes qui ont essayé de capturer cette fascination renouvelée des artistes contemporains pour « l'autre côté », pour reprendre le titre d'un roman célèbre d'Alfred Kubin. Ces expositions, et les publications qui les ont accompagnées, semblent toutefois n'être que la proverbiale pointe de l'iceberg.

Les réactions vis-à-vis de cette présence, que nous pourrions peut-être appeler une « mode culturelle », comme le fit jadis Mircea Eliade dans un

---

[5] Sixten Ringbom, "Art in the 'Epoch of the Great Spiritual': Occult Elements in the Early Theory of Abstract Painting," *Journal of the Warburg and Courtauld Institutes* 29 (1966): 386-418.

[6] Pour une étude panoramique intéressante sur cette tendance, cf. Lars Bang Larsen, "The Other Side," *Frieze* 106 (April 2007) : 114-119. Plus récemment, la revue *Frieze* a publié un numéro (135, Nov.-Déc. 2010) dont le thème principal est "Religion et Spiritualité", et qui présente une discussion de diverses formes d'art inspirée par l'occulte.

fameux essai[7], varient. Certains l'interprètent plutôt positivement comme une forme de « résistance à l'omnipotence du sens » , quoi que cela puisse vouloir dire[8]. D'autres en revanche, tenant à préserver les valeurs des Lumières et de l'engagement politique et social qui s'y rattachent, y voient un retour inquiétant de ce « sombre monde souterrain avec ses envoûtements, ses mystères, ses mirages et toutes sortes de magie 'naturelle'; un monde qui relève de cette imagination romantique résolument anti-urbaine, mais aussi de l'ancien régime, avec ses rituels et ses Raspoutines, ses opiacés et ses messes (non pas masses)[9]. »Curieusement, ici le ton et la rhétorique, voire les implications politiques, ne sont pas trop éloignés de celles de Robsjohn-Gibbings.

La variété des attitudes, souvent chargées d'émotion, semblent témoigner d'une relation ambiguë, qui peut sans doute être expliquée en tenant compte du visage double de l'occulte. On retrouve cette duplicité dans certains stéréotypes populaires aussi, mais elle a tout de même un fondement historique. D'une part l'occulte représente un monde de valeurs et de comportements qui est perçu comme « superstition », car il empêcherait l'émancipation politique, psychologique et sociale des êtres humains des illusions d'une réalité métaphysique qui n'aurait aucune existence concrète. En ce sens, il est perçu comme intrinsèquement *réactionnaire.* D'autre part l'occulte représenterait un monde de valeurs et de comportements qui s'oppose à l'*establishment*, aux autorités étatiques et religieuses, se fondant parfois sur des manières radicalement alternatives de concevoir la société, la politique, et la conscience individuelle. En ce sens, il est perçu comme intrinsèquement *révolutionnaire.*

Dans son aspect réactionnaire, l'occulte peut être facilement inséré dans la catégorie plus générale de la religion, et partager son sort malheureux dans l'art contemporain. En effet, la relation entre l'art contemporain et la religion est loin d'être simple[10]. Les religions

---

[7] Mircea Eliade, "Cultural Fashions and History of Religions," in *Occultism, Witchcraft, and Cultural Fashions: Essays in Comparative Religions* (Chicago: University of Chicago Press, 1978), 1-17.

[8] Michel Gauthier, « Le temps des nécromants », in *Fresh Théorie II.* Black Album (Paris: Editions Léo Scheer, 2006), 180.

[9] Dieter Roelstraete, "Great Transformations: On the Spiritual in Art, Again," *Afterall 20* (Spring 2009): 15. A la fin de la citation, la version originale anglaise contient un calembour, avec une référence évidente à la fameuse définition marxienne de religion, difficilement traduisible en français : " … its opiate and its masses: religious services, that is, not people".

[10] Cf. James Elkins, *On the Strange Place of Religion in Contemporary Art* (New York - London: Routledge, 2004); James Elkins and David Morgan, eds., *Re-Enchantment* (New York - London: Routledge, 2009); et Maria Hlavajova, Sven

majoritaires ne sont guère perçues aujourd'hui comme des sources d'inspiration intéressantes dans le discours dominant de l'art contemporain[11]. Si l'on considère les religions dominantes et l'occulte comme appartenant au même monde illusoire de réalité transcendante, qui s'éloignerait du rôle émancipateur attribué à l'art contemporain, il devient facile de les rejeter ensemble. Mais lorsque l'occulte est en mesure de se présenter sous son autre visage – le visage révolutionnaire – les choses changent. Dans ce cas, la distance entre l'occulte et les religions dominantes ou traditionnelles grandit, et une distinction nette entre les deux permet de percevoir l'occulte comme bien plus attrayant que la religion, sans qu'il y ait trop de dissonances cognitives. Il n'est pas trop difficile de comprendre pourquoi. L'art contemporain aime bien se présenter comme un lieu d'expérimentation, d'innovation permanente, de défi, de transgression des normes esthétiques - mais souvent aussi politiques et sociales - traditionnelles. Dans ce contexte, la religion est associée à la stabilité, à la tradition, au conservatisme et aux valeurs du passé.

L'occulte, d'autre part, semble offrir de nombreuses possibilités d'expérimentation (par exemple en tant qu'exploration du soi ou du corps), et partage avec l'art contemporain la réputation de vouloir contester toutes sortes de normes. En outre, il peut aussi se présenter (la question n'est pas de savoir ici si c'est à tort ou à raison) comme ayant été persécuté, à un moment ou à un autre, par la religion dogmatique et institutionnelle. Ainsi, l'attitude critique vis-à-vis de la culture et de la société que l'on retrouve tant dans l'art contemporain comme dans l'occulte pourrait conduire à une sorte d'étrange alliance. Dans cette perspective, l'occulte peut être utilisé par l'art pour son impact culturel très fort, sans que l'art ne partage nécessairement ses présupposés spirituels ou qu'il participe émotionnellement à ses croyances. Par ailleurs, il est vrai que la plupart des discours ésotériques de l'art contemporain peuvent être placés sous l'enseigne de l'ironie, de la provocation et du jeu (comme on peut le voir - pour ne citer qu'un exemple - chez le Center for Tactical Magic, basé en Californie). Il y a d'autres cas (un Joachim Koester ou une Maria Loboda par exemple) où il semble y avoir une

---

Lütticken, and Jill Winder eds., *The Return of Religion and Other Myths: A Critical Reader in Contemporary Art* (Utrecht - Rotterdam: BAK, basis voor actuele kunst - post editions, 2009).

[11] On pourrait donc se demander si le titre du numéro de *Frieze* que j'ai cité plus haut (cf. n. 6) a été vraiment bien choisi. « Spiritualité et Occulte » aurait peut-être été plus approprié. Il y a en effet peu de références à des formes institutionnelles ou majoritaires de religion dans les œuvres d'art présentées dans ce numéro.

attitude plus sérieuse et engagée, que même un cadre conceptuel particulièrement sophistiqué ne peut pas cacher entièrement. Dans ce cas, l'art contemporain semble offrir un espace pour l'exploration des dimensions cachées de la réalité, et surtout du soi, qui est parfaitement conforme aux objectifs traditionnels de l'ésotérisme et de la magie.

On pourrait donc se demander si l'art, dans ce contexte, n'est pas devenu l'un des débouchés possibles pour l'expression de pratiques spirituelles et d'idées que la sécularisation avait rendu peu accessibles pour l'homme moderne. Certes, l'ésotérisme avait déjà trouvé sa propre manière d'interagir positivement avec la modernité, par exemple à travers la psychologisation et la naturalisation de son discours, et avec d'autres stratégies culturelles analogues. Cependant, la ceinture de protection d'un discours fondé sur l'expérimentation illimité, qui se situe dans une dimension qui n'est pas forcement vérifiable par la pensée rationnelle et scientifique, a pu trouver dans l'art contemporain un environnement favorable.

Je vois quatre manières principales dans lesquelles la relation entre l'occulte et l'art (contemporain) peut s'établir. La première est la représentation de symboles ésotériques ou d'images associées à l'ésotérisme. Ceci a lieu bien sûr au niveau le plus explicite du langage visuel, et est basé sur le patrimoine d'images et de symboles de la tradition ésotérique, qui a évidemment une très longue et riche histoire. La deuxième est la production d'objets artistiques (une image, une sculpture, une installation, ou autre) qui peuvent être interprétés comme des talismans ou des fétiches, ou comme le résultat d'une manipulation de la matière qui peut être associée aux sciences occultes telles que l'alchimie ou la magie. Dans ce cas, l'objet artistique lui-même est le résultat d'un travail de type magique et/ou il est doté de pouvoirs magiques. La troisième est l'œuvre artistique comme moyen de provoquer des expériences extraordinaires, qui peuvent être interprétées comme ayant des qualités spirituelles, mystiques, initiatiques, chamaniques ou magiques. Cela a bien sûr une importance particulière dans le contexte du *body art* et des performances artistiques. Quelques-uns des ingrédients caractéristiques des performances artistiques extrêmes, tels que la douleur, la peur, ou les récits de mort et de retour à la vie, ont souvent été associés à l'initiation ésotérique et à l'expérience mystique. Enfin, la quatrième est le travail artistique comme le résultat d'une inspiration ou communication directe avec des entités spirituelles ou d'une expérience visionnaire ou mystique.

Je vais maintenant donner quelques exemples pour chacune de ces quatre catégories. Il faudrait pourtant garder à l'esprit que plusieurs d'entre eux pourraient appartenir à plusieurs catégories à la fois.

Pour l'art contemporain nous pouvons trouver des exemples du premier type dans les symboles maçonniques et les scènes d'initiation utilisés par Matthew Barney dans ses films (notamment *Cremaster 3*, 2002). On pourrait mentionner d'autres exemples: la référence subliminale de Banks Violette à la célèbre image du Baphomet de l'occultiste Eliphas Lévi dans son *Jägermeister Baphomet* (2003); ou la photographie de Joachim Koester du miroir magique de John Dee (2006). Dans certains cas, la référence n'est pas à un symbole ou à un objet, mais à une personne célèbre associée à l'occultisme, comme c'est le cas pour la « Madame Blavatsky » de Goshka Macuga (2007). Dans un autre cas, nous avons la reconstruction d'un espace ésotérique à travers l'affichage d'objets et de symboles liés à un courant particulier, tel que le spiritisme du XIXème siècle. C'est le cas de l'installation d'Olivia Plender, *The Medium and the Daybreak*.

Le deuxième type est la production d'objets artistiques qui peuvent se présenter comme des talismans ou des fétiches, qui sont produits à travers des procédés occultes, ou qui sont dotés de pouvoirs magiques. Un exemple intéressant pourrait être le « Beauty Mirror » de l'artiste letton Gints Gabrāns (2007) ou l'installation de Maria Loboda « The Grand Conjuration of Lucifuge Rofocale », qui a été préparé suivant les instructions pour l'évocation d'un démon tirées d'un grimoire magique (2006). Un autre exemple intéressant concerne un espace plutôt qu'un objet. Nous avons ici l'espace d'une galerie d'art qui a été envoûté par l'artiste Dane Mitchell.

Dans le troisième type, l'accent est mis sur le travail artistique comme un moyen d'induire des expériences extraordinaires. Quelques exemples intéressants peuvent être trouvés chez des artistes qui appartiennent à une génération déjà active dans les années 1960, tels que Joseph Beuys, avec ses performances chamaniques, et Hermann Nitsch, avec ses « actions », qui ont souvent été inspirés par des thèmes liés aux anciens cultes à mystères, les légendes du Graal, ou à d'autres traditions religieuses, spirituelles et mythiques. Mais d'autres exemples pourraient être trouvés aussi chez des artistes plus jeunes tels que Pawel Althamer (voir par exemple ses vidéos basées sur ses expériences avec des substances psychoactives, comme le LSD), ou Joachim Koester, avec ses films inspirés par la tradition populaire italienne du tarantisme ou par les « passes magiques » de Carlos Castaneda (2007 et 2009).

Enfin, le quatrième type, c'est-à-dire l'œuvre artistique qui se veut dérivée d'une inspiration ou communication directe avec des entités spirituelles ou d'une expérience visionnaire ou mystique. La ligne de partage entre le troisième et le quatrième type peut paraître floue, et elle l'est parfois véritablement, mais la distinction entre le deux me semble importante. Dans ce quatrième type, l'expérience extraordinaire n'est pas l'objectif final de la pratique artistique, mais la source initiale des œuvres à réaliser. L'œuvre artistique est donc l'expression ou la représentation de ce qui a été expérimenté par le biais, par exemple, d'une communication spirituelle, d'une vision mystique, ou, pour employer un langage plus neutre, d'un état modifié de conscience. Là encore, nous retrouvons des exemples où l'ironie, le jeu, et une certaine ambiguïté sophistiquée sont évidentes, comme chez le peintre allemand Sigmar Polke avec son fameux « Höhere Wesen befahlen » (1969). Mais il y a d'autres exemples où le travail artistique semble être fondé sur une véritable exploration du mystère de la mort, comme chez l'artiste suédois Carl Michael von Hausswolff, avec son utilisation des enregistrements des voix de personnes prétendument décédées recueillies par Carl Friedrich Jurgenson.

Il ne devrait pas être nécessaire de souligner ici que ces quatre catégories ne sont que des types idéaux, dont le seul but est de nous aider à classer le matériau offert par l'art contemporain. En réalité, les frontières sont beaucoup moins nettes, et il n'est pas difficile de trouver des travaux artistiques qui correspondent à plus d'une catégorie à la fois.

Je voudrais maintenant faire quelques remarques sur la dernière catégorie, qui me semble être la plus intéressante, et peut-être aussi la plus problématique. Au sein de la culture occidentale, l'interaction avec des entités spirituelles a été constamment perçue comme quelque chose de dangereux et d'illicite. Dans un contexte chrétien traditionnel cette interaction, sauf si elle était validée par la structure ecclésiastique après un examen attentif et délicat, pouvait facilement conduire à des accusations de sorcellerie et de satanisme. D'autre part, dans un contexte sécularisé, l'existence de ces entités est tout simplement niée, et la revendication d'une interaction avec elles ne peut être interprétée que comme le résultat d'une hallucination ou d'un dérangement mental. Il y a donc dans la culture occidentale une certaine continuité dans le rejet et la marginalisation de ce genre d'expériences. C'est toutefois cette situation même qui rend l'expérience attrayante pour ceux qui développent des formes de critique et de résistance aux valeurs culturelles dominantes. Pour ce qui est du contexte artistique, l'un des phénomènes les plus intéressants est lié aux débuts de l'art abstrait. Robsjohn-Gibbings n'en tenait pas compte, pour la simple raison que cela était encore ignoré ou négligé à son époque. On savait depuis un certain temps que des artistes comme Kandinsky et Mondrian avaient été motivés plus par des préoccupations de type spirituel que par des intentions purement formalistes.

Mais au cours des vingt dernières années, le récit rationaliste et formaliste sur les débuts de l'art abstrait est devenu encore plus problématique lorsqu'on s'est rendu compte que les premiers exemples d'abstraction avaient été créés par l'artiste suédoise Hilma af Klint (1862-1944) tout au début du XXème siècle. Or, elle affirmait que son art avait été inspiré directement par des entités spirituelles[12]. Que devrions-nous donc penser du fait que l'un des événements les plus marquants de l'histoire de l'art moderne a été le résultat non pas de conceptualisations purement formelles de problèmes esthétiques, mais de la communication directe avec des esprits célestes? Il semble évident que l'accès à une réalité alternative (il n'est pas question ici de déterminer s'il s'agit d'une réalité purement psychique ou autre), accessible à travers des formes d'altération de la conscience, saurait pousser ces artistes à créer un langage artistique nouveau, qui se développe dans des directions particulièrement novatrices et originales. Ce langage est très proche de « l'ambiguïté » et de « l'indétermination » décrites par Dario Gamboni, dans un livre important sur les « images potentielles », considérées par cet auteur comme l'un des phénomènes récurrents de l'art moderne[13]. Tout comme les occultistes à la fin du XXème siècle ont été attirés par les théories philosophiques idéalistes, parce qu'elles leur permettaient de concevoir une matière plus souple et maniable pour leurs opérations magiques, les artistes spiritualistes abstraits ont essayé de dissoudre la rigidité des images figuratives afin de les rendre accessibles à la puissance imaginative.

Comme Marina Warner l'a montré, la sécularisation n'a pas tué les fantômes, qui imprègnent encore, ou plutôt « hantent », la culture

---

[12] A partir de 1986, lorsque ses peintures furent exposées pour la première fois depuis sa mort, il y a eu un intérêt grandissant pour af Klint, qui a donné lieu à un certain nombre d'expositions et de publications. Cf., entre autres, Åke Fant, "The Case of Hilma af Klint", in *The Spiritual in Art: Abstract Painting 1890-1985* (New York - London - Paris: Los Angeles County Museum of Art - Abbeville Press Publishers, 1986), 155-163; Gurli Lindén, *I Describe the Way and Meanwhile I Am Proceeding along It. A Short Introduction on Method and Intention in Hilma af Klint's Work from an Esoteric Perspective* (Hölö: Rosengårdens Förlag, 1998); *3 x Abstraction: New Methods of Drawing* by Hilma af Klint, Emma Kunz, Agnes Martin, eds, Catherine de Zegher and Hendel Teicher (New York - New Haven - London: The Drawing Center - Yale University Press, 2005); *Hilma af Klint (1862-1944). Une modernité révélée* (Paris: Centre Culturel Suédois, 2008); *De geheime schilderijen van Hilma af Klint. The Secret Paintings of Hilma af Klint,* eds, Hedwig Saam and Miriam Windhausen (Arnhem: Museum voor Modern Kunst Arnhem, 2010).

[13] Cf. Dario Gamboni, *Potential Images. Ambiguity and Indeterminacy in Modern Art* (London: Reaktion Books, 2002).

contemporaine[14]. En effet, l'art de communiquer avec les esprits a gardé une présence dans l'art contemporain, et il est encore bien vivant aujourd'hui, comme on peut le voir chez des artistes tels que Michael von Hausswolff (que nous avons déjà cité), Raimundas Malašauskas, Nico Dockx, et John Roach. Mais dans l'art contemporain il ne s'agit plus nécessairement de la réception de messages spirituels qui concerneraient l'avenir de l'humanité, comme c'était le cas pour Hilma af Klint au début du XXe siècle, mais plutôt de l'exploration de territoires inconnus de l'esprit, faite parfois avec un sourire ironique de détachement. Parfois.

Jusqu'à des temps récents, à quelques exceptions près et avec quelques hésitations, un certain discours laïciste a donc été prédominant dans l'art contemporain. Et pourtant, cet accent nouveau mis sur les dimensions spirituelles et occultes semble maintenant indiquer un développement possible dans d'autres directions. De toute évidence, cette tendance est loin d'être représentative de l'art contemporain dans son ensemble, même si elle ne peut pas non plus être considérée comme marginale ou insignifiante. Est-ce que cette tendance est le signe d'une nouvelle perception de l'art comme possédant un but religieux ou spirituel? Est-ce qu'il est perçu comme une alternative possible au religieux? Après tout, Max Weber l'avait déjà compris il y a un siècle, lorsqu'il écrit ceci :

> L'art de son côté se constitue en un *cosmos de valeurs propres*, toujours plus conscientes, plus cohérentes, plus autonomes. Il revêt une fonction de libération *intramondaine*, quelque soit la définition qu'on donne à ce terme : libération du quotidien et aussi, avant tout, libération de la pression croissante du rationalisme théorique et pratique. Prétention qui le place en situation de concurrence directe avec les religions de salut[15].

Certains artistes des avant-gardes du début du XXème siècle, c'est-à-dire à la même époque où ces mots ont été écrits, avaient la même idée de ce que le rôle de l'art devait être dans la société moderne, et se considéraient parfois comme des prêtres d'une nouvelle religion. Ces sentiments et ces motivations semblent animer encore, peut-être de manière moins explicite, l'art contemporain. Je ne crois pas que l'art contemporain soit en train de se transformer en un nouveau mouvement religieux. Toutefois, si nous voulons le comprendre en tant que phénomène culturel complexe, je pense qu'il est important aussi de ne pas

[14] Cf. Marina Warner, *Phantasmagoria. Spirit Visions, Metaphors, and Media into the Twenty-first Century* (Oxford - New York: Oxford University Press, 2006).

[15] Max Weber, « Parenthèse théorique, » *Archives de sciences sociales des religions* 61, no. 1 (1986) : 20-21.

négliger les dimensions religieuses, spirituelles et ésotériques, qui ont pu se manifester ces dernières années.

Œuvres citées

Adorno, Theodor W. "Theses against Occultism." *The Stars down to Earth and Other Essays on the Irrational in Culture*. London - New York: Routledge, 1994.

Bang Larsen, Lars. "The Other Side." *Frieze*, 106 (April 2007):114-119.

Eliade, Mircea. "Cultural Fashions and History of Religions." *Occultism, Witchcraft, and Cultural Fashions: Essays in Comparative Religions*, Chicago: University of Chicago Press, (1978): 1-17.

Elkins, James and David Morgan, eds. *Re-Enchantment*. New York - London: Routledge, 2009.

Elkins, James. *On the Strange Place of Religion in Contemporary Art*. New York - London: Routledge, 2004.

Fant, Åke. "The Case of Hilma af Klint." In *The Spiritual in Art: Abstract Painting 1890-1985*, 155-163. New York - London - Paris: Los Angeles County Museum of Art - Abbeville Press Publishers, 1986.

Gamboni, Dario. Potential Images. Ambiguity and Indeterminacy in Modern Art, London: Reaktion Books, 2002.

Gauthier, Michel. « Le temps des nécromants ». In *Fresh Théorie II. Black Album,*180. Paris: Editions Léo Scheer, 2006.

Ginzburg, Carlo. "Mircea Eliade's Ambivalent Legacy." In C.K. *Hermeneutics, Politics, and the History of Religions. The Contested Legacies of Joachim Wach and Mircea Eliade*, edited by Wedemeyer and W. Doniger. Oxford - New York: Oxford University Press, 2010.

Hlavajova, Maria, Lütticken, Sven, and Winder, Jill (eds.). *The Return of Religion and Other Myths: A Critical Reader in Contemporary Art*. Utrecht - Rotterdam: BAK, basis voor actuele kunst - post editions, 2009.

Bang Larsen, Lars. "The Other Side," *Frieze* 106 (April 2007) : 114-119.

Lindén, Gurli. *I Describe the Way and Meanwhile I Am Proceeding along It. A Short Introduction on Method and Intention in Hilma af Klint's Work from an Esoteric Perspective*. Hölö: Rosengårdens Förlag, 1998.

Lukács, György. *Die Zerstörung der Vernunft*. Berlin: Aufbau-Verlag, 1954.

Pasi, Marco. "The Modernity of Occultism: Reflections on Some Crucial Aspects." In *Hermes in the Academy: Ten Years' Study of Western Esotericism at the University of Amsterdam*, edited by Wouter J.

Hanegraaff and Joyce Pijnenburg. Amsterdam: Amsterdam University Press, 2009:59-74.

Ringbom, Sixten. "Art in the 'Epoch of the Great Spiritual': Occult Elements in the Early Theory of Abstract Painting." *Journal of the Warburg and Courtauld Institutes* 29 (1966): 386-418.

Robsjohn-Gibbings, T.H. *Mona Lisa's Mustache. A Dissection of Modern Art*. New York: Alfred A. Knopf, 1947.

Roelstraete, Dieter. "Great Transformations: On the Spiritual in Art."*Again. Afterall* 20 (Spring 2009).

Warner, Marina. *Phantasmagoria. Spirit Visions, Metaphors, and Media into the Twenty-first Century*, Oxford - New York: Oxford University Press, 2006.

Webb, James. *The Occult Establishment*. La Salle: Open Court, 1985.

Weber, Max. "Parenthèse théorique."*Archives de sciences sociales des religions* 61, no. 1 (1986) .

THE CHIMERICAL ENCOUNTER OF WESTERN IDEALISM AND EASTERN MYSTICISM: NEW THOUGHT

PHILIPPE MURILLO

Uniting science and religion has always been a multi-faceted adventure for American people trying to embark on a chimerical perspective of combining supposedly irreconcilable opposites. Much experimentation has been conducted in American history, particularly regarding American metaphysical religion in the early 19th century. Indeed, for many years, research in American religion has been dominated by major theories such as the "evangelical thesis" by William G. McLoughlin[1] or Jon Butler's emphasis on the role of denominations.[2] However, with the recent interest and specific research in new religions and New Religious Movements, many scholars on both sides of the Atlantic such as J. Stillson Judah, Bryan Wilson, Gordon Melton, Dell DeChant and particularly Catherine Albanese have challenged the marginalization of metaphysical religion, thus paving the way towards a redefinition of the influence of non-mainstream spirituality within the science/religion dialogue. Away from the evangelical or denominational theses, Catherine Albanese has specifically stressed the importance of American metaphysical religion and the fostering influence of European esotericism into American popular culture, by replacing Transcendentalism as a significant tradition in American history (Albanese, 2007).

---

[1] William McLoughlin, *Revivals, Awakenings and Reform* (Chicago: University of Chicago Pres, 1978). This 'evangelical thesis' was continued by Nathan Hatch, Mark Noll and George Marsden.

[2] Away from the 'evangelical thesis', historian Jon Butler started scrutinizing the role of mainline denominations (state/church thesis), while playing down the importance of the 'magical' which he saw as a competing element to Christianity.

> For it was in America, in what many Europeans called the 'new world,' that the old worlds came together in combination to produce a new form of metaphysical religion. [3]

How was the relationship possible between popular metaphysics and science? How do religions or spiritualities use science as an authoritative foundation for their belief system: chimera or reality?

This paper addresses a number of issues starting with the historical and theological importance of this encounter between science and American metaphysical religion or New Thought; before considering the present and future evolution of this chimerical encounter in the development of a 'renewal'.

A chimerical experiment?

Interestingly enough, the pre-Darwinian and antebellum America period is to be remembered as a time of peaceful coexistence of science and religion, even to the extent of a symbiotic relationship (James R. Moore, David C. Lindberg, Ronald L. Numbers). Historian Herbert Hovenkamp[4] refers to it as a "honeymoon" period. The domination of natural science and evangelical Protestantism led to the emergence of modern alternative religious movements such as New Age, Transcendental Meditation or Unification Church and many more; where spiritualism, phrenology and mesmerism coexisted among other schools of thought. There is only one current surviving movement from that period: the metaphysical religious movement or metaphysical religion, which serves as a counter-example of the various academic theses of examining science and religion either as a conflictual or even as a warfare phenomenon. It all comes down to the question of popular philosophies and their relationships with science, where science has become an authoritative foundation for the construction of a belief system.

To begin with, 'ex uno plura' (*from one, many*) could be the motto of American metaphysical religion. Indeed, if we consider the history of this religious tradition in the United States we are struck by the fragmented building of a common ground. Considered as a global New Religious Movement (NRM), the American-born New Thought (movement), which

---

[3] Catherine L. Albanese, *A Republic of Mind and Spirit: A Cultural History of Metaphysical Religion* (New Haven: Yale University Press, 2007), 61.

[4] Herbert Hovenkamp, *Science and Religion in America, 1800-1860* (Philadelphia : University of Pennsylvania Press, 1978), 49.

initiated this metaphysical religious tradition, dates back to the early nineteenth century and began as a popular religion with emphasis on idealism and popular science. In addition, New Thought embodies the very pluralism of the American religious experience with a blend of Hindu mysticism, Esotericism, Transcendentalism, Christianity, Protestant liberalism (Unitarianism) and the influence of science over religion. Indeed, metaphysical tradition borrowed from the Hindu concept of the "multiple": Hinduism is not one but multiple, so is New Thought. As New Thought theologian Alan Anderson[5] noted, New Thought is "a link between East and West".[6] Indeed, the mystical conception of a unity between the "self" and the divine is at the crossroads of Western idealism and Eastern mysticism. This conception directly refers to Hinduism and the Western idea of a real universe featuring real and permanent beings. The aim of this personal transformation is a complete metamorphosis of the presence of the divine into a living reality. To enhance the Western influence on New Thought, Anderson reminds us that

> the central philosophical foundation for the New Thought movement as well as for any other metaphysical religion is idealism (…).[7]

Within the same perspective, Dell DeChant[8] confirms that the idealistic model of the movement has played a major role in its popularity, particularly in the lower and middle classes. However, controversy has

---

[5] Alan C. Anderson was Emeritus professor of philosophy and religion at Curry College, Milton, Mass. (1996-2000) and is considered as one of the most influential contemporary members of the New Thought, giving the movement an international recognition. He initiated the "Process New Thought" perspective in the late 1990s.

[6] Alan C. Anderson, "The New Thought Movement: A link between East and West", (speech delivered at the Parliament of the World's Religions, Chicago, September 3, 1993). http://websyte.com/alan/parl.htm (accessed March, 20, 2012)

[7] Alan C. Anderson, "Pluralistic Idealism : only mind, many minds," *Journal of the Society for the Study of Metaphysical Religion*, Clearwater, FL., 4, no.1 (Spring 1998) : 23.

[8] Professor Dell DeChant is the editor of the *Journal of the Society for the Study of Metaphysical Religion* and Chair of the Department of Religious Studies at University of South Florida. See "The American New Thought Movement" in *Introduction to New and Alternative Religions in America,* Volume 5, edited by Eugene V. Gallagher, W. Michael Ashcraft, Eugene V. Gallagher, Wesport, CT., Greenwood Publishing (2006): 67-89.

risen within the movement since Warren Felt Evans[9] introduced eastern perceptions in the philosophy of New Thought in the 1880's, such as eastern mysticism, even coining the expression *Christian pantheism*.

This brief outline of how American metaphysical religion became a reality highlights the complexity of the phenomenon and appears as a chimerical experiment in which syncretism emerges as a creed. It is now time to explore both the ideological and historical elements from which this religion inherited.

As a distinct heritage to metaphysical tradition, Esotericism has been one of the determining features in the development of this religion. First, the major influence comes from Emmanuel Swedenborg (1688-1772), a Swedish theologian and philosopher who experienced mystical visions and a dialogue with the spirits. His experience had a deep impact on the spiritualist movements in the United States. It placed mysticism and the power of the mind at the core of any theological consideration. It praises the idea of correspondence in the universe where these correspondences appear as terrestrial forms meant to be replicas of entities belonging to the kingdom of the spirits. Later, the influence of Frantz Anton Mesmer's (1734-1815) mesmerism and his animal magnetism emphasized the therapeutic domination in the negotiation of science and religion, and therefore rooted metaphysical religion into a clear scientific perspective.

Besides, Transcendentalism emerged as another fostering tenet of metaphysical tradition. To put it briefly, transcendentalism results from tensions between two theological conceptions within liberal Protestantism. In the nineteenth century, there was a pressing demand to depart from evangelical Christianity, to engage instead into a mystic experience. At that time, New Thought was developing along with Transcendentalists, Unitarians, progressive Quakers, Spiritualists, Theosophists, all claiming the interiority of meditation, universal brotherhood, and a redefinition of spirituality. The period must be understood as a transition from the Enlightenment to the modern era. From oriental theology, transcendentalism borrowed the divinity of mankind, and individualism –both concepts to be included in the construction of metaphysical religion. Following the works of Henry David Thoreau and Ralph Waldo Emerson, feminist activist Elisabeth Cady Stenton (1815-1901) awakened the population to a new era of emancipation with a view to "spiritualizing" religions, to access the realm of the "experiential". New Thought also took much of the influence of a humanistic vision of the world, showing a predominance of the spirit

[9] Warren Felt Evans (1817-1889) was a pioneer writer within the New Thought and was healed by P.P Quimby. Evans started teaching these healing principles and wrote his first book on this matter in 1889, *The Mental Cure.*

over the body and nature. And like Transcendentalists, New Thought emphasized the omnipresence of (Kantian) idealism.[10]

Another significant heritage element is the Protestant heritage, working as a positive and stabilizing factor for New Thought. From Protestantism, New Thought borrowed the concept of 'individualism', which has tended to become the hallmark of this religious movement. It is to be recalled that this individualism resulted from tensions between theologians and metaphysicians in the eighteenth century and later, which established a new personal and intimate relationship between the believer and the divine (as did John Edwards, for example).

Along with (Protestant) individualism, the popular dimension is another significant characteristic of this religion. Since the late nineteenth century, there has been an enhanced development of the "self" ("self-religion", "self-divinization") popularized by American romanticism and transcendentalism, and rooted mainly within low and middle classes looking for life betterment. In this respect, New Thought is a quintessentially American religion as William James[11] had already noticed, since it praises the sacralization of a positive worldview that emphasizes optimism, happiness and peace of mind (all these concepts are contained in the American Declaration of Independence: "Life, Liberty, and the pursuit of Happiness" [12]). The believer is in a process of acquiring power, and *New Thought,* within American metaphysical religion, is to be taken as a popular movement which gives the power from the ministers or the elite back to the people or the church, expressing a democratization process. To widen the scope of this popular impact, American metaphysical religion has even been using popular metaphysics to define the relationship between the believer and the divine. When it comes to metaphysics, J. Gordon Melton divides the metaphysical movements into two categories or "families": the "Christian Science-Metaphysical Family",

---

[10] "Kant's transcendental idealism distinguishes between things in themselves (*Dinge an sich*) and things as they appear to us or appearances (*Erscheinungen*), and makes a claim with respect to each side of this distinction. With respect to things as they are in themselves, Kant claims that we can have no cognition (*Erkenntnis*). Things as they appear to us, Kant argues, are mind-dependent, in some sense, and to some extent, they are empirically real and transcendentally ideal." (Lucy Allais, "Kant's Idealism and the Secondary Quality Analogy," *Journal of the History of Philosophy* 45, no. 3, (July 2007): 459.

[11] William James, *The Varieties of Religious Experience (*New York, Longman, Green & Co., 1902), 380.

[12] « Declaration of Independence » as cited in *Religion and American Cultures,* vol. 2, edited by Gary Maderman and Luis Léon (Santa Barbara, Calif. : ABC CLIO, 2003), 723.

and the "Spiritualist, Psychic, and New Age Family".[13] New Thought belongs to the former where there is no official use of occultism or magic. On this exclusion of magic, J. Gordon Melton differs from Catherine L. Albanese who believes in the presence of a specific magic:

> metaphysical practice is about what may be called magic, and magic (...) lies at the heart of American metaphysics.[14] (including mental and material magic).

If we consider the historical background, although there is still some controversy on the origins of New Thought itself, it is accepted that the influence of metaphysical ideas on this movement can be traced back to the scientific and spiritual experiments of Phineas Parkhurst Quimby in Boston, who developed a blend of mesmerism, idealism and healing techniques between 1838 and the late nineteenth century (Braden, 1963). However, the paternity of the 'organized' American metaphysical movement called New Thought is to be granted to the female leader Emma Curtis Hopkins (1849-1925), a New Thought pioneer, mystic healer and teacher who seceded from Christian Science in 1885. By the early twentieth century, New Thought was well-established in the United States and many subgroups even emerged. Undergoing a rapid growth and a significant global influence, the religious movement developed into more numerous denominations in more than 60 countries and totalled up to 632,000 members around the world.[15] Hence, diversity and pluralism do characterize the very nature of American metaphysical religion.

Indeed, this religion has always favored its inherent diversity toward a constantly renewed pluralism. In 1915, after the influence of Quimby's ideas and the still much-debated 'maternity/paternity' of Emma Curtis Hopkins, the New Thought movement joined an umbrella organisation or federation called International New Thought Alliance (INTA) including various denominations founded by Quimby's healed disciples: among the main ones are Divine Science, Unity, Religious Science, and Christian Science (which in the end seceded from New Thought). Each one retains its own diversity and originality, both in its membership policy and in its theology, while spreading the same metaphysical message and the same healing practices: the use of visualization, positive thinking (from Norman

---

[13] J. Gordon Melton, *The Encyclopedia of American Religions* (8th edition, Michigan, Gale Cengage, 2009), 613.

[14] Albanese, A Republic of Mind and Spirit: A Cultural History of American Metaphysical Religion, 6-7.

[15] According to figures available at http://www.adherents.com (accessed on November 21, 2011)

Vincent Peale,) Pantheism – all is God – and Western idealism, mixed with eastern mysticism.

Divine Science, also called The Christ Method of Healing, can be defined as a combination of Emma Curtis Hopkins's teaching and Quakers' tenets. Nona Brooks and Melinda Cramer were the two founders of this church: both had experienced unconventional therapy and agreed to create this church where healing and metaphysics combine, thus revealing a spiritual movement that is a token of the power of mind over matter. Its prominent female leadership is also characteristic of the church.

Another church is Unity, founded shortly after Divine Science by Charles (1854-1948) and Myrtle Fillmore (1845-1931) in Kansas City, Ka. With a definitely Christian perspective, Unity has always resented doctrinal conformism and reaffirmed the metaphysical reading of the Scriptures. The church has been active through various publications since 1889 such as *Daily Word* or *Unity Magazine*, which took part in spreading the metaphysical message, and are still available today. The message is one of individual health and both spiritual and material prosperity.

Also, Religious Science or Science of Mind is among the major founding churches, and has now the largest number of members. Ernest S. Holmes (1887-1960), one of Emma C. Hopkins's disciples, founded this church. In 1926 he wrote *Science of Mind* which became the founding charter of Religious Science. The main philosophy is focused on mysticism and "spiritual mind treatment".

The whole complexity of this spiritual phenomenon addresses many issues regarding first the coexistence of science and religion, and second, the pressing social longing for this kind of embodiment that New Thought represented. American metaphysical religion has proved that a chimerical combination is possible, where spirituality meets science and where they interact with each other.

Science, philosophy and spirituality: transcending chimeras

Science, philosophy and spirituality are transcended within the metaphysical religion to reach a popular religion where chimerical encounters between scientific experiment and spiritual power push the envelope of harmony beyond common grounds.

Religiously speaking, New Thought might be considered partly as an expression of primitive Christianity, advocating a return to a thorough, not literal, reading of the New Testament. New Thought members praise love, healing through prayer and educating the poor to achieve both spiritual and material prosperity, thus enhancing personal and individual relationships with the divine, thereby allowing a higher responsibility of the believer within his quest for happiness. The use of primitive Christianity within the movement is expressed through particular attention toward the popular sphere, by the chartering of a *practical spirituality*.

Within this practical spirituality, Phineas Quimby unveiled a new spiritual conception of "empowering people" in their quest for Truth. Historian Nathan Hatch already underlined this phenomenon by calling "insurgent movements" [such as the New Thought] phenomena that

> empowered ordinary people by taking their deepest spiritual impulses at face value rather than subjecting them to the scrutiny of orthodox doctrine and the frowns of respectable clergymen. [16]

Quimby developed his metaphysical apprenticeship by emphasizing a particular sense of the spiritual regarding mystical experience and the prevalence of the mind over the body. Such a unique metaphysical practice attracted many patients who used his healing technique to be cured. Although his theories could be questioned, his healing theology was soon recognized as a founding structural fabric for a new metaphysical religion. For him, disease is what follows the disturbance of the mind or spiritual matter. He believed there is something he called "First cause", "Wisdom", or even "the Christ" that lies beyond any spiritual matter.

Idealism was the foremost philosophy he built upon. Hence, he chose to adopt a metaphysical reading of the Scriptures by merging harmony with nature and stressing the self-control mind. This development led to the construction of a "self-religion" which became very popular thanks to American Transcendentalism. Also, with the spreading of Emmanuel Swedenborg's revived and reinterpreted doctrine of divine flux, which made available the mingled, spirit-matter dimension of his thought, Quimby engaged in a healing *experientialism*, marrying nature to mind, creating all the elements for the birth of America's metaphysical religion. Finally, Spiritualism has always been at the heart of the movement, as an expression originating from popular culture, that insists on the scientific aspect and intellectual liberation (that is how the New Thought has developed close links with feminism and the feminist movements since 1848).

---

[16] Nathan O. Hatch , *The Democratization of American Christianity* (New Haven, CT.: Yale University Press, 1989), 10.

Nevertheless, one of New Thought's originalities is its theological marginality, which makes it such a complex and unique phenomenon. Because ever since its emergence, the leaders of the movement have privileged a marginal orientation: the use of a metaphysical reading of the Bible as opposed to a literary understanding, symbolizing the integration into institutional religion. Charles Fillmore, one of Quimby's disciples and co-founder of Unity, chartered the authoritative *Metaphysical Bible* which has been used by all New Thought denominations. Between popular and traditional philosophical meaning, metaphysics in metaphysical religion deals with philosophical metaphysics in connection with the popular movement that is defined as American religious tradition, thinking of God in impersonal terms and stressing the power of mind to change the world. It has been practiced since the nineteenth century in Spiritualism, Theosophy, New Thought, and most recently, New Age[17]. The originality or marginality of the New Thought is its construction of a near metaphysics, bordering on myths and often substituted by the word "science" understood in its popular meaning.

By and large, Americans have always put their premium on "facts" that could be supported by evidence, discarding the "metaphysical" speculations of the Catholic Middle Ages. And an enlightened American citizenry, formed and shaped in the Protestant mold supports this pragmatic leaning advocated by this metaphysical religion.

Clearly, New Thought has its emphasis not on religiosity, but on spirituality which involves intensely personal, devotional and life-transforming aspects of religion. It is both spiritual and practical. In a way, it synthesizes the apparent opposites of practicality and spirituality. Its aim is to bridge the gap between science and religion.

> New Thought was an influence rather than an organization, an attitude of mind and not a religion, though it naturally organized a church and just as naturally in Boston. It was called the Church of the Higher Life. [18]
>
> The science I try to practise is the science that was taught 18,000 years ago and has never had a place in the hearts of man since…[19]

---

[17] Wadsworth Publishing Dictionary online, http://religion.wadsworth.com/intro/american/glossary.html#anchor2366583 (accessed April 2008).

[18] Gaius Glenn Atkins, *Religion in Our Times* (New York : Round Table Press, Inc., 1932), 296.

[19] Phineas Parkhurst Quimby, *The Quimby's Manuscripts* (novembre 1859), edited by Horatio W. Dresser (Thomas Y.Crowell Cy, 1921) 33.

An additional significant feature in this religion is the quite recent introduction of a 'process thought' perspective introduced by New Thought philosopher and leader Alan C. Anderson – a perspective which contributed to the return of the New Thought to the foreground of the American NRMs. [20] And in the renewal of popular metaphysics, this may be a hint to identify a transcending process of chimerical combinations: an advanced level of reconciliation between science, philosophy and religion. As Leigh Eric Schmidt from Princeton University pointed out in his book *In Restless Souls: the making of American Spirituality from Emerson to Oprah* [21] there are new elements in the national conversion about religion and public life (popular culture) in the United States. The new orientation of the New Thought movement called Process New Thought combines science and religion into a moderate perspective for a twenty-first century American religious identity.

To understand this renewal, it is necessary to expound the transition from "substance" New Thought to "process" New Thought. In the 19th century, the non substantialist paradigm was at the core of theological preoccupations. Substantialism has it that every phenomenon can be supported by constant realities. In substantialist theory, any location that may be the residence of a god triggers a feeling of the sacred since the object itself holds a sacred property. To go further, we will take the example of Huston Smith in *Beyond the Post-Modern Mind* [22], who examines three angles to perceive the reality of our world: the Christian angle, the modern angle, and lastly the post-modern one. In the Christian perspective, reality focuses on the person, while the mechanisms of the physical world are beyond our understanding: the path to salvation does not imply conquering nature, but following God's commandments as he revealed them to us. In contrast, from the modern angle, this reality is less certain and less important than its place in the universe; human reason may perceive this order when it is manifested in the laws of Nature: the path to human salvation goes through the discovery of the laws of Nature, and their uses, whenever possible. Finally, Smith concludes by showing that the post-modern perspective holds that people do not have certainties any longer concerning the order of reality, their perception of the cosmos has been disturbed by encyclopaedic skepticism; if reality is ordered, most people are not sure that the human mind may be able to apprehend this

---

[20] NRM : New Religious Movement.

[21] Leigh Eric Schmidt, *Restless Souls, The Making of American Spirituality* (New York : HarperCollins, 2005),11.

[22] Huston Smith, *Beyond the Post-Modern Mind* (Wheaton, Ill. : Theosophical Publishing House, 1989), 46.

order; maybe, salvation is not possible through self-realisation, except for our own satisfaction in a world where intellectualism has been destroyed. This pessimistic view has been rejected by New Thought and replaced by a positive thinking approach.

Another solution is the return to the primordial tradition (or perennial philosophy), also known as "ancient wisdom", inherited from Thomas Aquinas and Agostino Steuco, which holds that there is a set of universal values and truths common to all the peoples in the world. This perspective was taken over by Aldous Huxley in 1945 in his book *The Perennial Philosophy*.[23] In this tradition, there is another reality beyond the physical one, apprehended only by Man's senses and mind; this reality insists on the predominance of "intuition". Originally, New Thought has always favored this perspective since its creation. In primordialism, which is a kind of metaphysics, reality has many layers and gradations; it proclaims the similarity of the individual soul with that of divine Reality (each individual has some divine essence, even without being aware of it). The ultimate objective is to transform oneself like a metaphor of the presence of God in a living reality. There comes the problem of dissensions between the personal and impersonal paradigms. Suffice it to say that New Thought has privileged an evolution towards a philosophical personalism, a more intimate relationship with the presupposed divine, which is true of this new "process" approach.

Indeed, process thought or process theology, derives from the works of the American mathematician and philosopher Alfred North Whitehead, and later taken over by David Ray Griffin's Process Thought School in Chicago. To put it briefly, it holds that each unit of existence has "experiential" characteristics. Within this framework, God is both transcendent and personal, within a panentheistic perspective. In the late 1990s, Alan Anderson coined the expression Process New Thought that is both addressing a high perspective in theology and philosophy, and offering a practical opportunity to improve everyday life. Process thought or process theology has a different position from traditional theology in that there is no substantial soul or body. It contends instead that there is a vast matrix of occasions (experiences) that are structured within a complex organization into societies. One of these is the flow of human experience through time and is called the person or soul. In each of these occasions, we can find an element of self-determination and uniqueness. We are then all members of one another, we are in Christ and Christ is in us (pan*en*theism).

---

[23] Aldous Huxley, *The Perennial Philosophy* (London : Perennial, 1990).

On one hand, what Process New Thought does, is to address a high perspective in theology and philosophy by inserting "process thought" into the debate, while on the other hand, offering a practical opportunity to improve everyday life. Process New Thought holds that each unit of existence has "experiential" characteristics, thus offering a third way between dualism and materialism. Within this new theological and philosophical framework, God is both transcendent and personal. Inspired by Alfred North Whitehead in the early twentieth century, and taken over particularly by David Ray Griffin from Claremont College, the "process" thought and process theology—a more naturalistic form of theology has emerged as a true "third way" in New Thought. It develops one main idea which is the process of co-creation combining three necessary elements forming an equation: past (experiences) + divine offer + choice. Practical applications of it rely on two major pillars which are creativity and prehension. The latter is the idea that existence is made of various entities (actualities) that add up to become one. It eventually leads to a unification of a new experience. Each decision or change of thought produces a new creation. The new perspective develops a process analysis of creativity while applying it to practical everyday life. The concept of "co-creation" insists on an intense individualism partly inherited from mainline Protestantism. It also insists on dynamism, action and constant change, belief in abundance in economy, though abundance is an idea which is not usually associated with religion. In (Process) New Thought, the belief is also that there is only one power in the universe and that that power is good.

What is more, Process New Thought is pan*en*theistic, meaning that all is *in* God, rather than pantheistic (all is God), which is a total identification of the world and God, often with the idea that the world is an illusion. The individual is like in Oprah Winfrey's show, empowered, but to what extent? Process Theology is understood as a "renewal" of American Christian thought. It is the most advanced form of neo-liberal theology in the United States. This effort to unite science and religion is in tune with a desire to legitimize religion so as to reach a new mainstream status for the twenty-first century, all the while affirming each individual's freedom of belief.

Unlike conventional religions, American new religions (spiritualities) such as Process New Thought offer a dual perspective: a commodity for the spiritual consumer, and a genuine spiritual awakening. Throughout American history, liberal theology has been defined and influenced by three major schools of thought: Evangelical liberalism, personalistic idealism and the natural empiricism. In this case, we are talking of an "experientialist" perspective of liberal theology. By the end of the twentieth century, Friedrich D.E Schleimacher and Walter Rauschenbusch embodied the return to theological liberalism. At that time, mathematician and philosopher Alfred North Whitehead founded the new 'process thought' school in Chicago, which offered a fundamental basis for the spreading of American liberal theology. The main objective was to face the challenges and problems affecting the meaning of liberalism by suggesting a progressive Christian alternative. Clearly, the idea was dominated by a spirit of reform and not of revolution, with a re-conceptualization of the meaning of Christianity through the prism of modern knowledge and ethical values (a personal god, a loving, eternal mind/spirit). The hallmark of this new liberalism is the process of divine creativity, relativity, holism, ecological respect and divine benevolence. Metaphorically speaking, it is the assertion that "becoming" is more essential than "being"; where becoming implies a process of transformations, mutations or transgressions such as in the case of chimeras. Following the process tradition of Heraclitus, Henri Bergson, John Dewey, Samuel Alexander, Charles Hartshorne or Pierre Teilhard de Chardin, Alfred N. Whitehead identifies "process thought" as a redefinition of all the theologies and philosophies that encompass becoming, events and relativity as necessary values for knowledge.

Within Process New Thought, Alan Anderson's new orientation leads to a concrete application of liberal and process theology which can be viewed as a *reactivation* of liberal Protestantism in the sense that all these new concepts are more part of a reform than a revolution. From substantiality to process, the mutation of New Thought into Process New Thought is quite a revealing phenomenon of a shifting boundary within NRMs between marginalization and mainstreaming. Establishing a more intimate relationship with the divine, Anderson goes beyond a classical pantheism into a *panentheism* better fit for a metaphysical interpretation —all is *in* God. It shows that intuition is more powerful and reaches beyond physical reality by asserting that each individual has a divine essence, without necessarily being aware of it.

Conclusion

The challenging and even chimerical encounter between Western idealism and Eastern mysticism can be interpreted as a breakthrough towards the establishment of an 'interspiritual' and 'scientific' bridge connecting the mind and the body. We have shown that American metaphysical religion has blurred the frontier(s) between science and spirituality as a whole by creating a breach into chimerical perspectives on human spiritual empowerment. Unfathomable possibilities are offered to the human brain and spirit to open up new paths into the realm of unorthodox beliefs and creeds.

These findings suggest that, since the beginning, this "open at the top" religion was meant to bring together opposites and to demonstrate that chimerical encounters are possible within the negotiation of science, spirituality and religion. New Thought and now Process New Thought are live expressions of this pressing demand of post-contemporary human aspirations. Process Thought and liberal theology work side by side towards establishing a new charter of American intellectual and interspiritual identity. The new brand of New Thought using process philosophy reinterprets God as 'experience' as opposed to 'mind' or 'consciousness'. The introduction of Whitehead's optimistic and sustainable philosophy to New Thought ideology opens new practical perspectives for American metaphysical religion in this twenty-first century. This combination is an expression of enjoyment; it conveys a sense of happiness, the sacralization of a positive worldview, while stressing peace of mind.

Beyond this, practical applications of New Thought and Process New Thought are numerous encompassing for instance optimistic behaviour in everyday life (science has recently come up with support for such beliefs), visualization (in healing idealism), prosperity consciousness. Everyone creates their own world through their responses to God's initiatives in life, for those responses are largely experienced as thoughts. The Process New Thought perspectives embody the awakening of a metaphysical consciousness against the deadlock represented by traditional theological tenets. The paradigm shift is still one of a religion of 'exploration' not revelation. This exploration is that of a God that proceeds by attraction and not by domination. The challenge of this chimerical association of science, philosophy and spirituality transcends all the standards of religious studies to come up with a new pattern to attempt to connect religion with reality. It is an attractive vision of the need for a true engagement with the wonderful world around us.

Works Cited

Agar, Herbert and others. *The City of Man: A Declaration on World Democracy*. New York: The Viking Press, 1940.

Albanese, Catherine L. "The Subtle Energies of Spirit: Explorations in Metaphysical and New Age Spirituality" in *Journal of the American Academy of Religion* 67, no. 2 (June 1999): 305-325.

—. *A Republic of Mind and Spirit: a History of Metaphysical Religion.* New Haven: Yale University Press, 2007.

Anderson, Alan C. *Healing Hypotheses: Horatio W. Dresser and the Philosophy of New Thought.* New York and London: Garland Publishing, 1993.

Barker, Eileen, Beckford James A. and Karel Dobbelaere, eds. *Secularization, Rationalism, and Sectarianism: Essays in Honour of Bryan R. Wilson.* Oxford: Clarendon Press, 1993.

Braden, Charles S. *Spirits in Rebellion: The Rise and Development of New Thought.* Dallas: Southern Methodist University Press, 1963.

Bromley, David G. and J. Gordon Melton, eds. *Cults, Religion, and Violence.* Cambridge, England: Cambridge University Press, 2002.

Butler, Jon. *Awash in a Sea of Faith: Christianizing the American People.* New York: Harvard University Press, 1992.

Dechant, Dell. "The American New Thought Movement." In *Introduction to New and Alternative Religions in America,* vol 5, edited by Eugene V. Gallagher, W. Michael Ashcraft, Eugene V. Gallagher, 67-89. Wesport, CT. : Greenwood Publishing, 2006.

Dobson, Alan P., and Steve Marsh. *Us Foreign Policy Since 1945.* London: Routledge, 2001.

Dresser, Horatio W. *A History of the New Thought Movement.* New York: Thomas Y. Crowell Company, 1919.

Fillmore, Charles S. *Metaphysical Bible Dictionary.* Kansas City, Missouri, Unity School of Christianity, (1931) 1995.

Gahr, Evan. "Carter Can't Cloister His Faith in Religion." *Insight on the News*, June 20, (1994): 10.

Griffith, David R. *Religion and Scientific Naturalism, Overcoming the Conflicts.* Albany, N.Y.: State University of New York Press, 2000.

Harley, Gail M. *Emma Curtis Hopkins: Forgotten Founder of New Thought.* Syracuse, New York: Syracuse University Press, 2002.

Hartshorne, Charles. *Reality as Social Process: Studies in Metaphysics and Religion.* Glencoe, Illinois: The Free Press, 1953.

Hatch ,Nathan O. *The Democratization of American Christianity.* New Haven, CT.: Yale University Press, 1989.

Heinz, Holly A. "New Spirituality for a New Life." Accessed June 2010. http://www.1-spirit.net/pages/New_Thought/welcome.htm

Herbert Hovenkamp. *Science and Religion in America, 1800-1860.* Philadelphia : University of Pennsylvania Press, 1978.

Holmes, Ernest. *The Science of Mind,* New York: Dod Mead and Company, 1926.

James, William. *The Varieties of Religious Experience.* New York: Longman, Green & Co., 1902.

Judah, J. Stillson. "Metaphysics Midwestern America." *The Journal of Popular Culture* 17, no. 4 (2004): 134.

Lewis, James R. *Legitimating New Religions*. New Brunswick, NJ.:Rutgers University Press, 2003.

Marshall, William P. "The Culture of Belief and the Politics of Religion." *Law and Contemporary Problems*, 63, no. 1-2, 2000, 453.

Mcloughlin, William G. *Revivals, Awakenings, and Reform: An Essay on Religion and Social Change in America*, 1607-1977. Chicago: University of Chicago Press, 1978.

—. Bellah Robert N. *Religion in America.* New York: Beacon Press, 1968.

Melton, J. Gordon. "A talk given at the annual meeting of the Evangelical Ministries to New Religions." Lecture, Southern Baptist Theological Seminary, Louisville, Kentucky, February 22-23, 2002. http://www.cornerstonemag.com (accessed March 21, 2007).

—. *New Thought: A Reader*. Santa Barbara, California: Institute for the Study of American Religion, 1990.

—. *The Encyclopedia of American Religions*. 8th edition, Michigan: Gale Cengage, 2009.

Neuhaus, Richard John. "Metaphysical America." *First Things: A Monthly Journal of Religion and Public Life*, (March 2007): 26.

Niebuhr, Reinhold. *Does Civilization Need Religion? A Study in the Social Resources and Limitations of Religion in Modern Life*. New York: The Macmillan company, 1927.

—. *Moral Man and Immoral Society: A Study in Ethics and Politics*. New York: Scribner, 1960.

Noll, Mark A., ed. *Religion and American Politics: From the Colonial Period to the 1980s*. New York: Oxford University Press, 1990.

Roof, Wade Clark. "Religion and Spirituality: Toward an Integrated Analysis." In *A Handbook of the Sociology of Religion*, edited by Michele Dillon, 137-150. Cambridge, England: Cambridge University Press, 2003.

Satter, Beryl. *Each Mind A Kingdom*. Los Angeles: University Press of California, 1999.

Schmidt, Leigh Eric. *Restless Souls, The Making of American Spirituality*. New York: HarperCollins, 2005.

Stark, Rodney and Bainbridge, William Sims. *The Future of Religion.* Berkeley: University of California Press, 1985.

Stearns, Peter N. "Fear and Contemporary History: A Review Essay." *Journal of Social History* 40, no. 2, 2006.

Versluis, Arthur. *The Esoteric Origins of the American Renaissance*. New York: Oxford University Press, 2001.

Wadsworth Publishing Dictionary online, http://religion.wadsworth.com/intro/american/glossary.html#anchor2366583 (accessed April 2008).

Whitehead Alfred North. *Process and Reality: An Essay in Cosmology*. New York: The Macmillan Company, 1929.

SECTION II

DE L'IMAGINAIRE SCIENTIFIQUE AUX SCIENCES IMAGINAIRES

DANIEL S. LARANGE

> La magie réunit donc, dans une même science, ce que la philosophie peut avoir de plus certain et ce que la religion a d'infaillible et d'éternel. Elle concilie parfaitement et incontestablement ces deux termes, qui semblent d'abord si opposés : foi et raison, science et croyance, autorité et liberté.
>
> Elle donne à l'esprit humain un instrument de certitude philosophique et religieuse exacte comme les mathématiques, et rendant raison à l'infaillibilité des mathématiques elles-mêmes [...].
>
> La science a ses nuits et ses aurores, parce qu'elle donne au monde intellectuel une vie qui a ses mouvements réglés et ses phases progressives. Il en est des vérités comme des rayons lumineux ; rien de ce qui est caché n'est perdu, mais aussi rien de ce qu'on trouve n'est absolument nouveau.[1]

L'une des sources occultes du romantisme plonge dans les recherches scientifiques d'Isaac Newton, notamment dans la théorie de la gravitation et la théorie corpusculaire de la lumière qui deviennent des paradigmes récurrents dans le discours social des romantiques, tant chez les saint-simoniens, les fouriéristes, les positivistes que dans l'ésotérisme qui en découle[2].

---

[1] Éliphas Lévi, *Histoire de la magie* (Paris: Germer Baillière, 1860), 2-3.

[2] Daniel S. Larangé, « Les Lumières, métaphore maçonnique aux sources du romantisme français», *Études et recherches en philologie. Série langues romanes, Editura Universitatii din Pitesti* (2011) : 28-49.

L'abbé Alphonse-Louis Constant de Beaucourt (1810-1875) en retient l'articulation entre le monde naturel et l'imaginaire culturel ; il oriente sa recherche en direction d'une synthèse des sciences que son double, Éliphas Lévi, réalise en inaugurant la grande tradition de l'ésotérisme français du XIXème siècle. Poète, conteur, fabuliste, théologien, l'abbé Constant trouve dans l'épistémologie les outils nécessaires à la transmutation des savoirs et se transforme, tel l'alchimiste qui, par la découverte du Grand Œuvre comprend la poésie des éléments chimiques, en Éliphas Lévi, l'ésotériste qui synthétise le savoir scientifique en magie universelle. Comment l'imaginaire scientifique parvient-il à constituer une science de l'imaginaire?

Le destin proprement romanesque de la transformation du petit romantique abbé Constant en un grand mage et Grand-Maître maçonnique Éliphas Lévi se lit paradoxalement comme l'histoire du passage du désenchantement religieux d'une génération postrévolutionnaire au réenchantement pour une foi plus rationnelle et sociale. L'exploration de l'œuvre poétique, littéraire, politique et théologique, comme celle des traités ésotériques, va montrer l'usage réservé à l'argumentation scientifique et sa mythologisation moderne dans l'ésotérisme.

Du désenchantement social au réenchantement scientifique

La vie d'Alphonse-Louis Constant est marquée par une soif insatiable de savoir : il ne cesse de s'instruire alors même que le sort s'acharne contre lui et le décourage. Il naît dans une famille de modestes cordonniers et commence son instruction primaire sous la direction de l'abbé Jean-Baptiste Hubault Malmaison (1765-1864), qui organise dans sa paroisse un collège gratuit. Après le décès prématuré de son père, cet élève studieux et doué entre en 1825 au petit séminaire Saint-Nicolas du Chardonnet et devient le condisciple d'Alphonse Esquiros (1812-1876) dont l'amitié demeure indéfectible. Le supérieur du séminaire, l'abbé Antoine-Philippe Frère-Colonna (1786-1858), polytechnicien et théologien, laisse sur lui une empreinte indélébile : il lui inculque la thèse selon laquelle l'humanité, expulsée du sein de Dieu au moment du péché originel, y retourne peu à peu à travers l'histoire en se libérant de la matière, s'inspirant principalement de Giambattista Vico et des philosophes allemands comme Herder, Kant, Lessing et Schlegel.[3] Il

[3] Anne-Marie Amiot, *Baudelaire et l'illuminisme* (Paris: A.G. Nizet, 1982), 23-24.

l'initie également au magnétisme animal et à un joachimisme qui articule l'histoire chrétienne comme un cycle de répétitions de plus en plus évoluées.[4] Enfant, il se passionne déjà pour les grandes lois de la physique qui semblent organiser l'univers comme une horloge ou un automate et dont les mécanismes seraient tous liés entre eux dans un jeu d'interdépendance.

Une fois sa rhétorique terminée en 1830, le jeune homme passe deux années de philosophie au séminaire d'Issy[5] puis poursuit en théologie au séminaire de Saint-Sulpice où il est ordonné sous-diacre et tonsuré. Persuadé d'appartenir à une nouvelle génération postrévolutionnaire, précurseur d'importants changements sociaux et politiques, il cultive des idées millénaristes et se prépare à l'approche du siècle de l'amour caractérisé par « la grande réconciliation entre le loup et l'agneau»[6].

En 1835, il s'éprend d'une des catéchumènes dont il a la charge, Adèle Allenbach, l'une des plus belles femmes de Paris[7], et voit en elle la Sainte Vierge. Il lui voue alors une passion ardente et quitte sur un coup de tête le séminaire pour la rejoindre. Trop tard : la jeune fille l'a délaissé pour devenir actrice aux Délassements comiques et se faire courtiser par des artistes. Profondément abattue par la déchéance de son fils, la vieille mère du jeune homme se suicide.

Après ce drame, le jeune théologien se lie d'amitié avec la socialiste Flora Tristan (1803-1844) et collabore avec l'écrivain Alphonse Esquiros (1812-1876) à la revue *Les Belles Femmes de Paris*. Il révèle ainsi au public ses dons de dessinateur tout en s'introduisant dans le monde artistique parisien. Alors qu'il court de salon en salon, il rencontre et fait la connaissance, chez Delphine de Girardin, d'Honoré de Balzac, de Charles

---

[4] Henri de Lubac, *La postérité spirituelle de Joachim de Flore* (Paris/Namur: P. Lethielleux, 1979). Ernest Renan, « Joachim de Flore et l'Évangile éternel » in *Nouvelles études d'histoire religieuse* (Paris: Calmann-Lévy, 1899), 219 sqq. Karl Löwith, *Meaning in History* (Chicago: The University of Chicago, 1949), 145-149.

[5] Charles Grandidier, *Le Séminaire d'Issy : ancien château de Marguerite de Valois* (Paris: Chez l'auteur, 1853), 17.

[6] *Romanticism and Millenarianism*, ed. Tim Fulford (New York: Palgrave, 2002). Ann Taylor, *Visions of Harmony: A Study in 19th Century Millenarianism* (Oxford: Clarendon Press, 1987). Michael Barkun, *Disaster and the Millenium* (New Haven: Yale University Press, 1974). Ernest Lee Tuveson, *Millennium and Utopia: a study in the background of the idea of progress* [1st ed. 1949] (New York: Harper & Row, 1964), 207-219.

[7] *Les Belles femmes de Paris par des hommes de lettres et des hommes du monde* – 1re série (Paris : Au bureau Rue Christine, 1839), 266-269. Voir aussi : Charles Baudelaire, *Œuvres complètes* – tome 2, éds. Claude Pichois et Jean Ziegler (Paris: Gallimard, 1961), 1005.

Nodier, de Théophile Gautier, d'Alfred de Musset, d'Alphonse de Lamartine, d'Alexandre Dumas et de Jules Janin. Personnage hors norme, à la verve séduisante, il fascine son entourage par son érudition et sa vision mystique de l'Histoire, toute teintée de progressisme et d'enthousiasme pour les découvertes scientifiques.

Il y fait aussi la connaissance du Mapah Jean-Simon Ganneau (1800-1853)[8], fils de chapelier, qui, après des études pour devenir officier de santé, s'est intéressé à la phrénologie, théorie qui lui permet de définir les caractères de ses patients à partir de leur physionomie et des mensurations de leur crâne. Celui-ci défend aussi la doctrine de l'Évadisme, basée sur l'égalité des sexes, et a pour disciples le polytechnicien fouriériste Victor Considérant (1808-1893), le journaliste et critique d'art Théophile Thoré-Burger (1807-1869), l'éditeur et écrivain (P.J. Stahl) Pierre-Jules Hetzel (1814-1886), le journaliste, dramaturge et homme politique Félix Pyat (1810-1889). L'abbé Constant, admirateur de Franz Joseph Gall (1757-1828) et Johann Gaspar Spurzheim (1766-1832), découvre grâce à lui le lien de dépendance entre la physiologie et la cognition, lui confirmant ainsi le « prodige » de l'Incarnation, mettant en rapport l'humain et le mondain, la phrénologie et l'astrologie[9].

Ses supérieurs au collège de Juilly le maltraitent au point qu'il perd toute confiance dans l'institution humaine de l'Église et que, par écœurement, il compose, au grand dam du clergé, *La Bible de la Liberté* où il prétend que Dieu, étant le Mystère par excellence, échappe à l'entendement, y compris à celui de l'Église visible, et que seule une alliance modeste de l'Intelligence à l'Amour permettrait d'en révéler les traces : il faut cesser de Le chercher dans le monde mais Le rechercher en l'homme[10]. L'ouvrage est saisi à Versailles. L'auteur est arrêté, jugé pour hérésie, condamné et incarcéré à la prison de Sainte-Pélagie, où un grand nombre d'intellectuels peuvent méditer sur leurs écrits. Il y retrouve son ami Esquiros et fait la connaissance de l'abbé ultramontain Félicité de Lamennais (1782-1854), grand précurseur du catholicisme social et libéral[11].

---

[8] Éliphas Lévi, *Histoire de la magie avec une exposition claire et précise de ses procédés, de ses rites et de ses mystères* (Paris: Germer Baillère, 1860), 519-525.

[9] Éliphas Lévi, *Dogme et rituel de la haute magie* - tome 1er (Paris: Germer Baillère, 1861), 314-315 et *La Clef des grands mystères suivant Hénoch, Abraham, Hermès Trismégiste, et Salomon* (Paris: Germer Baillère, 1861), 226-227.

[10] Abbé Constant, *La Bible de la Liberté* (Paris: Le Gallois, 1841), 3.

[11] Éliphas Lévi, *Dogme et rituel de la haute magie* - tome 1er (Paris: Chacornac Frères, 1930), 47.

Son amie Flora Tristan lui confie, peu de temps avant sa mort, le manuscrit intégral d'une œuvre (qu'il édite sous le titre *L'Émancipation de la femme ou le Testament de la paria*), car elle voit en lui un grand défenseur de la cause féministe et le promoteur d'une mariologie moderne[12]. Il publie également un manifeste pacifique inspiré de Silvio Pellico (1789-1854) avec qui il correspond : *La Fête-Dieu ou le Triomphe de la paix religieuse* (1845). La science justement guidée par la foi est la voie de lumière vers la réconciliation entre le Créateur et ses créatures, et vers la restauration d'un âge d'or.

Les idées utopistes et humanistes absorbent l'abbé Constant. Deux mouvements suscitent chez lui de longues méditations : le saint-simonisme et le fouriérisme. Cela le conduit à vouloir démontrer par le biais des sciences le lien de parenté entre christianisme et socialisme. Selon lui, ce dernier serait la forme sécularisée d'une religion enfin libérée d'institutions archaïques.

> Le socialisme n'est déjà plus un système ; c'est la religion universelle de toutes les intelligences actives et de tous les cœurs jeunes et vivants.
> Le Christianisme va enfin réaliser ses promesses ; et la philosophie, en arrivant à l'unité par la synthèse, devient essentiellement religieuse. La raison va ainsi se réconcilier pour jamais avec la Foi.
> Le temps des superstitions est passé. On n'abusera plus les hommes avec des images mystérieuses, on ne les fera plus trembler par d'inexplicables énigmes.[13]

En 1845, dans le *Livre des larmes ou Le Christ consolateur : essai de conciliation entre l'Église catholique et la philosophie moderne*, il développe une philosophie religieuse qui rattache le cœur à l'intelligence : l'institution religieuse doit s'actualiser et se défaire de traditions sclérosées.

Encouragé par le chansonnier Pierre-Jean de Béranger (1780-1857), il compose aussi des chansons rassemblées dans un recueil empreint de mysticisme : *Les Trois Harmonies : chansons & Poésies* (1845). La poésie habille toute une réflexion poétique et sociale fondée sur la théorie des Harmonies qu'il se réapproprie. Le thème est en vogue, notamment grâce à Antoine Fabre d'Olivet (1767-1825), Alphonse de Lamartine (1790-

---

[12] Daniel S. Larangé, « Théologie mariale et discours féministe : la foi romantique en l'avenir du pouvoir féminin selon l'abbé Alphonse-Louis Constant, » *Tangence 94* (2010) : 113-134.

[13] Abbé Constant, *La Dernière Incarnation : légendes évangéliques du XIXe siècle* (Paris : Librairie sociétaire, 1846), 119.

1869)[14] et l'abbé Paul Lacuria (1806-1890)[15] D'après Jacques Crépet, Charles Baudelaire (1821-1867) s'en serait inspiré pour ses *Correspondances.*[16]

> Rien n'est muet dans la nature
> Pour qui sait en suivre les lois :
> Les astres ont une écriture,
> Les fleurs des champs ont une voix,
> Verbe éclatant dans les nuits sombres,
> Mot rigoureux comme des nombres [...][17]

Cette théorie des correspondances se fonde sur les principes de l'analogie et de l'allégorie. L'harmonie doit aboutir à la pacification des passions qui déchirent le siècle et instaurer une nouvelle ère de concorde, d'abondance et de félicité, tant individuelle que sociale. Le savoir et les sciences en particulier ont un rôle important à jouer pour une organisation rationnelle de la société. Sciences physiques et sciences naturelles viennent à se compléter et trouvent dans les mathématiques leur expression linguistique la plus adaptée, selon l'enseignement même du Philosophe Inconnu[18].

> Traitera-t-on avec dédain de mysticisme et d'illuminisme ces mathématiques vivantes, ces proportions des idées et des formes, cette révélation permanente dans la raison universelle, cet affranchissement de l'esprit, cette base inébranlable donnée à la foi, cette toute-puissance révélée à la volonté ?[19]

---

[14] *Harmonies poétiques et religieuses* (1830). Voir Aurélie Loiseleur, *L'Harmonie selon Lamartine : utopie d'un lieu commun* (Paris : Honoré Champion), 2005.

[15] Paul Lacuria, *Les Harmonies de l'Être exprimées par les Nombres* [1re éd. 1844] (New York : Arno Press), 1976.

[16] Charles Baudelaire, *Œuvres complètes – tome 1*, ed. Claude Pichois (Paris: Gallimard, 1975), 840. Jean Pommier, *La Mystique de Baudelaire* (Paris: Les Belles Lettres, 1932). Frank Paul Bowman, *Éliphas Lévi visionnaire romantique* (Paris : Puf, 1969), 6-7.

[17] M.A. Constant de Baucour, *Les Trois Harmonies : Chansons & Poésies* (Paris: Fellens et Dufour, 1845), 300.

[18] Louis-Claude de Saint-Martin, *Les Nombres*, ed. Robert Amadou (Paris: Documents martinistes, 1983).

[19] Éliphas Lévi, *Dogme et rituel de la haute magie - tome 2* (Paris: Germer Baillère, 1856), 310.

Cette vision pétrie d'un platonisme assez courant et diffus[20] parmi les mystiques de son époque récupère l'organon aristotélicien, prônant l'usage systématique des formes de la pensée et de la démonstration comme condition scientifique, afin de persuader de la validité de ses énoncés. Dieu est une nécessité. Son christianisme, foncièrement christocentrique, est spéculatif : la religion est un système imaginaire profondément inscrit dans chaque homme et indispensable à la cohésion sociale. Ce paradoxe doit être dépassé par rapport à une position trop souvent mal comprise, tant à son époque qu'aujourd'hui. Ainsi répond-t-il à la question de l'existence historique de Jésus en distinguant la réponse scientifique de la réponse spirituelle et morale :

> La conception de Jésus-Christ est définie par le symbole : il a été conçu du Saint-Esprit, est né de la Vierge Marie. C'est donc une réalité symbolique et dogmatique. Vous voulez de plus savoir, je pense, si c'est une réalité historique et scientifique ? Certainement non, d'après les règles si simples que j'ai établies dans mes livres.
>
> Tout ce qui est de foi échappe à la science.
> Tout ce qui est de science échappe à la foi.
> Il n'existe aucun document historique sur la personne de Jésus.[21]

Très bien introduit dans les cercles saint-simonien et fouriériste et dans les organes, il y traite à sa guise de la question sociale en cherchant des réponses scientifiques. L'industrialisme et le socialisme finissent par le passionner. La notion théologique de Providence est remplacée par celle, plus sécularisée, de Progrès[22], et le Peuple élu le conduit au messianisme. Sa déception vis-à-vis de l'Église catholique et sa défection de toute forme d'institution le conduisent à repenser la religion en termes spirituels, voire à défendre une mystique sociale.

D'ailleurs, il fréquente assidûment Charles Fauvety (1813-1894), un philosophe maçonnique d'origine protestante, qui fonde la Religion laïque en recherchant une harmonie entre la foi et la raison, concevant Dieu à travers un panthéisme et l'immortalité de l'âme comme une probabilité. Les deux hommes fondent en 1845 la revue mensuelle *La Vérité sur toutes choses*. Ils partagent une certaine conception mystique du politique, selon

---

[20] Michel Brix, *Le Romantisme français : esthétique platonicienne et modernité littéraire* (Leuven: Peeters, 1999).

[21] Éliphas Lévi, *Cours de philosophie occulte : lettres au baron Spédalieri : de la kabbale et de la science des nombres*, ed. Christiane Buisset (Paris: Guy Trédaniel, 1988), 145.

[22] Karl Löwith, *Meaning in History* (Chicago: University of Chicago Press, 1957), 60-103.

laquelle la réconciliation nationale serait la pierre angulaire de l'édifice d'une nouvelle société universelle annonçant le rétablissement de la Jérusalem céleste sur terre[23].

Science et foi

L'articulation entre science et foi doit être nuancée. L'imaginaire scientifique ouvre la voie aux sciences de l'imaginaire. Le terme même d'imaginaire doit être ici interprété en tant qu'élément constitutif de l'identité de chaque individu, dans sa dimension psychologique et sociale. Le discours magique, si prisé dans le romantisme,[24] témoigne aussi d'aspirations sociales et politiques plus profondes, d'espoirs et de craintes, de toute une représentation d'un réel configuré où le platonisme se terre derrière la parade de l'aristotélisme, et réciproquement. Selon Éliphas Lévi, quand la science montre ses limites, la passion et l'amour prennent le relais, stimulant à leur tour la quête des savoirs.

> Où la science s'arrête, la foi commence, et la foi puise ses révélations hypothétiques dans les aspirations du cœur toujours plus insatiable et plus courageux que l'esprit. Or, le cœur humain peut s'appuyer sur une force ou se laisser égarer par une faiblesse […].
>
> Pour suppléer à l'insuffisance de la science, on peut donc faire appel ou à l'exaltation des sentiments, ou à la surexcitation des instincts lâches.[25]

L'amour reste le grand principe de la science romantique : de même que l'amour est une voie de connaissance traditionnelle, la science moderne progresse par amour du savoir. Le christianisme, incarné par le Fils de Dieu, l'Amour poussé au sacrifice, est donc bien l'expression la plus adéquate pour l'épanouissement des savoirs, même si les institutions humaines la pervertissent.

---

[23] Charles Fauvety, *Le Représentant du Peuple* (Paris :1848) ; *Programme politique* (Paris: Braux, 1849) ; *Catéchisme d'économie politique et industrielle : question du libre-échange* (Paris: Moque, 1849).

[24] Yves Vadé, *L'Enchantement littéraire : écriture et magie de Chateaubriand à Rimbaud* (Paris: Gallimard, 1990).

[25] Éliphas Lévi, *La Science des esprits : révélation du dogme secret des kabbalistes, esprit occulte des évangiles, appréciations des doctrines et des phénomènes spirites* (Paris: Germer Baillière, 1865), 291.

> [C]omme si la foi, qui est la confiance de l'amour, n'était pas aussi durable et souvent plus invincible que la raison ; comme si l'amour, qui est la raison de la foi, n'était pas aussi la raison de l'existence des êtres soumis aux recherches de la science. Quand elle ne sait plus, il commence à croire, et quand la raison épuisée s'arrête et tombe au seuil de l'infini, la foi ouvre ses ailes, s'élance, déchire les nuages, fait descendre jusqu'à terre l'échelle lumineuse de Jacob et sourit doucement en tendant la main à sa sœur.[26]

Il courtise à l'Institution Chandeau de Choisy-le-Roy la sous-maîtresse Eugénie Chenevier lors de ses promenades en compagnie de sa jeune élève Marie-Noémi Cadiot (1832-1888). Or c'est cette dernière qui, après lui avoir envoyé des lettres enflammées, fugue pour le rejoindre. Le père de la jeune fille exige alors le mariage, sous peine de le dénoncer pour détournement de mineure : Constant se résigne.

La révolution de février 1848 lui donnant plus de liberté, il fonde avec ses amis Esquiros et Le Gallois un cercle politique, le *Club de la montagne*, composé de travailleurs. Lors des journées de juin, l'insurrection des classes laborieuses conduite par la réaction tente de faire périr la République naissante. Le 23 juin 1848, un marchand de vin qui lui ressemble est fusillé à sa place.

À cette époque, il lit la *Kabbala Denudata* de Christian Knorr de Rosenroth (1636-1689), étudie les écrits de Jacob Böhme (1575-1624), de Louis-Claude de Saint-Martin (1743-1803), d'Emanuel Swedenborg (1688-1772), d'Antoine Fabre d'Olivet (1767-1825), d'Augustin Xaho (1811-1858), et de Johann Joseph von Görres (1776-1848). Il rencontre aussi, parmi les polytechniciens, Auguste Comte (1798-1857) avec qui il partage un positivisme teinté de mysticisme. Il reconnaît dans la loi des trois états (théologique, métaphysique et positif)[27] le cycle historique décrit par de Joachim de Flore. Il lui emprunte aussi la classification rationnelle et positive des sciences en fonction de leurs interdépendances et interactions. Au sommet du savoir absolu, il place la « science des esprits ».

Fin 1850, il fait la connaissance de l'abbé Jacques-Paul Migne (1800-1875), fondateur et directeur de la librairie ecclésiastique de Montrouge, qui lui commande pour sa collection un *Dictionnaire de la littérature chrétienne*. L'ouvrage, de plus de mille pages, devient rapidement une référence et confirme son adhérence orthodoxe au catholicisme, dont jamais il ne se défera.

---

[26] Ibid., 42.

[27] Auguste Comte, *Plan des travaux scientifiques nécessaires pour réorganiser la société*, imprimé pour la première fois par Claude Henri de Saint-Simon dans *Du Contrat social* (Paris: Chez les marchands de nouveautés, 1822).

En même temps, il commence la rédaction du *Dogme et rituel de la haute magie*, sous le pseudonyme hébraïsant d'Éliphas Lévi Zahed. Suivant l'exemple de Claude Henri de Rourvoy de Saint-Simon (1760-1825) et d'Auguste Comte, il cherche à concilier les principes de rationalité scientifique et d'amour[28] qui sustente sa conception de la magie. Éliphas Lévi renoue ainsi avec la poésie scientifique de l'Humanisme, bien avant René Ghil (1862-1925) qu'il inspire, et ses conversations avec Camille Flammarion (1842-1925) inspirent sa contemplation :

> L'amour s'est révélé dans la nature par l'harmonie, le triangle sacré fait briller sa lumière, le nombre trois complète la divinité soit dans ton idéal soit dans la connaissance transcendante de toi-même. Ton intelligence est devenue mère parce qu'elle a été fécondée par le génie de la foi [...].
>
> Lève maintenant les yeux et contemple le ciel. Vois la splendeur et la régularité des astres. Prends le compas et le télescope de l'astronome et monte de prodige en prodige, calcule le retour des comètes et la distance des soleils, tout cela se meut selon les lois d'une hiérarchie admirable. Toute cette immensité pleine de mondes absorbe et surpasse tous les efforts de l'intelligence humaine.[29]

Son discours magique s'inspire explicitement du positivisme religieux, surtout en ce qui concerne le magnétisme (la loi des attractions/répulsions) et le spiritisme (les esprits).

> La foi n'est qu'une superstition et une folie si elle n'a la raison pour base, et l'on ne peut supposer ce qu'on ignore que par analogie avec ce qu'on sait. Définir ce qu'on ne sait pas, c'est une ignorance présomptueuse; affirmer positivement ce qu'on ignore, c'est mentir.[30]

Le principe d'objectivité est fondé sur un théologisme consistant à placer Dieu au centre de tous les savoirs, car l'esprit humain est conçu comme un reflet de l'Esprit divin : « La science suppose nécessairement Dieu, étudie l'esprit de l'homme dans ses aspirations les plus hautes, examine les hypothèses relatives aux esprits inconnus et rejette les fantômes ».[31] La science positive est devenue indispensable à la réalisation de l'humanité. Elle seule pourra apporter la paix au monde :

---

[28] Georges Dumas, *Psychologie de deux messies positivistes : Saint-Simon et Auguste Comte* (Paris: Félix Alcan, 1905).

[29] Éliphas Lévi, *Le Grand Arcane ou l'occultisme dévoilé* [1re éd. 1868-1869], Paris, Guy Trédaniel, 1980, p. 299-300.

[30] Éliphas Lévi, *Dogme et rituel de la haute magie - tome 1er*, 377.

[31] Éliphas Lévi, *La Science des esprits*, 2.

> La science est calme, elle est douce parce que l'éternité lui appartient, elle est le soleil des esprits et la chaleur des âmes, elle est le temple du culte universel, car elle change le monde entier en un temple où l'on adore l'inaccessible raison et la justice indéfectible.[32]

La priorité est paradoxalement accordée dans ses ouvrages au savoir scientifique qui valide toute profession de foi.

> La vraie foi ne saurait être en contradiction avec la vraie science. Aussi, toute explication du dogme qui démontrerait la fausseté doit-elle être réprouvée par la foi.
>
> Nous ne sommes plus au temps où l'on disait : je crois parce que c'est absurde. Nous devons dire maintenant : je crois parce qu'il serait absurde de ne pas croire ;
>
> Credo quia absurdum non credere.
>
> La science et la foi ne sont plus deux machines de guerre prêtes à s'entrechoquer, ce sont les deux colonnes destinées à soutenir le fronton du peuple de la paix. Il faut nettoyer l'or du sanctuaire si souvent terni par la crasse sacerdotale.[33]

Or il ne s'agit surtout pas de confondre la science avec la foi qu'un dialogue permanent rassemble[34] :

> Si jusqu'à présent la science et la foi ont paru inconciliables, c'est qu'on a toujours essayé en vain de les mêler ensemble et de les confondre. Il n'y a qu'un moyen de les concilier, c'est de les distinguer et de les séparer l'une de l'autre d'une manière complète et absolue. Consulter le pape lorsqu'il s'agit de la démonstration d'un théorème, soumettre à un mathématicien une distinction théologique, ce seraient deux absurdités équivalentes. L'immaculée conception de la Vierge n'est pas une question d'embryologie et la table des logarithmes n'a rien de commun avec les tables de la Loi. La science est forcée d'admettre ce qui est démontré et la foi quand elle est réglée par une autorité qu'il est raisonnable et nécessaire d'admettre, ne peut rien rejeter de ce qui est article de foi.

---

[32] Éliphas Lévi, *Les Mystères de la Kabbale ou L'Harmonie occulte des Deux Testaments* [1re éd. 1861] (Paris: Guy Trédaniel, 1991), 256.

[33] Éliphas Lévi, *Le Grand Arcane ou l'occultisme dévoilé* [1re éd. 1868-1869] (Paris: Guy Trédaniel, 1980), 49.

[34] Éliphas Lévi, *La Clé des grands mystères suivant Hénoch, Abraham, Hermès Trismégiste, et Salomon* (Paris: Germer Baillère, 1861), 96-104.

> La science ne démontrera jamais que Dieu et l'âme n'existent pas et l'Église a été forcée de se dédire devant la démonstration des systèmes de Copernic et de Galilée. Cela prouve-t-il qu'elle peut se tromper en matière de foi ? Non, mais qu'elle doit rester dans son domaine. Elle-même ne prétend pas que Dieu lui ait révélé les théorèmes de la science universelle[35].

Ce principe est au cœur de la méthode kabbalistique qu'il enseigne par correspondance :

> Sans la foi, la science conduit au doute ; sans la science, la foi mène à la superstition. Les deux réunies donnent la certitude et pour les unir il ne faut jamais les confondre. L'objet de la foi, c'est l'hypothèse et elle devient une certitude quand l'hypothèse est nécessitée par l'évidence ou par les démonstrations de la science.
>
> La science constate des faits. De la répétition des faits elle préjuge des lois. La généralité des faits en présence de telle ou telle force démontre l'existence des lois. Les lois sont nécessairement voulues et dirigées par l'intelligence. L'unité dans les lois fait supposer l'unité de l'intelligence législative. Cette intelligence que nous sommes forcés de supposer d'après les œuvres manifestes, mais qu'il nous est impossible de définir, est ce que nous appelons Dieu ![36]

Le messianisme du mathématicien, ingénieur et philosophe polonais Jozef Maria Hoénë-Wroński (1776-1853)[37] confirme son eschatologie de type apocalyptique. Il faut à tout prix retrouver l'Universalité à travers le particularisme car « l'humanité n'a qu'une grande âme qui passe de génération en génération »[38]. L'histoire est fondamentalement gnostique (la quête du Savoir Absolu) et commence par l'éclatement de l'Unité primordiale, la séparation et la chute dans le Multiple. Elle s'oriente vers la réunion de tous les symboles car la matière se contracte et le retour à l'Unité marquera le Plérôme. Il en appelle ainsi à une réforme de la société : « La forme de notre société est usée, et son âme veut passer à des peuples nouveaux. Alors tout se constituera selon la loi de l'unité.[39] »

---

[35] Éliphas Lévi, *Le Grand Arcane ou l'occultisme dévoilé* [1re éd. 1868-1869], 358-359.

[36] Éliphas Lévi, *Le Livre des splendeurs : études sur les origines de la kabbale* [1re éd. 1869-1870] (Paris: Guy Trédaniel, 1979), 233.

[37] Hoënë Wroński, *Messianisme ou réforme absolue du savoir humain ; nommément : réforme des mathématiques comme prototype de l'accomplissement final des sciences, et réforme de la philosophie comme base de l'accomplissement final de la religion* (Paris: Firmin Didot, 1847).

[38] Abbé Constant, *La Bible de la Liberté* (Paris: Le Gallois, 1841), 13.

[39] Ibid., 24.

Cette quête le conduit à penser la complémentarité entre les savoirs, par conséquent à développer une gnose qui s'articule sur des triades :

L'âme → sciences des Lettres
La conscience → sciences des proportions, des rapports mathématico-géométriques et de la musique
Le corps → sciences naturelles et techniques

Le mal a été nécessaire pour éprouver l'homme et le purifier afin qu'il atteigne sa maturité par l'expérience. La triple répartition joachimite se retrouve dans l'expérimentation des trois principes issus de la Révolution : Liberté – Égalité – Fraternité. Afin que le Plérôme se réalise, il faut libérer l'âme, établir une égalité entre les consciences et une fraternité entre les êtres.

Pour y parvenir, il faut un gouvernement de sages qui soient moraux, philosophes et instruits afin d'élire un roi, prêt à sacrifier son individualité à la cause collective, véritable Christ réhabilité au sein d'un peuple d'élus.

Hermétisme et herméneutique

L'hermétisme de Constant relève du travail herméneutique. L'exégèse biblique lui donne accès à l'encyclopédie chrétienne des fables et des récits. Son art mantique est appuyé par le Prognomètre de Wroński, appareil complexe qui lui permet de lire les signes dispersés dans la nature car la réalité est un tissu de symboles dont le contenu change en fonction du temps et de l'espace[40]. Ces signes, images et symboles sont des représentations « polarisées », configurées au sein du champ magnétique diffus dans l'univers. Ils subissent le jeu d'attraction-répulsion :

> La science constate les phénomènes de l'aimant astral et de l'aimant minéral. L'aimant animal se manifeste tous les jours par des faits que la science observe avec défiance, mais qu'elle ne peut déjà plus nier, bien qu'elle attende avec raison pour les admettre qu'on en puisse terminer l'analyse par une synthèse incontestable.

---

[40] Pierre Laurant, *L'Ésotérisme* (Paris: Le Cerf et Fides, 1993), 108 and *L'Ésotérisme chrétien en France* (Lausanne: L'Âge d'homme, 1992), 62-63. Voir aussi : Christiane Buisset, *Éliphas Lévi : sa vie, son œuvre, ses pensées* (Paris: Guy Trédaniel, 1984), 159 and Auguste Viatte, *Victor Hugo et les illuminés de son temps* (Montréal: L'Arbre, 1942), 82. Alain Mercier, *Éliphas Lévi et la pensée magique du XIXème siècle*. (Paris: Seghers, 1974), 105-106.

> On sait que l'aimantation produite par le magnétisme animal détermine un sommeil extraordinaire pendant lequel l'âme du magnétisé tombe sous la dépendance du magnétiseur avec cette particularité que la personne endormie semble laisser oisive sa vie propre et particulière pour manifester uniquement les phénomènes de la vie universelle. Elle reflète la pensée des autres, voit autrement que par les yeux, se rend présente partout sans avoir conscience de l'espace, perçoit les formes bien mieux que les couleurs, supprime ou confond les périodes du temps, parle de l'avenir comme s'il était passé et du passé comme s'il était à venir, explique au magnétiseur ses propres pensées et jusqu'aux reproches secrets de la conscience, évoque dans son souvenir les personnes auxquelles il pense et les décrit de la manière la plus exacte sans que le somnambule ou la somnambule les ait jamais vues, parle le langage de la science avec le savant et celui de l'imagination avec le poète, découvre les maladies et en devine les remèdes, donne souvent de sages conseils, souffre avec celui qui souffre et pousse parfois d'avance un cri douloureux en vous annonçant des tourments qui doivent venir.
>
> Ces faits étranges mais incontestables nous entraînent nécessairement à conclure qu'il existe une vie commune pour toutes les âmes, ou du moins une sorte de réflecteur commun de toutes les imaginations et de toutes les mémoires où nous pouvons nous voir les uns les autres, comme il arrive dans une foule qui passe devant un miroir.[41]

Le flux magnétique est conçu comme un ensemble de lumières vives et mortes qui se combinent différemment. Le discours mystique plonge finalement dans des variations néoplatoniciennes chargées de réminiscences de l'hermétisme médiéval, plus ou moins connues. Le véritable magnétiseur est un voyant divin, au contraire du fou qui exerce une fascination diabolique. Il communique son inspiration de la même manière que le poète enthousiaste dans le dialogue platonicien intitulé *Ion*. Le mesmérisme finit ainsi par se confondre avec le magnétisme. Constant fait une lecture platonisante de la kabbale juive, ce qui lui permet de jongler avec les harmonies divines, les analogies et les oppositions. Ces harmonies se configurent en fonction d'un jeu d'attraction-répulsion qui rappelle le discours mystico-social de Charles Fourier, fondé notamment sur les principes de la physique newtonienne[42].

---

[41] Éliphas Lévi, *Le Grand arcane* [1re éd. 1898] (Paris: La Maisnie/Guy Trédaniel, 1980), 7.

[42] Charles Fourier, *Traité de l'unité universelle* (Paris: Société pour la propagation et la réalisation de la théorie de Fourier, 1822-1823, tome 1, xcviii, 4-5, 20, 26, 32-35, 37, 44, 56, 101, 105, tome 2, 64, 71, 114-115, 276, 348 ; tome 3, 266-267 ; tome 4, 273, 285, 375).

> [Dieu] maintient l'harmonie entre les astres par les lois de l'attraction et c'est par les mêmes lois qu'il a réglé d'avance les destinés des hommes. Les attractions sont donc proportionnelles aux destinées.
>
> Or, les attractions diverses ont toutes les volontés dans différents cercles magnifiquement coordonnées entre eux. Une immense chaîne d'harmonie rattache à Dieu toutes ses œuvres, et de série en série il distribue la vie à tous les êtres.
>
> La série distribue les harmonies. Des relations analogiques existent entre les séries et sont comme les degrés de l'échelle de la science, de cette échelle d'or qui vit autrefois un prophète pendant son sommeil, et qui aidait les esprits à monter de la terre au ciel et à redescendre du ciel à la terre.
>
> Voilà les bases de la science nouvelle ; elles sont appuyées sur toutes les traditions philosophiques et religieuses ; et nous pouvons dire que ce ne sont pas les principes d'une École, mais les théorèmes de la Science la plus achevée et les dogmes incontestables de la vraie Religion universelle.[43]

Dans un registre plus prosaïque, Mme Constant a entamé une liaison avec le marquis de Montferrier, le beau-frère de Wroński, et décide de partir s'installer définitivement à Lausanne. Profondément blessé, Constant se remet au travail pour tenter d'oublier son chagrin.

Au printemps 1854, il se rend à Londres, y rencontre le Dr John Ashburner (1793-1878), célèbre pour ses travaux sur le rêve[44] et le magnétisme animal[45], membre du *Charing Cross Spirit Power Circle*, et l'écrivain Sir Edward Bulwer-Lytton (1803-1873) qui l'introduit dans les cercles rosicruciens. Il y tente une série d'évocations au cours desquelles le fantôme du néopythagoricien Apollonius de Tyane[46] apparaît pour lui indiquer l'endroit où se trouve le *Nycthemeron*[47]. Ce qui paradoxalement va à l'encontre des préceptes d'Éliphas Lévi, opposé aux expériences de magie. Celui-ci fit même promettre à ses disciples de ne jamais tenter la moindre expérience, et de ne s'occuper que de la partie spéculative de la philosophie occulte.

Revenu en France en août 1854, Éliphas Lévi loge quelque temps dans l'atelier de peintre de son ami chiromancien Adolphe Desbarolles (1804-1886). Il achève *Dogme et rituel de la haute magie*, qui enfin lui fait

---

[43] Constant, *La Dernière Incarnation*, 119-120.

[44] John Ashburner, *On a Theory of Sleep* (London: Walton & Mitchell, 1848).

[45] John Ashburner, *Facts in clairvoyance: extracted from the Zoist, with observations on mesmerism, and its application to the philosophy of medicine, and to the cure of disease* (New York: H. Baillière, 1848).

[46] Jan van Rijckenborgh, *Le Nycthemeron d'Apollonius de Tyane* (Haarlem: Rozekruis, 1994).

[47] Éliphas Lévi, *Dogme et rituel de la haute magie - tome 1er*, 132-135.

connaître le succès, sans la fortune. Ses manuels de magie forment des recueils éclectiques où son savoir encyclopédique sert une réflexion philosophico-mystique sur le monde, la société et l'homme, et ce, par une écriture où prose poétique, fable, dialogue et essai se combinent, renouant ainsi avec les grands traités de théologie et d'alchimie médiévaux. Le genre littéraire réussit de sorte qu'il fonde, en 1855, avec Charles Fauvety la *Revue philosophique et religieuse*. En 1859, la publication de l'*Histoire de la magie* lui attire définitivement la sympathie des ésotéristes français férus de sciences, tels Henri Delaage (1825-1882), Luc Desages (1820-1903), Paul Auguez (1792-1864), Jean-Marie Ragon (1781-1862), Jean-Henri Favre (1823-1915) et le Dr Fernand Rozier (1839-1922). La publication de *la Clef des grands mystères* en 1861 clôt la trilogie commencée avec *Histoire de la magie* et *Dogme et rituel de la haute magie*. Il travaille beaucoup, initiant par correspondance à l'occultisme des érudits appartenant à la plus haute aristocratie, jusqu'à l'évêque d'Évreux, Mgr Devoucoux, à qui il enseigne la kabbale. Grâce à l'argent perçu en rémunération de ses leçons, il vit dans un relatif confort matériel, enrichissant sans cesse sa bibliothèque. Avec le comte Alexandre Branicki (1821-1877), hermétiste, il mène des expériences du Grand-Œuvre dans un laboratoire installé au château de Beauregard, à Villeneuve-Saint-Georges, propriété de la veuve d'Honoré de Balzac, Ewelyna Hańska (1801-1882). Le gendre de son hôtesse, le comte Jerzy Wandalin-Mniszech (1822-1881), les rejoindra quelque temps plus tard.

> L'exemple de l'extraordinaire transformation de l'abbé Alphonse-Louis Constant en Éliphas Lévi témoigne de l'alchimie qui s'opère au XIXème siècle entre le catholicisme social et l'occultisme scientiste. Il illustre également la montée du positivisme à travers un discours mystique où la religion, la question sociale et les sciences se mélangent afin de produire de nouvelles idéologies qui vont en se « totalisant ».[48]

Le penseur Constant-Lévi est bien le fruit du romantisme qui se construit sur des aspirations paradoxales et dans un effort continu de séculariser la foi, afin d'en extraire l'essence du savoir. La saveur qui ressort de cette opération se retrouvera et perdurera dans toutes les tentatives esthétiques, voire éthiques, de son siècle et des siècles

[48] Ernest Seillière, *Le Mal romantique : essai sur l'impérialisme irrationnel* (Paris: Plon-Nourrit et Cie, 1908) ; Hannah Arendt, *The Origins of Totalitarism* (New York: Harcourt, Brace, 1951). Norman Cohn, *The Pursuit of the Millenium: revolutionary messianism in medieval and Reformation Europe and its bearing on modern totalitarian movements* (New York: Oxford University Press, 1970).

suivants[49]. L'homme romantique est appelé à conjuguer sa conscience de la science avec la science de sa conscience, autrement dit sa foi dans le savoir avec la connaissance de soi. Ainsi, la science moderne, la sociologie et la psychanalyse sont-elles les fruits de leur temps.

À travers sa recherche de l'universel, le romantique aura finalement ouvert une voie vers l'individualisme social (généralisé) qui caractérise nos sociétés occidentales si fières de leur altruisme d'apparat, et vers la croyance en une solide rationalité assurée par une science institutionnalisée. Or, si la science prétend chasser l'imaginaire au nom du profane, elle s'en nourrit secrètement. Le discours magique d'Éliphas Lévi apparaît alors aussi comme un discours social sur une forme de société convaincue, et bercée ici encore par l'illusion de la croyance (opposée à la connaissance du cœur), que seule la science est à même de la guérir des maux dont elle souffre. À ce titre, le discours magique est un discours de l'imaginaire social sur certains aspects irrationnels de la science, en particulier l'usage néfaste qui peut en être fait.

Œuvres citées

Les Belles femmes de Paris par des hommes de lettres et des hommes du monde – 1re série. Paris: Au bureau Rue Christine, 1839.

Amiot, Marie. *Baudelaire et l'illuminisme.* Paris: A.G. Nizet, 1982.

Arendt, Hannah. *The Origins of Totalitarism*. New York: Harcourt, Brace, 1951.

Ashburner, John. *Facts in clairvoyance: extracted from the Zoist, with observations on mesmerism, and its application to the philosophy of medicine, and to the cure of disease*. New York: H. Baillière, 1848.

—. *On a Theory of Sleep*. London: Walton & Mitchell, 1848.

Barkun, Michael. *Disaster and the Millenium.* New Haven: Yale University Press, 1974.

Baudelaire, Charles. *Œuvres complètes – tome 1*, édité par Claude Pichois. Paris: Gallimard, 1975.

—. *Œuvres complètes – tome 2*, édité par Pichois et Jean Ziegler. Paris: Gallimard, 1976.

Bowman, Frank Paul. *Éliphas Lévi visionnaire romantique*. Paris: Puf, 1969.

[49] Philippe Muray, *Le XIXème siècle à travers les âges* (Paris: Denoël, 1984). Alain Mercier, *Les Sources ésotériques et occultes de la poésie symboliste* (Paris: Nizet, 1969 et 1974).

Brix, Michel. *Le Romantisme français : esthétique platonicienne et modernité littéraire*. Leuven: Peeters, 1999.

Buisset, Christiane. *Éliphas Lévi : sa vie, son œuvre, ses pensées*. Paris: Guy Trédaniel, 1984.

Chacornac, Paul. *Éliphas Lévi (1810-1875), rénovateur de l'occultisme en France* (1re éd. 1926). Paris: Éditions traditionnelles, 1989.

Cohn, Norman. *The Pursuit of the Millenium: revolutionary messianism in medieval and Reformation Europe and its bearing on modern totalitarian movements*. New York: Oxford University Press, 1970.

Comte, Auguste. *Plan des travaux scientifiques nécessaires pour réorganiser la société*. Paris: Chez les marchands de nouveautés, 1822.

Constant de Baucour, M.A. *Les Trois Harmonies : Chansons & Poésies*. Paris: Fellens et Dufour, 1845.

Constant, Abbé. *Harmonies poétiques et religieuses,* Paris: Librairie sociétaire, 1830.

—. *La Bible de la Liberté*. Paris: Le Gallois, 1841.

—. *La Dernière Incarnation : légendes évangéliques du XIXe siècle*. Paris: Librairie sociétaire, 1846.

De Lubac, Henri. *La postérité spirituelle de Joachim de Flore*. Paris/Namur: P. Lethielleux, 1979.

Dumas, Georges. *Psychologie de deux messies positivistes : Saint-Simon et Auguste Comte*. Paris: Félix Alcan, 1905.

Fauvety, Charles. *Catéchisme d'économie politique et industrielle : question du libre-échange*. Paris: Moque, 1849.

—. *Programme politique*. Paris: Braux, 1849.

Fourier, Charles. *Traité de l'unité universelle*. Paris: Société pour la propagation et la réalisation de la théorie de Fourier, 1822-1823, tome 1-4.

Grandidier, Charles. *Le Séminaire d'Issy : ancien château de Marguerite de Valois*. Paris: Chez l'auteur, 1853.

Hoëné, Wroński. *Messianisme ou réforme absolue du savoir humain ; nommément : réforme des mathématiques comme prototype de l'accomplissement final des sciences, et réforme de la philosophie comme base de l'accomplissement final de la religion*. Paris: Firmin Didot, 1847.

Lacuria, Paul. *Les Harmonies de l'Être exprimées par les Nombres* [1re éd. 1844]. New York: Arno Press, 1976.

Larangé, Daniel S. « Théologie mariale et discours féministe : la foi romantique en l'avenir du pouvoir féminin selon l'abbé Alphonse-Louis Constant. » *Tangence* 94 (2010) : 113-134.

—. « Les Lumières, métaphore maçonnique aux sources du romantisme français » *Études et recherches en philologie : Série langues romanes : Editura Universitatii din Pitesti* (2011) : 28-49.

Laurant, Pierre. *L'Ésotérisme chrétien en France.* Lausanne : L'Âge d'homme, 1992.

—. *L'Ésotérisme.* Paris: Le Cerf et Fides, 1993 ;

Lévi, Éliphas. *Histoire de la magie avec une exposition claire et précise de ses procédés, de ses rites et de ses mystères*. Paris: Germer Baillère, 1860.

—. *Dogme et rituel de la haute magie - tome 1er*. Paris: Guiraudet et Jouaust, 1854.

—. *Dogme et rituel de la haute magie - tome 2*. Paris: Guiraudet et Jouaust, 1856.

—. *La Clé des grands mystères suivant Hénoch, Abraham, Hermès Trismégiste, et Salomon*. Paris: Germer Baillère, 1861.

—. *La Science des esprits : révélation du dogme secret des kabbalistes, esprit occulte des évangiles, appréciations des doctrines et des phénomènes spirites*. Paris: Germer Baillière, 1865.

—. *Le Livre des splendeurs : études sur les origines de la kabbale* [1re éd. 1869-1870]. Paris: Guy Trédaniel, 1979.

—. *Le Grand Arcane ou l'occultisme dévoilé* [1re éd. 1868-1869]. Paris: Guy Trédaniel, 1980.

—. *Cours de philosophie occulte : lettres au baron Spédalieri : de la kabbale et de la science des nombres*, édité par Christiane Buisset, 145. Paris: Guy Trédaniel, 1988.

—. *Les Mystères de la Kabbale ou L'Harmonie occulte des Deux Testaments* [1re éd. 1861]. Paris: Guy Trédaniel, 1991.

Loiseleur, Aurélie. *L'Harmonie selon Lamartine : utopie d'un lieu commun*. Paris: Honoré Champion, 2005.

Löwith, Karl. *Meaning in History*. Chicago: The University of Chicago, 1949.

Mercier, Alain. *Éliphas Lévi et la pensée magique au XIXème siècle*. Paris: Seghers, 1974.

Muray, Philippe. *Le XIXème siècle à travers les âges*. Paris: Denoël, 1984.

Tristan, Flora. *L'Émancipation de la femme ou le Testament de la paria.* Paris: Au bureau de la direction de la vérité, 1846.

Pommier, Jean. *La Mystique de Baudelaire*. Paris: Les Belles Lettres, 1932.

Renan, Ernest. "Joachim de Flore et l'Évangile éternel. In *Nouvelles études d'histoire religieuse*, 219 sqq. Paris: Calmann-Lévy, 1899.

Rijckenborgh, Jan van. *Le Nycthemeron d'Apollonius de Tyane*. Haarlem: Rozekruis, 1994.

—. *Romanticism and Millenarianism*, edited by Tim Fulford. New York: Palgrave, 2002.

Seillière, Ernest. Le Mal romantique : *essai sur l'impérialisme irrationnel*. Paris: Plon-Nourrit et Cie, 1908.

Taylor, Ann. *Visions of Harmony: A Study in 19th Century Millenarianism*. Oxford: Clarendon Press, 1987.

Tuveson, Ernest Lee. *Millennium and Utopia: a study in the background of the idea of progress* [1re éd. 1949]. New York: Harper & Row, 1964.

Vadé, Yves. *L'Enchantement littéraire : écriture et magie de Chateaubriand à Rimbaud*. Paris: Gallimard, 1990.

Viatte, Auguste. *Victor Hugo et les illuminés de son temps*. Montréal : L'Arbre, 1942.

LES NOCES ALCHIMERIQUES DE CHRISTIAN ROSE-CROIX

SEBASTIEN GREGOV

> Dévoilés, les mystères s'avilissent,
> Et profanés, ils perdent leur force.
> Ne jette donc pas de perles aux pourceaux
> Et ne fais point aux ânes une litière de roses.
> —Incipit des *Noces Alchimiques de Christian Rose-Croix, anno 1459*

En l'an 1604, Halley traverse le ciel du Wurtemberg, en Allemagne du sud : les âmes s'échauffent, on croit la fin du monde arrivée, les astrologues dament le pion aux astronomes – quand ce ne sont pas les mêmes qui troquent juste leur robe (pensons à Newton !). De 1614 à 1616 paraissent les trois manifestes rosicruciens, qui défrayent un temps la chronique: blague de potaches à Tübingen, ou vaste complot ourdi pour ébranler les fondements de la vieille Europe ? En 1623, on s'inquiète en haut lieu de l'apparition d'une secte rose-croix à Paris. Une chimère ? Dans le même temps encore, on construit outre-Rhin un globe aux frais d'un noble allemand, de l'intérieur duquel on pourra contempler les étoiles et le mouvement des sphères – le *minutus mundus* – le microcosme – n'aura jamais été plus à portée d'imagination…

Toutes ces découvertes scientifiques de l'époque se retrouvent épinglées dans les *Noces Alchimiques de Christian Rose-Croix, anno 1459*[1] ou dans la *Confessio* et la *Fama Fraternitatis* : mais elles s'y mêlent aux licornes, aux griffons, devisent avec les faiseurs d'or, rencontrent le

[1] Pour lire le texte en français, voir deux éditions au choix, l'une universitaire, érudite et neutre, l'autre plus engagée et partiale, mais dont les commentaires regorgent d'originalité: Bernard Gorceix, *La Bible des Rose-Croix* (Paris: PUF, 1970). Jan Van Rijckenborg, *Les mystères de la Fraternité des Rose-Croix. L'appel de la Fraternité de la Rose-Croix* (Fama Fraternitatis). *Le témoignage de la Fraternité de la Rose-Croix* (Confessio Fraternitatis). *Les noces alchimiques de Christian Rose-Croix anno 1459*, 4 vol. (Haarlem: Rozekruis Pers, 1989).

géant Atlas, Cupidon ou la déesse Vénus endormie… et entraînent dans leurs rêves de démesure et de perfection les chercheurs de l'époque, mi-alchimistes, mi-ermites, qui graviront les étages de la tour de l'Olympe à la recherche de la sagesse pérenne. Les hermétistes d'alors la nomme *philosophia naturalis.* Aujourd'hui, on parlerait plutôt de sciences dures ou de recherche appliquée !

On tentera ici de démêler le scientifique du merveilleux, l'imaginaire de l'expérientiel, et surtout de voir pourquoi ils s'entremêlent, et ainsi seulement - peut-être - de percer plus avant le secret (encore en grande partie intact !) des *Noces alchimériques* de Christian Rose-Croix… Pour ce faire, on commencera par replacer le décor politique, religieux et idéologique dans lequel les personnages des *Noces* évoluent. L'histoire des idées scientifiques sera ensuite le prolongement de cette histoire événementielle, notamment celle de l'alchimie, matière de synthèse par définition - et par excellence. Enfin, la forme littéraire du conte (entre journal intime et récit de voyage) expliquera sa modernité, à savoir ce mélange subtil de sciences et de mystères. On partira d'une exigence de méthode: lire les *Noces Alchimiques* tel que l'aurait fait un lecteur du premier quart du XVIIIème. Pour saisir la manière de penser propre à une époque, il faut appliquer ce qu'Hélène Metzger appelle "la méthode philosophique en histoire des sciences", selon laquelle l'historien doit "se faire le contemporain des savants dont il parle" et expliquer en quoi leurs conceptions étaient *pensables à leur époque et non impensables à la nôtre*.

L'esprit occidental ne se meut que par soubresauts, il ne connaît l'exaltation que *dans et par* la crise, comme l'ont montré Paul Hazard[2] et Roland Edighoffer[3]. La première moitié du XVIIème est ainsi marquée par une triple crise de la civilisation occidentale : scientifique et religieuse, avant de devenir politique. En effet, Descartes tente d'opposer à la poétique mythique du cosmos à la Newton, le grand livre de la Nature écrit par le doigt de Dieu, un « univers-machine », presque un « cadavre cosmique »[4]. Cette crise philosophique devient ontologique, et les clivages dogmatiques, essaimant d'Allemagne avec la Réforme, débouchent bientôt sur la Guerre de Trente Ans (1618-1648). L'absolutisme naissant se servira à dessein de toutes ces peurs pour asseoir son pouvoir. L'on sait

[2] Paul Hazard, *La crise de la conscience européenne. 1715-1680* (Paris : Boivin, 1935).

[3] Roland Edighoffer, *Les Rose-Croix et la crise de la conscience européenne au XVIIème siècle* (Paris : Dervy, 1998).

[4] Jean Borella, *La crise du symbolisme religieux* (Lausanne : L'Age d'Homme1990).

que les *Noces* circulaient sous le manteau dès 1605, et qu'elles sont très vraisemblablement le premier des trois manifestes fondateurs du rosicrucisme, par ordre d'écriture, malgré leur parution en troisième position. Les *Noces* se font l'écho des préoccupations métaphysiques de l'époque et deux panacées sont proposées par les rose-croix pour redresser les mœurs et les habitudes de penser de leurs contemporains. Aucune d'elles n'est *stricto sensu* rationnelle, mais en même temps, elles correspondent à ce que le siècle met sous le terme générique de sciences: en réalité un patchwork mêlant religion et sciences, mais dont la quintessence est plus une pensée magique qu'autre chose. La première réponse est d'ordre mythique : les rose-croix prétendent éduquer pour sauver le monde de ses déboires métaphysiques ; la seconde est pseudo-scientifique, et c'est l'endroit où naissent toutes les chimères…

La déclaration de principe des rose-croix de 1614 à l'endroit des sciences se lit dans la *Confessio*, visant à rallier à sa cause les « savants d'Europe »: les rose-croix se présentent en bons protestants, détenteurs d'une connaissance secrète.[5] Elle commence par répondre à l'interrogation suivante : pourquoi les scientifiques de l'époque auraient-ils besoin des services des rose-croix ? Suit une analyse des sciences du temps. Le constat de départ est cinglant : la philosophie est "en train de rendre l'âme[6].

Fidèle en cela au modèle médiéval, classant arts majeurs et mineurs, les rose-croix voient dans la rhétorique, et donc la pensée discursive, la reine des disciplines scientifiques. Et ils lui intiment de se régénérer. Plus loin, ils la détaillent ainsi : « elle comprend surtout la Théologie, la Médecine, et avant tout la Science du Droit : c'est une philosophie qui sonde le ciel et la terre à l'aide d'un excellent art d'analyse, ou qui, en un mot, exprime essentiellement que chaque homme est un microcosme ». Ici, un élément radicalement nouveau, en revanche, fait son apparition. Comment amender cette philosophie ?

Il faut d'abord partir du postulat que le monde est déchu, et que la sagesse primordiale ou gnose, la connaissance du père Créateur, connaissance de première main, a disparu. Se référant à Salomon, qui serait, aux yeux des auteurs, l'un des derniers sages, ils listent les sciences

---

[5] « Il est cependant dans notre intention, par amour pour les savants, de donner de plus amples éclaircissements sur toute partie de la *Fama* qui pourrait paraître trop concentrée, ou qui ne saurait être exprimée en d'autres langues. Nous espérons qu'ainsi, ceux-ci nous accueilleront plus favorablement, et qu'ils admettront plus aisément notre point de vue. » in Jan van Rijckenborgh, *Le témoignage de la Fraternité de la Rose-Croix* (Haarlem, Pays-Bas: Rozekruis Pers, 1983) 9 et sq.

[6] Ibid., 10.

où le roi excellait, puis délaissant l'époque vétérotestamentaire, ils reviennent à l'époque contemporaine, s'adressant aux « véritables disciples de la sagesse et [aux] réels adeptes de l'art spagirique »[7]. Le but est désormais de percer la moitié inconnue du monde. Cette sagesse recouvrée doit agir, telle une "panacée parfaite pour tous les corps [médecine], qui peut transmuter les métaux vils en l'or le plus pur [alchimie], et qui peut éloigner des hommes toute maladie [médecine], toute inquiétude, toute détresse et toute affliction [religion]" : bref prêcher et guérir les malades, selon l'adage biblique bien connu. Pour ce faire, la Fama propose d'entamer l'écriture de trois livres majeurs N., M. et T. sur la nature, le monde et Dieu (*Liber Naturae*, *Mundi* et *Theos*): un *Liber Naturae* ou "règle de tous les arts" ; pour l'écrire, le personnage hautement mythique de Frère C.R.C. (Christian Rose-Croix), fondateur de la prétendue fraternité, Allemand, « Père spirituel très illuminé », vit une vie d'ascèse dès son plus jeune âge, et voyage, se formant à toutes les sagesses et tous les savoirs scientifiques : Christian a 5 ans lorsqu'il part pour Jérusalem, qu'il n'atteindra jamais. Il s'arrête en effet à Damas, où il apprend la science des plantes médicinales, puis à Damcar, en Arabie Heureuse (à 16 ans), où il est rompu aux mathématiques et à la physique ; il traduit le *Liber M.* (*Mundi*) de l'arabe en latin ; il revient en Egypte (à 19 ans), puis à Fez où il apprend la magie africaine (la science des élémentaux, c'est-à-dire des esprits de la nature) et les *periodis seculorum* [8], que maîtrisait Salomon cité plus haut : en d'autres termes, il acquiert le don de l'histoire, mais aussi celui de prophétie. A 22 ans, il revient en Europe, via la péninsule ibérique, pour faire profiter les savants de ses savoirs.[9] Il est très mal accueilli, personne ne veut de sa sagesse, car les savants ont trop peur de perdre leur réputation… Mais l'idée est lancée d'une axiomatique, d'un condensé encyclopédique, bien avant l'Encyclopédie de D'Alembert et Diderot, d'une somme théologico-

[7]Terme alchimique dérivé de *spao*, je sépare, et de *ageiro*, j'unifie ; c'est l'équivalent de la formule latine : *solve et coagula*.

[8] Pour les Rosicruciens classiques, le temps est cyclique; d'ailleurs, quand ils ont une décision importante à prendre, ils consultent leur « rota mundi », et sont alors en mesure de prévoir l'avenir...

[9] Ibid., XXV: « Il s'entretint avec les savants d'Espagne, quant à ce qui manquait de nos arts et quant à la façon de les aider ; d'où l'on pouvait tirer des indices certains sur les siècles suivants [prophétie] et en quoi ils devaient concorder avec les siècles passés [histoire] ; comment réformer les défauts de l'Ecclésia [politique] et toute la philosophie morale [éthique]. Il leur montra de nouvelles plantes [botanique], de nouveaux fruits et animaux [entomologie] qui ne suivaient pas les lois de l'ancienne philosophie et il leur communiqua de nouveaux axiomes qui pouvaient tout résoudre parfaitement [logique] »

politique qui contiendrait toutes les connaissances humaines ; c'est à cette fin que C.R.C. crée bientôt une fondation, puis un laboratoire de recherche (avec à terme le projet d'une université). L'image du microcosme préside encore et toujours à cette construction:

> C'est là que prit son origine la belle association où, de même qu'en tout noyau est enfermé un arbre ou un fruit entier, de même l'ensemble du vaste monde serait présent dans un 'petit homme' dont la religion, la politique, la santé, les membres, la nature, les paroles et les œuvres suivraient à l'unisson la mélodie de Dieu, du ciel et de la terre. Tout ce qui serait en dissonance avec cela serait erreur, falsification et œuvre du diable, qui seul est le moyen premier et la cause dernière de la dissonance du monde, de son aveuglement et de son obscurantisme.[10]

Les rose-croix se réclament également d'Haselmayer et de Paracelse. Mais pour que l'œuvre au service de l'humanité n'échoue pas cette fois, sous les coups des délateurs et des détracteurs, Frère C.R.C. utilise la transmutation des métaux pour s'allouer une fortune personnelle et créer un laboratoire, puis un "collège doctoral" : il se donne cinq années, période durant laquelle il construit en effet une école (qu'il nomme la demeure *Sancti Spiritus*), entame la rédaction des livres susdits, et se consacre majoritairement à la mathématique et à l'ingénierie, « fabriquant *ex omnibus hujus artis partibus* maints beaux instruments »[11]. C.R.C a donc désormais 27 ans, et son pèlerinage véritable commence : il va créer un réseau savant européen en toute clandestinité: La *Fraternitas Rosae Crucis*. Il engage d'abord quatre assistants, qui doivent coucher scrupuleusement par écrit ses instructions.[12] Le cénacle s'étend bientôt à trois autres membres, car parallèlement, C.R.C. continue d'officier en qualité de médecin (paracelsien).

La Caritas, ou l'art de soigner le monde, est le but ultime des rose-croix « car nos *rotae* ont pris leur commencement au jour où Dieu prononça le *fiat* et se termineront lorsqu'il dira *pereat.* »[13] Mais ils restent

---

[10] Ibid., XXIV.

[11] « … dont cependant il ne nous est parvenu que peu de choses, ainsi que nous le verrons », Ibid, p. XXVII. La suite est en fait à lire lors de la découverte fortuite du tombeau de C.R.C., son "compendium de l'univers", par ses collaborateurs, 120 ans après sa mort!

[12] « … afin que les membres futurs, qui devraient être admis à cause d'une révélation spéciale, ne soient trompés par aucune syllabe ni aucune lettre. » Ibid., XXVII.

[13] Bernard Gorceix, *La Bible des Rose-Croix* (Paris : PUF, 1988), 9. Gorceix traduit par « cycles », mais ajoute une note de bas de page où il est question des roues de la fortune, ou bien des révolutions ou périodes astrales (Ibid., no. 2, 6)

en prise directe avec le monde contemporain, aussi décident-ils de se séparer et de voyager dans le monde entier pour y répandre leur savoir, et continuer d'engranger les connaissances de tous "en cachette des savants". Ils jurent fidélité sur six articles d'une confession de foi, un contrat sextuple, qui est la charte de fondation de l'université nomade/ virtuelle des rose-croix[14]. Le mythe fondateur est posé. La première réponse au défi du temps est contractuelle. Vient ensuite la réponse pratique, magique, derrière le paravent de la science : comment pallier l'illogisme d'une époque troublée? Le tiraillement entre tradition religieuse et modernité scientifique se ressent dans les *Noces*, où le merveilleux fantastique le dispute au merveilleux mécaniste. Autrement dit, le merveilleux va puiser aux sources du conte et de la fable, comme au balbutiement de la modernité. Ainsi se laissent appréhender trois sortes de merveilleux: le merveilleux technicien: les portes s'ouvrent toutes seules, le trône du roi est à autopropulsion, un globe tournant géant fait office de planétarium et se meut de lui-même; les tableaux sont de vivants automates; dans le domaine maritime, les navires se passent de timonier, et l'horlogerie n'est pas en reste, puisque le mouvement perpétuel a été inventé !

Cette première source d'enchantement se double d'un merveilleux (sur)naturel, notamment minéralogique, assez similaire. Ainsi des lumières perpétuelles éclairent-elles le caveau de Dame Vénus: il y a des escarboucles, comme dans le planétarium; des feux follets s'agitent au sommet des mâts des navires entre le palais et l'île d'Olympe... enfin, la dernière source d'inspiration du merveilleux est, elle, purement chimérique: lors du banquet, des serviteurs, invisibles aux yeux des commensaux non initiés, se dévoilent aux initiés, tels des anges; le bestiaire médiéval, qui puise à la tradition de l'héraldique, reprend certains *topoi*, tels la Licorne, figure de pureté – et accessoirement dans son sens alchimique, du Mercure recréateur – le lion, le griffon, le géant, les putti, en la personne de Cupidon... Lors de la traversée en nacelle (au sixième jour), nymphes et sirènes offrent aux alchimistes une perle. A aucun moment cependant, ces divers aspects du merveilleux n'entrent en conflit les uns avec les autres, ils sont simplement l'expression bigarrée d'une nature facétieuse, que la raison n'a pas encore domptée, et qui ajoute foi tout autant à la véracité des nombres qu'à la force de prédiction de la fée imagination.

---

[14] 1. Seule profession autorisée: le traitement des maladies, et encore gratuitement; 2. Vivre selon l'usage du pays; 3. Assister une fois par an à la conférence internationale des recteurs de l'université virtuelle; 4. Assurer sa succession; 5. Avoir pour sceau le sigle R.C.; 6. Garder le silence sur la Fraternité pendant 100 ans.

Dans les deux manifestes, en outre, la perspective des Rosicruciens est éminemment pratique: ils ont vocation à former des *Ingenia* « hautement éclairés » qui ont à remettre « à l'honneur les arts dégradés et imparfaits afin que l'homme comprenne enfin sa noblesse et sa majesté »[15]. Ces ingénieurs sont d'autant plus nécessaires que le monde est devenu, sous l'influence des sciences mécanistes, une "Machina céleste", que la compétence que confèrent les sciences appliquées aide à décrypter correctement. Mais en même temps, il faut des hommes disposés à entrevoir le lien entre céleste et terrestre, capables aussi bien de lire le livre M. que le livre T.

En résumé, les premiers Rosicruciens désirent qu'on dépasse le plan ordinaire de la philosophie et de la pensée pour accéder à celui d'une véritable Gnose, c'est-à-dire d'une appréhension 'suprarationelle' du réel, transcendant le dualisme du sujet et de l'objet pour devenir une co-naissance à l'harmonie entre l'homme et l'univers, entre microcosme et macrocosme. Cette *metanoïa* est non seulement nécessaire, mais elle est encore possible, à condition que l'homme lui-même se transforme et se trouve transfiguré. Il change alors de visage, en d'autres termes, il lui pousse des ailes d'ange, bref, il devient un hybride, une sorte de Dieu mortel ou d'Homme immortel, en un mot: une chimère. En résumé, les rose-croix veulent, à première vue, former une génération d'alchimistes opératifs et spirituels pour sauver la planète en désordre... Mais plus en profondeur, ils entrevoient la possibilité de transformer le plomb de la personnalité égotique en or d'une personnalité renée, transmutée, transfigurée. L'alchimie, au niveau microcosmique (individuel), est ce qu'est l'astrologie au niveau macrocosmique (planétaire): une panacée, doublée d'une prophétie eschatologique.

Nous allons maintenant nous focaliser sur une histoire en particulier, celle de l'alchimie, car c'est la discipline par excellence qui, pour reprendre les termes de Marc Fumaroli[16], apparaît comme un foyer de rencontre « entre philosophie morale, philosophie naturelle, médecine, philosophie du langage et philologie, dans une même aspiration sotériologique à l'unité des savoirs ». Bref, elle est viscéralement hostile à toute frontière entre les disciplines, voilà pourquoi, dans le même souffle, les *Noces* vont convoquer les mathématiques, la naométrie[17], l'astrologie, l'architecture des jardins, la mécanique, l'optique, la musique, la peinture,

[15] Ibid., XXI.
[16] Cité par Didier Kahn, *Paracelsisme et alchimie en France à la fin de la Renaissance 1567/1625,* Thèse de doctorat, Paris IV – La Sorbonne, 1998.
[17] Science de la mesure du temple dans la Bible.

la littérature et, *last but not least*, la religion. De plus, elle exprime une posture philosophique chère à l'homme du XVIIème: à monde instable, science de la transformation ! Devant tant d'horreur, d'incertitude et de guerre, rien ne vaut, pour survivre, la métamorphose intégrale et constante, pour s'adapter ou pour passer inaperçu, c'est-à-dire pour échapper au pire, qui peut toujours advenir. Et c'est là qu'elle « vire au théâtre », et qu'une distance se crée par rapport à la matière originale (l'art transmutatoire) pour devenir réflexion philosophique sur la vérité et ses limites. Finalement, l'entreprise paie, car ce n'est bientôt plus l'alchimie qui est au centre des *Noces*, mais bien le recul sur son temps, dont a pu faire preuve Andreae, et qui lui confère encore aujourd'hui sa modernité et sa permanence mythique, ce qui en fait donc une chimère littéraire !

L'Alchimie au XVIIème est le point d'orgue d'une longue tradition.. L'Europe médiévale connaît les deux principes Soufre et Mercure, auquel s'ajoute un troisième dès 1260: l'Arsenic. En 1330, Petrus Bonus est le premier à comparer la pierre philosophale au Christ. Dès lors, l'alchimie devient chrétienne. Ainsi, le Pseudo-Lulle[18] :

> De même que Jésus-Christ a pris la nature humaine pour la délivrance et la rédemption du genre humain, prisonnier du péché par la suite de la désobéissance d'Adam, de même, dans notre art, ce qui est souillé criminellement par une chose est relevé, lavé et racheté de cette souillure autrement, et par la chose opposée.

Toujours à la même époque (1350), Jean de Roquetaillade établit le lien entre Grand Œuvre et Passion du Christ. L'alchimie durant la Renaissance s'épure et se conceptualise. C'est l'époque de la *Table d'émeraude* (attribuée au théologien luthérien Osiander en 1541). Denis Zachaire déclare avoir réussi à transmuter du mercure en or le jour de Pâques 1550 (également le premier jour dans les *Noces Alchimiques*). Quand Rodolphe II de Habsbourg est empereur (1576-1612), la capitale de l'alchimie est Prague. Les adeptes de l'époque y convergent : Heinrich Khunrath (auteur d'un admirable *Amphitheatrum sapientiae aeternae*, 1602), Oswald Croll, Michael Maier[19]. L'écrivain François Béroalde de Verville (1558-1612), dont s'inspire Andreae pour écrire ses *Noces*[20], écrit

---

[18] Pseudo-Raymond Lulle, *Le Testament*, trad. (du latin) Kasteel, Hans van, préf. Kahn Didier, Grez-Doiceau (Belgique: Beya, Paris: diff. Edidit, 2006).

[19] Michel Maier est l'auteur, entre autres, des *Arcanes très Secrets*, 1613, et de *l'Atalante fugitive*, 1618.

[20] Il s'inspire notamment du *Songe de* Polyphile, interprété par Béroalde dans son *Tableau des riches inventions* paru en 1600.

vraisemblablement le *livre des figures hiéroglyphiques*[21] vers 1590. Il n'est édité qu'en 1612. Il développe la notion d'*ars magna*, une mutuelle délivrance de la matière et de l'esprit par la réalisation de l'œuvre, à la fois spirituelle et physique. Arrive Paracelse, qui vient brouiller les cartes. Il est mentionné par la *Fama Fraternitatis* et tenu en haute estime par les Frères rose-croix. Le médecin suisse n'a rien écrit d'alchimique au sens courant du terme (transmutation des métaux, production d'or), puisqu'il se concentre sur son utilisation médicale. Dans son *Opus paragranum* (1533), il substitue aux quatre Éléments les trois Substances (tria prima) que sont le Soufre, le Mercure et (c'est Paracelse qui l'ajoute) le Sel.[22]A la tradition médiévale et au paracelsisme s'ajoute, avec Pic et Ficin, la doctrine du *spiritus mundi*, qui fait son entrée dans le champ des concepts alchimiques, avec la quintessence et l'élixir.

Rosicrucisme, paracelsisme et alchimie ont ainsi travaillé à se conforter mutuellement. Il convient d'en étudier les rapports réciproques, les passerelles, et ainsi de voir quelle place occupe l'art spagirique dans les *Noces*. Deux hauts lieux de l'alchimie se distinguent dans le récit (une fontaine et une tour), ainsi qu'une figure emblématique: la "noble et vertueuse Demoiselle" Alchemia. C'est d'abord la théorie des "signatures" que les *Noces* empruntent au paracelsisme : Paracelse entrevoit l'univers comme traversé d'influences, de « sympathies » qui scellent les correspondances entre macrocosme et microcosme. La deuxième notion clé est celle des deux lumières : celle de la nature et celle de la grâce. La première suffit à mener à bien l'*Opus* alchimique, la transmutation des métaux. En revanche, la seconde nécessite l'intercession divine. Les deux éléments conjugués amènent à l'épisode de la fontaine hermésienne dans les *Noces*. Au-dessus d'une fontaine se dresse un lion. Une licorne vient s'agenouiller devant lui ! Le lion, sur ces entrefaites, brise une épée et la jette dans l'eau.

Puis il rugit, dévore une colombe blanche venue l'honorer et tenant un rameau d'olivier dans son bec, ce qui réjouit fort la licorne blanche au collier d'or ! La licorne peut représenter la volonté pure, le lion la force illimitée et la colombe le comportement dévoué de l'alchimiste qui voue sa vie à instaurer la paix à l'intérieur de lui-même: les trois côtés d'un

---

[21] Interprétation alchimique de l'arche du cimetière des Innocents à Paris , antidaté et faussement attribué à Nicolas Flamel.

[22] «Parmi toutes les substances, il en est trois qui donnent à chaque chose leur corps, c'est-à-dire que tout corps consiste en trois choses. Les noms de celles-ci sont : Soufre, Mercure, Sel. Si ces trois choses sont réunies, alors elles forment un corps (...). La vision des choses intérieures, qui est le secret, appartient aux médecins.»

triangle parfait, ailleurs appelé le « triangle de feu ». La licorne est le Mercure volatile, le lion le Soufre, la colombe le Sel, et leur mariage rend l'eau de la fontaine vivifiante, au sens où elle donnera désormais la vie éternelle ; c'est l'eau de la vie : les alchimistes doivent se laver les mains et le visage avec, avant d'être présentés au roi, autrement dit, ils doivent faire peau neuve, se purifier pour le Grand Œuvre. Un peu plus loin, au quatrième jour, le mystère hermétique se dévoile un peu plus : le premier jour du voyage initiatique des *Noces* a été le jour de l'appel ; le second fut le jour du jugement – les candidats indignes furent chassés ou punis ; le troisième jour, le Temple du jugement (le château royal) fait place au Temple de l'initiation. Tous les éléments nécessaires aux Noces alchimiques sont réunis et tous les préparatifs terminés. Le Grand Œuvre peut commencer. On retrouve ainsi les alchimistes au pied de la fontaine. Cette fois, le lion tient une plaque où figure l'inscription :

> Hermès prime,/ Après tant de malheurs/ qui frappèrent le genre humain,/ Sur le conseil de Dieu par décret divin / et avec l'assistance de l'art / Médecine dont la fin est la santé,/ Je coule ici./ Que celui qui le peut y boive./ Qui le veut s'y lave ;/ Que celui qui l'ose me trouble./ Frères, buvez et vivez.[23]

L'épée du jugement a disparu. L'alchimiste *bonafide* se voit offrir – par décret divin – l'accès à la sagesse pérenne, qui grâce à l'"Art pur", peut effacer les erreurs du passé et conduire à la félicité présente : par la guérison des malades. On marche toujours bel et bien dans les pas de Paracelse, et les deux lumières ont ici fusionné. Le deuxième grand moment alchimique est, bien entendu, l'ascension de la tour de l'Olympe. La tour est conçue comme un athanor. Chaque étage y correspond à une étape du Grand Œuvre ; on ne les détaillera pas toutes ici, mais qu'il soit juste mentionné que les artifices littéraires prévalent parfois sur la rectitude du procédé alchimique : l'œuf qui éclot pour donner un oiseau, que l'on sacrifie pour récolter son sang, tout ceci est souvent plus allégorique que pratique.

Mais n'est-ce pas ce que, paradoxalement, les *Noces* veulent nous dire : l'alchimie n'a jamais servi que l'intelligence du texte.L'artifice est donc ici d'ordre théâtral. Si le mythe et la science sont si imbriqués l'un dans l'autre, c'est parce que précisément, il s'agit alors de les fondre ou mieux, de les transcender : l'architecture magique et initiatique de la pièce en sept actes (ou sept jours) veut convaincre du bon droit de la religion face aux frasques des hétérodoxes et autres hérétiques : l'illusion finale de

---

[23] Bernard Gorceix, *La Bible des Rose-Croix*(Paris : PUF, 1998), 82.

la pièce – sa fin est tronquée ou ouverte, selon le point de vue[24] – renvoie dos à dos les deux lectures, mystique et scientifique et consacre peut-être une troisième voie, une voie de conciliation ou de révolution : la voie gnostique. La forme littéraire des *Noces*, baroque, voire un soupçon maniériste, s'allie à dessein parfaitement au fond alchimique, héraldique et symbolique, pour mieux le dépasser. En effet, la mise en abîme du monde, en littérature baroque, est un moyen de salut pour surmonter l'abîme du monde comme le souligne Shakespeare en faisant dire à l'un de ses personnages: « All the world's a stage, and all the men and women merely players: they have their exits and their entrances; and one man in his time plays many parts, his acts being seven ages. »[25]

De même, la fin tronquée des Noces s'achève sur deux "in quarto" manquants, soit le chiffre « huit » (8=2x4), autrement dit sur l'infini si bien que le récit des *Noces* se clôt sur l'infini et repart en boucle *ad libitum*.. C'est terriblement logique, dans la mesure où le maniériste abhorre la fin de toute chose (il court contre la montre, il fuit sans cesse le temps qui passe et se clôt par la mort). Toute fin est un éternel recommencement, et c'est là le talent finale du maniérisme : en préférant le temps cyclique, la boucle, au temps linéaire, l'artiste maniériste se confère à lui-même l'éternité.

Finalement, les trois livres que se proposait d'écrire CRC ont peut-être déjà été écrits : la *Fama* dévoile les lois naturelles selon la *philosophia naturalis*, la Naturphilosophie du XVIIème siècle ; la *Confessio* recentre le propos sur le monde des hommes, sur les faits de civilisation et les découvertes scientifiques ; mais il revient aux *Noces* de spiritualiser les deux premiers volets en replaçant au cœur des manifestes rosicruciens l'image de l'homme-microcosme, résumé de l'univers et créé à l'image du divin, seul capable de concilier les rêves les plus fous – nos chimères – et le pragmatisme le plus lucide : la connaissance de soi, en tant que science hermétique incluant toutes les autres – puisque je suis dans le monde, comme le monde est en moi. La critique larvée des faiseurs d'or, des « scientistes » matérialistes, est là en filigrane dans les trois manifestes: ne vous arrêtez pas à la richesse, ni même à l'illusion de devenir maître et possesseur de la nature, mais percez jusqu'à son secret le

---

[24] Lire, au sujet de l'ésotérumène du gardien, Frey-Jaun, Regine. *Die Berufung des Türhüters. Zur « Chymischen Hochzeit Christiani Rosenkreutz » von Johann Valentin Andreae (1586-1654)* (Bern: Peter Lang, 1989).

[25] William Shakespeare, dir. Brissenden, Alan. *The Oxford Shakespeare*, *As you like it*, (Oxford: Oxford University Press, 2008), II, 7.

plus intime. S'il transfigure, l'homme redevient un Dieu, et il emporte la création naturelle dans son sillage !

Aussi "l'esprit des *Noces*" plane-t-il jusqu'à nos jours, à l'instar des premières rencontres annuelles d'Eranos, à Ascona en Tessin suisse[26], d'intellectuels férus de sciences comme d'ésotérisme, tels C.G. Jung, Antoine Faivre, Henri Corbin, Mircea Eliade, Gilbert Durand. C'est l'esprit de Faust et des récits d'initiation d'Hermann Hesse, l'imaginaire d'un Corto Maltese (Hugo Pratt) ou d'un Umberto Eco, qui soufflent encore de nos jours, soit la quête d'unité entre science et conscience, religion et raison… A ceci près que, dans les *Noces*, une petite voix, à peine perceptible, dit : ce n'est pas dans ce monde que le salut peut se trouver, mais dans un tout autre, caché au-dedans, et pourtant si lointain. Un cœur de licorne doté d'un cerveau d'airain : n'est-ce pas ce que nous sommes, nous autres hommes du XXIème siècle ?

Œuvres citées

Borella, Jean. *La crise du symbolisme religieux.* Lausanne: L'Age d'Homme, 1990.

Edighoffer, Roland. *Les Rose-Croix et la crise de la conscience européenne au XVIIème siècle.* Paris: Dervy, 1998.

Frey-Jaun, Regine. *Die Berufung des Türhüters. Zur « Chymischen Hochzeit Christiani Rosenkreutz » von Johann Valentin Andreae* (1586-1654), Coll.: Deutsche Literatur von den Anfängen bis 1700 - volume 3/ Bern: Peter Lang, 1989.Gorceix, Bernard. *La Bible des Rose-Croix.* Paris: PUF, 1970.

Hazard, Paul. *La crise de la conscience européenne. 1715-1680*, Paris: Boivin, 1935.

Larue, Anne. « théâtre et alchimie, » In *Aspects de la tradition alchimique au XVIIème siècle*, actes du colloque international de l'université de Reims-Champagne-Ardenne, dir. Greiner, Franck. Paris: SEHA, 1998.

Maier, Michel. *Atalante fugitive*, trad. Etienne Perrot. Paris: Dervy, 1997.

Shakespeare, William, dir. Brissenden, Alan. *The Oxford Shakespeare, As you like it*, Oxford: Oxford University Press, 2008.

[26] Les rencontres eurent lieu de 1933 à 1988 (C.G. Jung les fréquente de 1933 à 1952), et reprirent de plus belle un an plus tard avec un nouveau cycle, et se poursuivent jusqu'à aujourd'hui.

Pseudo-Raymond Lulle. Le Testament, trad. (du latin) Kasteel, Hans van, préf. Kahn Didier, Grez-Doiceau: Beya, Paris: diff. Edidit, 2006.

Rijckenborgh, Jan van. *Les mystères de la Fraternité des Rose-Croix. L'appel de la Fraternité de la Rose-Croix (Fama Fraternitatis). Le témoignage de la Fraternité de la Rose-Croix (Confessio Fraternitatis). Les noces alchimiques de Christian Rose-Croix, anno 1459*, 4 vol. Haarlem: Rozekruis Pers, 1989.

WHERE WEBER WAS WRONG: THE EXAMPLE OF EMERGENCE THEORY

EGIL ASPREM

Introduction: Science and the Disenchantment of the World

At the end of 1918, while the November revolution swept a Germany exhausted by war, Max Weber told students at the University of Munich that the "disenchantment of the world" was "the fate of our times". Through an "intellectualist rationalization, created by science and by scientifically oriented technologies", modern man had come to hold an understanding of the world in which everything runs by perfectly explicable mechanisms. Although science had not, perhaps, uncovered all of them at the present stage in history, there was no doubt that exact knowledge was obtainable *in principle*. This was the disenchantment of the world: the universe was capable of being explained rationally, since "no mysterious incalculable forces … [came] into play."[1]

Weber's idea of a process of disenchantment has since been important for historians and sociologists interested in such things as the apparent disappearance of magic in the West, the relation between science and religion, and the influence of science on our modern view of the world and our place within it. In fact, Weber's thesis contained several assumptions concerning the operation of both science and religion in the history of Western civilisation. However widespread, these assumptions do not necessarily hold for the way science has *actually* been practiced, or how actual people have pictured the relation between science and religion. In the present essay, I shall look closer at some "anomalous" material that belongs to Weber's own days, connected especially with the philosophy of biology. But before advancing with this material, I would first like to make clear what exactly is being assumed by the disenchantment thesis concerning the operation of science in particular.

---

[1] Quotations from Weber, "Science as a Vocation," in *From Max Weber: Essays in Sociology*, translated and edited by H.H. Gerth and C. Wright Mills (New York: Oxford University Press), 129-156.

When we unpack the various claims that Weber made in connection to disenchantment by breaking the concept down into its composite parts, three major assumptions come to the fore (see Table 1). First, an *epistemological optimism* follows from the rejection of "mysterious incalculable forces", and the attendant claim that all facts of the world can be known by "empirical science" and calculations. Secondly, there is what could be called an *axiological scepticism*, or scepticism concerning questions of ethics, meaning, and value. This point revolves around Weber's insistence on the fact/value distinction: while science can know facts about the world, there is no meaning *inherent in* those facts that can tell us how to live our lives. In other words, science's strong claim to knowledge is restrained by stating that it can know nothing about questions of meaning. Thirdly, another kind of restriction is imposed by what may be referred to as *metaphysical scepticism*. Again, this means that science cannot go beyond "empiricism" and calculations, hence making metaphysics impossible.

<table>
<tr><th colspan="3">The Disenchantment of the World</th></tr>
<tr><td>Epistemological optimism</td><td>Axiological scepticism</td><td>Metaphysical scepticism</td></tr>
<tr><td>Rejection of mysterious, incalculable forces</td><td>Separation of facts and values</td><td>Science can know nothing beyond what is empirically given</td></tr>
<tr><td rowspan="2">Nature can in principle be understood by empiricism and reason</td><td>Science can know nothing of "meaning"</td><td>Metaphysics is impossible</td></tr>
<tr><td colspan="2">= Intellectual sacrifice necessary to possess religion</td></tr>
</table>

Figure 11-1 : Three dimensions of disenchantment

I will argue that these three assumptions, far from being taken for granted as guiding principles by scientists themselves, have, in fact, to various extents and degrees, constantly been challenged *inside* science. Both the promise of explaining the world completely and the shortcomings of meaning and metaphysics have more often been the dreams of certain philosophers than the working principles of scientific actors. This becomes particularly clear when we avoid the typical bias towards equating "science" with fundamental "physics". In this paper, I shall focus on some discussions in early 20th century philosophy of biology, which seem to centre directly on the fundamental problems of disenchantment. With reference to the three dimensions listed above, this case study will focus primarily on the *epistemological* questions, although I will show that there are sometimes strong links to the axiological and metaphysical dimensions as well.

The revolt against mechanism in 20th century biology

I will focus here on some fundamental theoretical struggles which were being fought in the life sciences during the first three decades of the 20th century. By the beginning of the century, a number of biologists, particularly working with embryology, were starting to see major shortcomings with the mechanistic-materialist paradigm that characterised the discipline in the 19th century. As science historian Anne Harrington has shown, a range of holistic and vitalistic alternatives were being proposed in strongly polemical terms against not only the mechanistic conception in science itself, but also against negative trends of "fragmentation" and "materialism" which some claimed to dominate modern culture and politics.[2] Harrington focused on Germany, where explicit vitalists such as Hans Driesch and Jakob von Uexküll were striking heavily against the mechanistic programme, sparking much heated debate. But the opposition against strict mechanistic conceptions of life was a much wider phenomenon, and other branches of it were developed in other countries.

In terms of disenchantment, we might identify three main types, or clusters of positions, in these debates. In ascending order from the most "disenchanted" to the most "enchanted", these are: mechanism, organicism/holism, and vitalism. These three positions should be distinguished clearly from one another, both because they represent different epistemological positions yielding quite different answers to the question "what is life?", and because they tended to represent different ideological positions as well, often entrenched in bitter polemics. Let me briefly introduce them.

Mechanism, or *philosophical* mechanism, is in this context, obviously, the view that the phenomenon of life can be understood and explained entirely in terms of the mechanical interaction of its constituent parts. The whole is *nothing more than* the sum of its parts. This, then, is entirely in holding with the disenchanted perspective, subsuming life under the wider umbrella of mechanistically explicable and calculable phenomena.

---

[2] Anne Harrington, *Reenchanted Science: Holism in German Culture from Wilhelm II to Hitler* (Princeton, NJ: Princeton University Press, 1996).

At the opposite end of the spectrum stands vitalism. The basic epistemological position of vitalism in early 20th century biology may aptly be summed up by one of its main spokespersons, Hans Driesch, who wrote, in 1908, that

> [n]o kind of causality based upon the constellations of single physical and chemical acts can account for organic individual development; this development is not to be explained by any hypothesis about configuration of physical and chemical agents. Therefore there must be *something else* which is to be regarded as the sufficient reason of individual form-production.[3]

The vitalist position goes on to claim that this "something else" is separate from ordinary matter. It represents, indeed, a mysterious and incalculable force, an irreducible *vital force* that drives the development of organisms and explains life as a phenomenon completely *sui generis*. In the case of Driesch, this vital force was identified with his principle of *entelechy*. Entelechy was described in decidedly anti-mechanistic, anti-materialistic, and anti-reductionistic terms; it was an immaterial organising and directing force, which consumed no energy while still possessing agency in the development of organisms. This position, then, in rejecting mechanism and holding that life must be understood with recourse to something non-material, stands in stark contrast to the disenchanted perspective.

While the distinction between these two positions is quite clear, I will now focus on certain positions that stand, so to speak, in the middle. This is the domain of holism and organicism – a cluster of positions that equally reject mechanism and vitalism. Taken as a whole, this category has in fact been very influential in biology. Highly esteemed biologists such as Hans Spemann, Joseph Needham, and Richard Goldschmidt defended organicist views. Indeed, the contemporary biologist Scott Gilbert and the philosopher of science Sahotra Sarkar have argued that "[m]any of the principles of organicism remain in contemporary developmental biology", but that they "are rarely defined as such."[4] Speculating about the reasons for this lack of acknowledgement of a living organicist heritage, Gilbert and Sarkar mention, among other things, that organicism has held rather bad company through the years, and has thus

---

[3] For a larger discussion of the "vitalistic or autonomous factor" of entelechy, see Hans Driesch, *The Science and Philosophy of the Organism*, Two Volumes (London: Adam and Charles Black, 1908), Vol. 1, 142-149; Vol. 2, 129-265.

[4] Scott Gilbert and Sahorta Sarkar, "Embracing Complexity: Organicism for the 21st Century." *Developmental Dynamics* 219 (2000): 1.

been "guilty by association". In addition to being associated with vitalism, which it in fact rejected, stances of organicism and holism have often been linked to political ideologies on the extreme right and left, as well as to various "spiritual" positions urging the reenchantment of the world. As Gilbert and Sarkar somewhat sardonically remark: "With ideological friends like these, who needs enemies? The association with such company as vitalism, fascism, communism, and New Age spirituality should be enough to bring down any philosophy."[5]

Nevertheless, its continued presence and – it would seem – active relevance in the development of modern biology merits a closer look. In terms of our interest in disenchantment, organicism may be said to occupy a midway position: while it rejects mysterious forces – that is, it usually follows an entirely materialistic ontology – it also rejects mechanism, and hence the absolute calculability of the world. To the organicist, life *is* indeed more than the sum of its parts. The question is how this ought to be explained and justified, philosophically as well as scientifically.

In what follows, I shall focus on one particularly central concept which was developed in this context, namely that of *emergence*. Varieties of the concept have been instrumental in distinguishing organicist thought from materialistic-mechanism on the one hand and explicit vitalism on the other. While organicism has continued to have a presence in developmental biology, it is also safe to say that the concept of emergence has enjoyed much attention in the philosophy of science.[6] However, turning our attention to the context in which emergentism was born we find that it was designed to do much more than merely accounting for some scientific-internal problems. In fact, we find in emergentism a strong engagement with religious motivations and questions of theology. This, I will suggest, furthermore draws into doubt the applicability of Weberian disenchantment's strict divide between scientific rationality and religious concerns.

---

[5] Ibid., 5.

[6] For an up-to-date overview of its relevance in contemporary philosophy of science, see e.g. the essays in *Emergence: Contemporary Readings in Philosophy of Science*, ed. Mark A. Bedau and Paul Humphreys (Cambridge, MA: The MIT Press, 2008).

The Emergence of Emergence

Some General Issues

Tracing the "emergence of emergence" we must move across the channel to Great Britain, where we find the "classic" phase of what is sometimes called "British emergentism" in the 1920s.[7] In this classic phase, three authors deserve particular mention. Listed in chronological order of when their major contributions were published, these are the philosopher Samuel Alexander (1859–1938), the psychologist Conway Lloyd Morgan (1852–1936), and the philosopher Charlie Dunbar Broad (1887–1971).[8] It will be noted from the outset that although the problems they were concerned with were connected to the question of life, none of these were biologists by profession. This actually points to a crucial feature of the controversies for defining and explaining life in the early decades of the 20th century: it was a highly interdisciplinary contest, with different disciplinary interests squaring off for authority over the field.[9]

---

[7] For overviews, see e.g. Brian P. McLaughlin, "The Rise and Fall of British Emergentism," in *Emergence*, *op.cit.* or Timothy O'Connor and Yu Wong, "Emergent Properties," *Stanford Encyclopedia of Philosophy*, 2006, http://plato.stanford.edu/entries/properties-emergent/index.html (accessed May 10, 2011).

[8] One should take note that the very general ideas may still be traced back to John Stuart Mill (who made a distinction between heteropathic and homeopathic effects), through George Henry Lewes (who coined the nouns emergent and resultant roughly corresponding to Mill's distinction). Here I will however focus on what I call the "classic" phase of emergentism, seeing that it was in the 1920s that this became seen as something of a philosophical field of speculation in itself with clear connections to the special sciences as well as to metaphysics and religion.

[9] A great number of books and articles about life as such were written in this period by non-biologists. Particularly interesting is the range of physicists writing about it, especially after the advent of quantum mechanics. Niels Bohr for example wrote articles with titles such as "Light and Life" (1933), "Biology and Atomic Physics" (1937), and "Physical Science and the Problem of Life" (1949/1957) as reproduced in Niels Bohr, *Atomic Physics and Human Knowledge* (New York: John Wiley & Sons, Inc., 1958). Erwin Schrödinger's *What Is Life? The Physical Aspect of the Living Cell* (Cambridge: Cambridge University Press, 1944) is a seminal example of that trend. Another perspective was given by Pascual Jordan, who included a chapter on 'Die biologischen Perskektiven der neuen Physik' in his *Physicalisches Denken in der neuen Zeit* (Hamburg: Hanseatische Verlagsanstalt, 1935) and later wrote a whole volume entitled *Die Physik und das Geheimnis des organischen Lebens* (Braunschweig: Friedr.Vieweg, 1941). The trend also

The question of reductionism and mechanism was loaded with connotations of which scientific discipline is most "fundamental"; on the entirely "disenchanted" perspective, life is capable of being reduced ultimately to chemistry, which is furthermore reducible to physics – and all can, theoretically, be explained from that level upwards given sufficient computational power. If this view could be challenged, one would have come a great deal further in claiming true autonomy and independence for the field of biology and the phenomenon of life more broadly.

To start with some more general issues let me illustrate how all the three above mentioned authors defined emergence in defence of a middle position between what we might see as the two extremes on the enchantment-disenchantment axis. Alexander, for example, stated that: "[l]ife is not an epiphenomenon of matter but an emergent from it. On the other hand, there seems to be no need for postulating … a new substance, a directing principle, or, as Prof. Hans Driesch calls it, an 'entelechy' or 'psychoid'."[10]

The content of this point becomes clearer if we turn to Lloyd Morgan. He stresses an important distinction in emergentist thinking between "emergents" and "resultants":[11] while emergents represent something entirely novel in nature, which supervenes on new kinds of relations between entities, resultants are the merely *additive* features of a system which can be calculated and predicted by adding up the parts making up the whole. This distinction is crucial for differentiating emergentism from mechanistic and vitalistic positions, as Morgan makes clear.

> The essential feature of a mechanical … interpretation is that it is in terms of resultant effects only, calculable by algebraic summation. It ignores the something more that must be accepted as emergent. It regards chemical compound as only a more complex mechanical mixture, without any new kind of relatedness of its constituents.
>
> It regards life as a regrouping of physico-chemical events with no new kind of relatedness expressed in an integration which seems, on the evidence, to mark a new departure in the passage of natural events.[12]

---

continues today, with a notable example being *Quantum Aspects of Life*, ed. Paul Derek, C. W. Davies and Arun K. Pati, *Quantum Aspects of Life* (London: Imperial College Press, 2008).

[10] Samuel Alexander, *Space, Time, and Deity* (London: Macmillan and Co., Limited, 1920), vol. 2, 64.

[11] These terms had already been introduced by Lewes. Cf. note 7 above.

[12] Conwy Lloyd Morgan, *Emergent Evolution* (London: Williams and Norgate, 1923), 8.

Morgan follows this up by simultaneously allying emergence with naturalism and the modesty of "proper science", while eschewing popular vitalistic notions as "questionable metaphysics":

> The gist of [emergent evolution] is that such an interpretation [i.e. the mechanistic one] is quite inadequate. Resultants there are; but there is emergence also. … That it cannot be mechanically interpreted in terms of resultants only, is just that for which it is our aim to contend with reiterated emphasis. But that it can only be explained by invoking some chemical force, some vital élan, some entelechy, in some sense extra-natural, appears to us to be questionable metaphysics.[13]

Similarly we find the distinctions being reiterated in somewhat new terms by C. D. Broad:

> I think that those who have accepted [Substantial Vitalism] have done so largely under a misapprehension. They have thought that there was no alternative between Biological Mechanism … and Substantial Vitalism. They found the former unsatisfactory, and so they felt obliged to accept the latter. … [H]owever … there is another alternative type of theory, which I will call "Emergent Vitalism", borrowing the adjective from Professors Alexander and Lloyd Morgan.[14]

Broad, being an eminent representative of analytic philosophy is also the one who presents the most concise technical definition of his own emergentist position, defined elegantly and concisely in contradistinction to mechanism. I will end this more general introduction of the idea of emergence by quoting from it:

> Put in abstract terms the emergent theory asserts that [1] there are certain wholes, composed (say) of constituents A, B, and C in a relation R to each other; [2] that all wholes composed of constituents of the same kind as A, B, and C in relations of the same kind as R have certain characteristic properties; [3] that A, B, and C are capable of occurring in other kinds of complex where the relation is not of the same kind as R; and [4] that the characteristic properties of the whole R(A, B, C) cannot, even in theory, be deduced from the most complete knowledge of the properties of A, B, and C in isolation or in other wholes which are not of the form R(A, B, C). The mechanistic theory rejects the last clause of this assertion.[15]

---

[13] Ibid.

[14] Charlie Dunbar Broad, *The Mind and Its Place in Nature* (London: Routledge & Kegan Paul Ltd, 1925), 58.

[15] Ibid., 61.

Emergence and religion

However, besides this purely epistemological difference on the existence of true "emergents" in nature, broader implications concerning worldviews were also very much present in these debates. In fact, all of the above authors discuss emergence in the context of broader concerns about worldview. For two of them – Alexander and Morgan – the question of religion was a major concern. In fact, Alexander's *Space, Time, and Deity* and Lloyd Morgan's *Emergent Evolution* were both given as Gifford Lectures in Natural Theology in Scotland, a culturally very important lecture platform dedicated to continuing the British strand of thinking about theology in terms of nature and reason. Since these two works were also the two first defining works of emergentism, it means that the whole field of speculation gained its identity in the context of natural theology.

Samuel Alexander's development of emergence as a technical term in his Gifford lectures in Glasgow between 1916 and 1918 (published as *Space, Time, and Deity* in 1920) is often mentioned as initiating the new wave of British emergentist philosophies in the 1920s. It is however more precise to say that the concept was created in a dialogue between Alexander and Morgan; Alexander himself states that the basic idea is taken from Morgan's *Instinct and Experience* from 1912.[16] Morgan then went on to spend the entire first lecture of his own Gifford in 1922[17] giving an interpretation of Alexander's version of emergent properties. Alexander was certainly in need of interpretation. His two-volume work was a highly speculative treatise of metaphysics, which tried to account for reality *in toto*, including extensive discussions of theology which were absolutely central to his whole project. It is intriguing to note that while Alexander is today often given reference and credit as a forefather of emergentism among contemporary philosophers of science, we seldom see anyone take note of the obvious religious motivation and foundation of his project.[18] In fact, the discussion of what Alexander calls 'the quality of

[16] Alexander, *Space, Time, and Deity*, 14. The relevant reference is the last chapter of Morgan's book ("Finalism and Mechanism: Body and Mind"), which discusses at length the conflict between mechanism and vitalism.

[17] The crucial first series of these lectures was published as *Emergent Evolution* in 1923. The second follow-up series was published as *Life, Mind, and Spirit* in 1926.

[18] Thus, for example, the entry on "Emergent Properties" in the *Stanford Encyclopedia of Philosophy* is ready to include Alexander as a major figure alongside, if not above, J. S. Mill and C. D. Broad, placing his philosophy of emergence as "very close in detail to a standard form of non-reductive physicalism" and comparing some other passages with views of Jerry Fodor. At the same time the authors humbly and bluntly admit that "Alexander's views are

deity' seems almost inseparable from the concept of emergence itself as it is used within his at times supremely confusing metaphysical system.[19] It will obviously be impossible to give it full justice in only a few pages; nevertheless, the "emergence of emergence" from this particular source is important enough from the point of view of our argument that I will at least give it the briefest of outlines to convey the general position. I will treat Alexander as my prime example of the religious impulse on British emergentism.

As with other such systems, Alexander uses emergence to paint a picture of a "layered reality" where "higher" levels of being emerge from more fundamental ones. In Alexander's scheme, these range from the most fundamental level of "Space-Time", up through the phenomena of physics and chemistry, to life and mind, all the way up to the "quality of deity". To picture this I have borrowed a diagram from Lloyd Morgan's exposition of Alexander, which neatly pictures the main concepts (Figure 11-2).

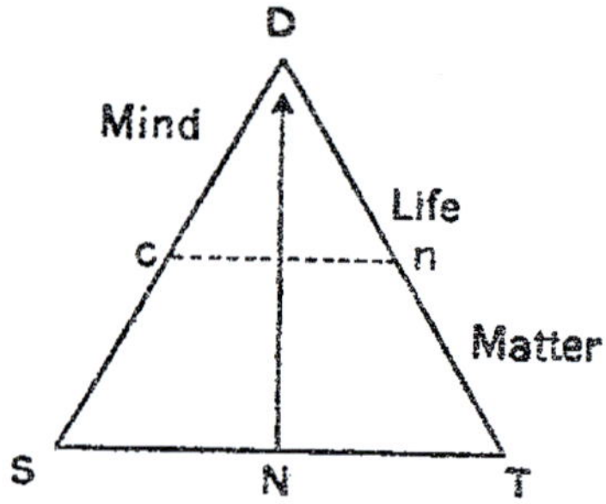

Figure 11-2 : Morgan's triangular depiction of Alexander's emergentist system.

The triangle pictures how, from the base line (S, T) of Space-Time, the higher orders of matter, life, and mind emerge. The triangular form is

embedded within a comprehensive metaphysics, some crucial aspects of which are, to these readers, obscure." Similarly, the author of a recent survey article of British Emergentism expressed doubts regarding Alexander, because 'to be frank, I find apparently conflicting passages in his texts and I am uncertain how to resolve the apparent conflicts.' He then found that he would spare the reader from these conflicts and instead "note a reading of Alexander's texts that [he] find[s] plausible". The concept of "deity", so central to Alexander as we shall see, is not mentioned in the article (or in the rest of the anthology for that matter). McLaughlin, "The Rise and Fall of British Emergentism", 31.

[19] See especially Alexander, *Space, Time, and Deity*, vol. 2, 341-372.

furthermore meant to suggest that Space-Time encompasses the whole of reality in its *extension*, but not in its *qualities*: along the upward axis of emergence we get a gradual narrowing down of new phenomena, which are nevertheless all included within Space-Time, as full and complete separate or finite parts of it. The central axis along which these novel qualities are represented to emerge is what Alexander called the *nisus* of the universe (N), which pushes reality onwards towards novelty. More specifically though, the aim of this nisus is the "quality of deity", shown here at the apex of the triangle (D).

The nisus is intimately connected with time, but here in a special understanding closer to Bergson's conception of *duration* than to the Einsteinian notion of a fixed space-time continuum. This, I should add, is only one out of many confusing double uses of the same term which makes Alexander's work particularly obscure. The concept of nisus is key to seeing how Alexander's emergentism is linked to his theology. Even though emergence as such is simply defined to suggest that genuinely novel "qualities" arise from increasing complexity in the arrangement and internal relations of entities already existing on the "lower" levels, the addition of a driving nisus more firmly weds this emergence of novelty and a notion of *progress* which may even seem to be of a teleological, directive type.

This becomes clearer when we try to unravel what is meant by the "quality of deity". Alexander discusses deity back and forth at length, in the process making several claims about it which do not entirely seem to add up. Deity is intimately connected with his attempts to define God on the one hand, and with a discussion of the relation between different levels of emerged existents on the other. We read, for example, that "deity is not so much the quality which belongs to God as God is the being which possesses deity".[20] But at the same time, deity is as we have seen just one step on the evolutionary ladder and hence simply a local event *within* space and time. Trying to grapple with these ideas we read further that "Deity is … the next higher empirical quality to mind, which the universe is engaged in bringing to birth."[21] At this point we see the contours of a bizarre theology in which God, or a being (or even several beings) with the quality of deity, is a certain hypothetical entity which will be created by the universe at some point in the future – a kind of God at the End of History.

This rather intriguing possibility, however, is suddenly cast aside when we read further the following description of the deity:

[20] Alexander, *Space, Time, and Deity*, vol. 2, 343.
[21] Ibid.

> For *any level of existence*, deity is the next higher empirical quality. It is therefore a variable quality, and as the world grows in time, deity changes with it. On each level a new quality looms ahead, awfully, which plays to it the part of deity.[22]

Now deity is defined as "what ever comes next" on a certain level of emergence. The elements of the periodic table were once the anticipated deity of the primordial plasma, and the first living cells the deity of non-living chemical compounds. Whatever will emerge at a later point in time with a greater order of existence than our consciousness can possibly fathom, then, is merely a manifestation of deity relative to our own limited existence. Whatever it may be, it is sure to come, as the universe is constantly generating deity through "that restless movement of Time, which is … the nisus towards a higher birth."[23]

At this point I will have to skip a number of complexities in Alexander's text, and go directly to what seems to be his final line concerning the relation between God, deity, nisus, and emergence. In the final instance, the following quotation seems representative: "God does not possess the quality of deity but *is the universe as tending to that quality*".[24] In other words, he now stresses the *activity* of God, understood as the universe itself with its progressive nisus, always at any level tending towards "deity". In this sense it is not the actual attainment of any specific level within the structure of Space-Time that is the point – these will always be transitory and obviously finite. Rather, the point is the immanent creative activity in itself.

We may deduce from this brief discussion that the concept of emerging "deity" becomes close to inseparable from the concept of emergence in general. This is particularly so since "God's deity is different for plants and men and angels, and varies with the lapse of time"[25] in the sense that it is always the "next level" before it is attained. Deity appears more as a perspective on emerging structures in the world – before they have in fact fully emerged. By using emergence in this way Alexander also distinguishes between God and "his deity": deity manifests itself in variable ways and through a certain lapse of time, but God "as such" is identified with the totality of the universe in its constant striving (nisus) *towards* deity.

---

[22] Ibid., 348. My emphasis.
[23] Ibid.
[24] Ibid., 361.
[25] Ibid., 366.

Conclusion

What does the emergence of emergence tell us about the so-called process of disenchantment? At least three things can be suggested. As part of the vitalism controversy, discussions of emergence testify to the actual complexity of the epistemic issues which disenchantment sets down as characteristic of modern science. While that controversy already in itself should indicate that the complete disappearance of "mysterious incalculable forces" was something which could still be debated within special sciences such as biology, emergence and the broader discourse on holism and organicism shows that serious intellectual attention was given to a plurality of responses to this problem. In terms of disenchantment, I have suggested that the emergentist position represents an embrace of incalculability, but *without* "mysterious forces". In the terminology I introduced at the beginning of this paper, this stands in opposition to the type of epistemic optimism that is characteristic of disenchantment.

Secondly, I have focused on the role of religion in formulating some of these ideas. I used Alexander to illustrate the point, but Morgan could also have served that point. In fact, while he claimed a thoroughly naturalistic attitude for his speculations, Morgan also explicitly stated that he wanted to challenge the notion that naturalism precluded theology. In the introduction to his lectures, he asked whether a "naturalistic interpretation suffices, or whether some further supra-naturalistic explanation is admissible…, *not as superseding* but as *supplementing* the outcome of scientific enquiry".[26] From the outset, his answer was positive: "I shall claim that it is admissible, and that there is nothing in emergent evolution, which purports to be strictly naturalistic, that precludes an acknowledgement of God".[27] Thus, a major objective for Morgan was to show that the emergentist interpretation made room for divine activity within nature.

Thirdly, both these major expressions of emergentism were formed in the context of the Gifford lectures in natural theology, and they bear a clear stamp thereof. Considering that emergentist and organicist thought has continued to wield an influence on modern biology as well as contemporary philosophy of science – sometimes in covert form, as Gilbert and Sarkar reminded us – this context of dialogue between religious, scientific and metaphysical concerns seems all the more important given what I termed the "pessimistic" strains laid on science by disenchantment. They seem, indeed, to indicate that metaphysical

---

[26] Morgan, *Emergent Evolution*, 2. Emphasis added.
[27] Ibid.

speculation and religious concerns have not been as separated from the production and direction of scientific research as the disenchantment thesis would suggest.

Works Cited

Abbott, Derek, Paul C. W. Davies and Arun K. Pati, eds. *Quantum Aspects of Life*. London: Imperial College Press, 2008.

Allen, G. E. "Mechanism, vitalism and organicism in late nineteenth and twentieth-century biology: The importance of historical context." *Studies in the History and Philosophy of Biological and Biomedical Science* 36 (2005): 261–283.

Alexander, Samuel. *Space, Time, and Deity*. Two volumes. London: Macmillan and Co., Limited, 1920.

Bedau, Mark A. and Paul Humphreys, eds. *Emergence: Contemporary Readings in Philosophy of Science*. Cambridge, MA: The MIT Press, 2008.

Bohr, Niels. *Atomic Physics and Human Knowledge*. New York: John Wiley & Sons, Inc., 1958.

Broad, Charlie Dunbar. *The Mind and Its Place in Nature*. London: Routledge & Kegan Paul Ltd, 1925.

Driesch, Hans. *The Science and Philosophy of the Organism*. Two Volumes. London: Adam and Charles Black, 1908.

Gilbert, Scott and Sahorta Sarkar. "Embracing Complexity: Organicism for the 21st Century." *Developmental Dynamics* 219 (2000), 1-9.

Harrington, Anne. *Reenchanted Science: Holism in German Culture from Wilhelm II to Hitler*. Princeton, NJ: Princeton University Press, 1996.

Jordan, Pascual. *Physicalisches Denken in der neuen Zeit*. Hamburg: Hanseatische Verlagsanstalt, 1935.

—. *Die Physik und das Geheimnis des organischen Lebens*. Braunschweig: Friedr. Vieweg, 1941.

McLaughlin, Brian P. "The Rise and Fall of British Emergentism." In *Emergence: Contemporary Readings in Philosophy of Science*, 19-60, edited by Mark A. Bedau & Paul Humphreys. Cambridge, MA: The MIT Press, 2008.

Morgan, Conwy Lloyd. *Emergent Evolution*. London: Williams and Norgate, 1923.

—. *Life, Mind, and Spirit*. London: Williams and Norgate, 1926.

O'Connor, Timothy and Hong Yu Wong. "Emergent Properties." *Stanford Encyclopedia of Philosophy*, 2006. Accessed May 10, 2011. http://plato.stanford.edu/entries/properties-emergent/index.html.

Schrödinger, Erwin. *What Is Life? The Physical Aspect of the Living Cell.* Cambridge: Cambridge University Press, 1944.

Weber, Max. 1946. "Science as a Vocation." In *From Max Weber: Essays in Sociology*, 129-156, translated and edited by H.H. Gerth and C. Wright Mills. New York: Oxford University Press: 129-156.

PART C:

GEMMATION

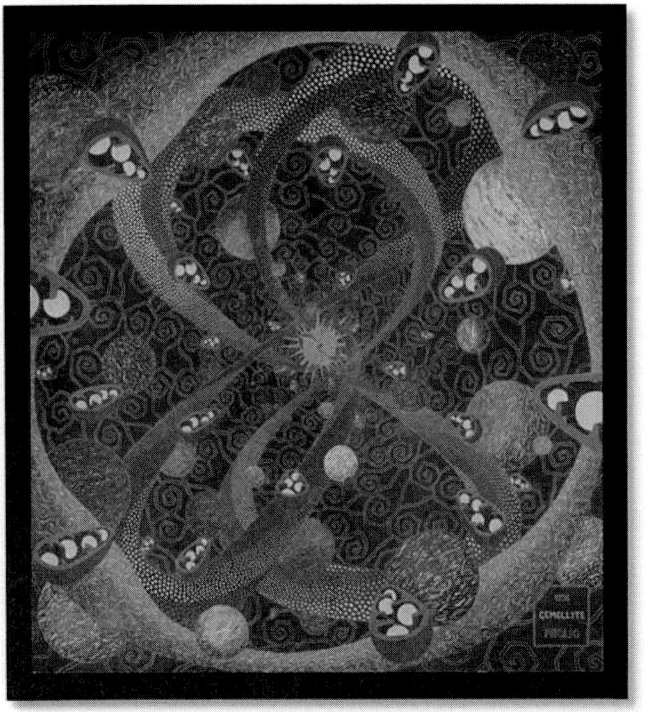

ILL. C-1 : François Miglio, *Gemmation*, detail from Triptyque EXPLORATION I (H 85cm x L 80cm)

SECTION I

ENJEUX DES INTERACTIONS ENTRE SCIENCES ET FABLE DE LA FONTAINE A AUBER

SABINE GRUFFAT

Le genre de la fable n'a pas pour vocation d'aborder l'animal comme un être naturel, qui aurait des caractéristiques biologiques propres. L'apologue, à visée pédagogique, incite par définition à considérer la bête comme la représentation allégorique de qualités morales, comme le « truchement » d'un univers poétique et didactique. C'est ce que rappelle La Fontaine au dauphin dans sa dédicace du premier recueil des Fables en 1668:

> Tout parle en mon ouvrage, et même les poissons.
> Ce qu'ils disent s'adresse à tous tant que nous sommes.
> Je me sers d'animaux pour instruire les hommes. (vers 4-6)

Adopter un point de vue scientifique pour aborder les fables où cohabitent des animaux réels et un bestiaire imaginaire revient donc à déplorer les aberrations d'Ésope qui signale que la nature des hyènes est d'être, selon les années, tantôt mâles, tantôt femelles ou les inexactitudes de La Fontaine qui fait manger à la cigale des mouches et des vermisseaux. Ces éventuels reproches ne présentent pas cependant grand intérêt puisque, encore une fois, ils ne tiennent pas compte des finalités du genre ; on trouve après tout des extravagances similaires dans l'histoire de la zoologie lorsque Pline décrit le regard mortifère et l'haleine vénéneuse du basilic.

Se réclamant d'un esprit plus rationnel, les fabulistes du XVIIIème siècle ont cependant cherché à naturaliser leurs personnages. Dans son *Discours sur la fable* (1719), Houdar de La Motte insiste sur la nécessaire justesse des allégories qui doivent rester « conformes aux idées qu'on a des choses »[1] et rendre exactement les caractères des animaux. Aussi critique-t-il par exemple l'improbable société de chasseurs que forment le lion, la génisse, la chèvre et la brebis. De même, Richer plaide en faveur

[1] « Discours sur la fable » dans La Fontaine. *Œuvres complètes*, éd. J.-P. Collinet (Paris: Gallimard, « Bibliothèque de la Pléiade », 1991) t. I, 939.

de la vraisemblance des images propres à favoriser la portée pédagogique d'un genre qui donne aux enfants les premières notions du monde : « rien n'y est contraire à l'instinct des animaux qui sont les personnages ordinaires de la fable[2] ». Mais il admet qu'il ne convient pas d'y regarder de trop près :

> Le renard sert à manger à la cigogne sur une assiette [...] Tout cela semble peu naturel [...] Cependant, lorsqu'on y fait réflexion, on doit convenir que la fable étant en possession de prêter la parole et même la raison aux animaux [...] elle peut bien avec autant de vraisemblance leur donner une batterie de cuisine.[3]

Pourtant, comme le souligne Patrick Dandrey à propos de La Fontaine, il est peu probable que cette littérature soit restée totalement fermée aux avancées de la biologie et à la modification du regard porté sur les animaux :

> Même dans un genre littéraire comme l'apologue, dont le génie procède de la confusion entre *naturalia*, *moralia* et *litteraria*, on distingue une discrète évolution, répercussion de ce mouvement de fond : la relation du géographe vient y supplanter çà et là la légende forgée ou transmise par le voyageur, le savoir théorique et l'expérience attestée y sont convoqués pour défendre la thèse de l'intelligence animale avec plus de poids que les plaisantes traditions des *moralia* de Plutarque et de Montaigne.[4]

C'est cette articulation complexe des savoirs zoologiques et de la fable que l'on se propose d'analyser sur la période menant des animaux-machines de Descartes à l'*Histoire naturelle* de Buffon, de La Fontaine à l'abbé Aubert.

Il semble en effet qu'à partir de La Fontaine, l'apologue se soit fait de plus en plus le véhicule des découvertes scientifiques, même si ces nouveaux savoirs valent rarement pour eux-mêmes et sont instrumentalisés à des fins morales, littéraires ou philosophiques. Ainsi, les XVIIème et XVIIIème siècles sont marqués par l'invention puis le perfectionnement du microscope qui contribuent au développement de l'entomologie : des savants comme Marcello Malpighi, Robert Hooke ou Swammerdam[5] exposent les résultats de leurs microdissections : le tube digestif, l'appareil

[2] Henri Richer, *Fables nouvelles mises en vers* (Paris, Etienne Ganeau, 1729), 9.
[3] Ibid., 10-11.
[4] Patrick Dandrey. *La fabrique des Fables : essai sur la poétique de La Fontaine* (Paris : Klincksieck, 1992), 149.
[5] Marcello Malpighi, *Dissertation epistolica de bombyce* (1669) ; Robert Hooke, *Micrographia* (1665) ; J. Swammerdam, *Histoire générale des insectes* (1669).

respiratoire de la chenille, l'anatomie de la puce, le système circulatoire de l'éphémère deviennent l'objet d'observations minutieuses et enthousiastes. Réaumur écrit une *Histoire des insectes* et se passionne pour les abeilles dont il affine la connaissance en les observant à travers des ruches vitrées.

Ces diverses recherches n'ont pas vraiment renouvelé le personnel des fables : le ciron, connu depuis l'Antiquité, est mentionné en tant qu'animal emblématique des êtres minuscules. Toutefois, un texte comme « La Besace » (I, 7) où La Fontaine inventorie les créatures de l'univers, depuis le singe jusqu'à l'homme en passant par l'éléphant, la baleine, la fourmi et le ciron, n'est pas sans rappeler les disproportions qu'évoquent Pascal ou Malebranche. Les emboîtements vertigineux que ce dernier imagine à partir des cirons[6] ne relèvent-ils pas de la même logique que la multiplication des « mondes infinis » qu'évoque le fabuliste dans « Démocrite et les Abdéritains » (VIII, 26) ? C'est alors tout un jeu d'oppositions entre l'infiniment grand et l'infiniment petit, une rêverie sur l'extension de l'univers et la segmentation des êtres que permettent les instruments d'optique[7]. De cet incommensurable physique à la démesure morale, il n'y a qu'un pas et l'on en revient à une lecture anthropomorphique : le ciron sert d'étalon à La Fontaine pour condamner la myopie des hommes envers leurs propres défauts[8] et à Houdar de La Motte pour dénoncer leur vanité :

> Tel s'imagine être un géant
> Qui n'a pas plus d'une coudée.
> Aristote n'a pas trouvé notre vrai nom.
> Orgueil et petitesse ensemble,
> Voilà tout l'homme ce me semble.

---

[6] Malebranche, *La Recherche de la vérité*, dans *Œuvres, I* (Paris: Gallimard, « Bibliothèque de la Pléiade », 1979), 55: « Les petits animaux dont nous venons de parler ont peut-être d'autres petits animaux qui les dévorent, et qui leur sont imperceptibles à cause de leur petitesse effroyable, de même que ces autres nous sont imperceptibles. Ce qu'un ciron est à notre égard, ces animaux le sont à un ciron ; et peut-être qu'il y en a dans la nature de plus petits et plus petits à l'infini, dans cette proportion si étrange d'un homme à un ciron. »

[7] Furetière note dans son Dictionnaire : « Le microscope nous a fait découvrir plusieurs parties en un ciron. » Antoine Furetière, *Dictionnaire universel contenant generalement tous les mots françois, tant vieux que modernes, & les termes de toutes les sciences et des arts*, (1690) édition de 1725, revue et augmentée par Henri Basnage de Beauval & Jean-Baptiste Brutel de La Rivière, article Microscope consultable en ligne : http://gallica.bnf.fr/ark:/12148/bpt6k50614b.

[8] « La Besace », I, 7 : « Lynx envers nos pareils, et taupes envers nous,
Nous nous pardonnons tout, et rien aux autres hommes :
On se voit d'un autre œil qu'on ne voit son prochain. » (vers 28-30)

Est-ce donc là ce qu'on nomme raison ?
Quoiqu'il en soit, voici quelqu'un qui nous ressemble ;
Au bon cœur près, tout homme est mon ciron. [9]

Richer caricature ces effets de contraste en rendant invisible un moucheron, autre « avorton » présomptueux :

Sur la corne d'un Bœuf qui paissait dans les champs,
Un Moucheron, jouet des vents,
Alla s'asseoir, atome imperceptible :
Sans microscope, il n'était pas visible.[10]

Mais ce monde minuscule peut aussi donner vie à des tableaux édifiants, à des spectacles fascinants. Fénelon, précepteur du duc de Bourgogne, soutient en effet dans ses traités pédagogiques qu'il convient d'exploiter l'imagination féconde des enfants en leur racontant des histoires agrémentées par des « images sensibles ».[11] Profitant peut-être de l'intérêt de Colbert pour l'élevage du bombyx, il imagine une fable, *Les Abeilles et les vers à soie,* où un conflit oppose devant Jupiter les « mouches à miel » et les « insectes rampants ». L'un d'eux prend la parole pour faire valoir les qualités de son peuple :

Chacun de nous montre les merveilles de la nature, et se consume dans un travail utile […] Nous avons la vertu de Protée pour changer de forme. Tantôt nous sommes de petits vers composés d'onze petits anneaux entrelacés avec la variété des plus vives couleurs qu'on admire dans les fleurs d'un parterre. Ensuite nous filons de quoi vêtir les hommes les plus magnifiques jusque sur le trône, et de quoi orner les temples des dieux […] Enfin nous nous transformons en fève, mais en fève qui sent, qui se meut, et qui montre toujours de la vie. Après ces prodiges, nous devenons tout à coup des papillons avec l'éclat des plus riches couleurs. [12]

Cette renaissance va de pair avec un éloge de la paix et du travail utile assurant la prospérité d'un pays. Mais au delà de cette politique chrétienne, la transfiguration de la chrysalide tient le milieu entre l'univers enchanté d'Ovide, peuplé de métamorphoses animales, et les phénomènes

[9] Antoine Houdar de La Motte, *Fables nouvelles* (Paris, 1719) « Le Bœuf et le Ciron », I, 13, vers 1-14.
[10] Henri Richer, *Fables nouvelles mises en vers*, « Le Bœuf et le Moucheron », II, 8 ; vers 1-4.
[11] Pierre P. Fénelon, *De l'éducation des filles*, dans *Œuvres*, ed. Jacques Le Brun, Gallimard, « Bibliothèque de la Pléiade », 1983, 124.
[12] Ibid., 225.

fabuleux de la nature. Le fabuliste fait surgir le papillon chatoyant de la chenille méprisée ; il met en scène une transformation magique qui ne doit rien à une quelconque formule mais qui n'est pas sans évoquer l'enthousiasme de Réaumur ou de Trembley devant les mystères insoupçonnés du monde réel. L'auteur de l'*Histoire des abeilles* suggère en effet qu'il s'agit désormais de passer des représentations mythologiques à l'examen scientifique des phénomènes jusque-là inconnus, de se fonder sur l'expérience pour ne rapporter que des faits avérés.[13] Il entend même profiter de l'engouement que manifeste le public mondain envers les contes et lui substituer une saine curiosité pour le « merveilleux vrai » des insectes[14]. Il faut faire « disparaître tout le faux merveilleux dont le nom de métamorphose et de transformation donnait des idées confuses, mais en même temps […] laisser bien du merveilleux réel à observer.[15] » Il convient donc de repenser les catégories du merveilleux pour qu'il ne soit plus en sciences naturelles synonyme de mensonge et de récit fantaisiste. L'inconcevable prend place dans une nature qui n'en finit pas de susciter la stupeur. C'est cette orientation plus scientifique qu'adopte Fénelon lorsqu'il écrit *l'Histoire naturelle du ver à soie*[16] : le texte s'ouvre sur un retour aux origines de l'humanité où les hommes, d'abord vêtus de feuilles, puis de peaux de bêtes, apprirent enfin à filer. Puis il entreprend de décrire au présent la lente métamorphose de la larve, « enfermé[e] dans une écorce transparente comme une perle », puis du vers dont « la tête est grosse à proportion du reste et par le microscope ressemble à celle d'un corbeau ».[17] Malgré la référence à l'instrument d'optique, l'animal hybride semble encore tenir des bêtes chimériques, impression qu'accentue le vif

[13] Réaumur, *L'Histoire des abeilles* (Paris: Imprimerie nationale, 1734), t.V, 7-8 : « je ne vous dirai rien que de bien vu, et de bien avéré ; il entrera beaucoup de merveilleux dans mon récit, mais rien de faux ; je détruirai des fables anciennes, dont je ne doute pas qu'on n'ait bercé votre enfance ; à la place, vous aurez des vérités qui ne vous surprendront pas moins, et vous contenteront davantage. »

[14] Réaumur, *Mémoires pour servir à l'histoire des insectes* (Paris: Imprimerie royale, 1734), t. I, 10 : « Le goût du merveilleux est un goût général, c'est ce goût qui fait lire plus volontiers des romans, des historiettes, des contes arabes, des contes persans, et même des contes de fées, que des histoires vraies. Il ne se trouve nulle part autant de merveilleux, et de merveilleux vrai que dans l'histoire des insectes. »

[15] Ibid., 351.

[16] Le texte est intégré à l'ensemble disparate, choisi par les éditeurs du XIXème siècle et repris par Jacques Le Brun, *Fables et opuscules pédagogiques*, *op. cit.*, 271-273.

[17] Pierre Fénelon, *Histoire naturelle du ver à soie*, XLV, op. cit., 272.

intérêt de Fénelon pour les mues successives qu'il peint avec minutie en insistant sur les couleurs changeantes d'un corps diaphane :

> La partie supérieure devient argentée [...] Son crâne prend la couleur d'agate. Il croît, a des taches rouges, devient transparent : on voit les feuilles à travers son corps. – Changement de peau blanche en pourprée: sa vieille peau se déchire, alors il se resserre, pousse entraille en haut, sa vieille peau se ride, et passe d'anneau en anneau ; cependant léthargie.[18]

Si le texte ne se réclame plus vraiment de la fable, il conserve bien quelque chose des contes dans ce goût des formes chantournées, dans cette thématique omniprésente de la métamorphose. La subjectivité du pédagogue se manifeste du reste dans le choix de cette palette trop vive pour être vraie et dans l'allusion à des pratiques superstitieuses qui voudraient que les œufs des papillons soient arrosés de vin et « couvés sous les aisselles des femmes[19] ». On voit cependant, toute proportion gardée, combien la fable et la science tendent à interférer : l'incroyable ne se cantonne plus au domaine du fabuleux. C'est pourquoi Houdar de La Motte revendique le droit de mêler « le naturalisme menteur » des « temps passés », qu'animaient les phénix, les sirènes et les sphinx, aux « vérités modernes »[20] :

> Qui peut le plus, ne peut-il pas le moins ?
> Les plines d'autrefois, ce sont les subalternes ;
> Ceux d'aujourd'hui, voilà les bons témoins.
> Ils savent rejeter l'opinion commune
> Qui n'a de fondement que la crédulité. (vers 13-17)

écrit-il dans *L'Écrevisse qui se rompt la jambe*. Se réclamant de l'esprit critique et de la démarche empirique des Lumières, l'auteur entend actualiser l'évocation des animaux en tirant parti des découvertes scientifiques. Ce nouveau savoir est aussi l'occasion pour le disciple de Fontenelle de condamner toute allégeance aveugle aux Anciens, de s'affranchir des déterminismes qui pèsent sur la pensée et sur la création

[18] Ibid.

[19] Ibid., 275. À ce propos, Jean Perrot se réfère à un document publié en 1602, *Bref discours sur la manière d'élever le ver à soie,* qui précise d'une part que le bombyx a en réalité des couleurs ternes et qui confirme d'autre part le rôle des femmes. Jean Perrot propose même de voir une symétrie entre ce maternage du ver et la relation du précepteur à son élève : « Contes et fables pour les enfants du XVIIe siècle à nos jours », *Diogène*, 2002/2 n° 198, 61.

[20] Antoine Houdar de La Motte, *L'Écrevisse qui se rompt la jambe*, II, 14.

littéraire. Autrement dit, le détour par la science serait surtout au service d'une réflexion métapoétique : La Motte y puiserait des sujets originaux lui permettant de faire valoir son indépendance littéraire et ses capacités d'invention. Il se fait alors l'écho des observations récentes de Réaumur sur la régénération des pinces de crustacés. L'apologue rapporte que l'écrevisse a la capacité de reproduire une patte amputée et le fabuliste se plaît même à souligner l'apparente invraisemblance de ce phénomène :

> « L'écrevisse a, dit-on, des jambes de relais.
> S'en rompt-elle une ? Il s'en trouve au passage
> Une autre que nature y substitue exprès.
> Une jambe est enfin un magasin de jambes.
> Vous riez ; vous prenez ceci
> Pour l'histoire des Sévarambes.
> N'en riez point. C'est un fait éclairci. » (vers 23-29)

Le savoir scientifique, pourtant établi sur un « scrupuleux témoignage », passe d'abord pour une extravagance de l'imagination, pour une fable et il appartient précisément à la fable de l'inverser en vérité. La Motte construit ainsi sa fiction sur la confrontation de deux points de vue qui signale la difficulté du projet : il commence par expliquer dans un long prologue la régénérescence animale à un lecteur incrédule dont il va s'agir de susciter l'adhésion. La leçon de choses précède la narration et en déroule les étapes pour instruire le destinataire réticent à admettre ce que l'opinion commune n'a pas encore attesté. Le récit redouble ensuite le prologue et oppose l'écrevisse boiteuse à une grenouille « raillant hors de saison », bientôt obligée d'admettre ses torts. La moralité invite alors à substituer à « l'instinct » de la bête l'usage de la « raison » en sachant discerner les « remèdes » derrière les « maux apparents ». L'expérimentation scientifique, en révélant la nécessité de dépasser une appréhension figée du monde, doit favoriser un mouvement d'innovation dont prétend profiter l'invention littéraire. En relayant les acquis inédits de la science, la fable se donne la possibilité de remettre en cause les références passées et de faire advenir le monde des Lumières.

Trembley opère le mouvement inverse : il s'amuse à attester la véracité du mythe de l'hydre en regardant repousser les membres coupés de ses polypes et prétend même renchérir sur la fable par ses expériences démiurgiques :

> J'ai coupé les têtes de celui qui en avait sept ; et, au bout de quelques jours, j'ai vu en lui un prodige qui ne le cède guères au prodige fabuleux de l'Hydre de Lerne. Il lui est venu sept nouvelles têtes ; et si j'avais continué à couper à mesure qu'elles poussaient, il n'y a pas à douter que je n'en eusse vu pousser d'autres. Mais, voici plus que la Fable n'a osé inventer. Les sept têtes, que j'ai coupées à cet Hydre, ayant été nourries, sont devenues des Animaux parfaits, de chacun desquels il ne tenait qu'à moi de faire une Hydre. [21]

Les fabulistes comme les savants puisent ainsi dans les nouvelles découvertes matière à enchanter leur public, à renouveler les sujets et, dans une certaine mesure, les prétentions du genre de l'apologue. La fable devient le laboratoire de petites expériences offertes au regard du lecteur ; lieu par excellence d'un merveilleux familier, elle peut prétendre vulgariser les savoirs les plus déroutants et prolonger l'entreprise des encyclopédistes. C'est ce qu'envisage en particulier l'abbé Aubert lorsque, publiant ses *Fables nouvelles* en 1773, il y joint

> des notes relatives à l'Histoire, à la Géographie, à la Mythologie et à l'Histoire naturelle. Je me suis principalement étendu sur ces dernières, comme étant les plus nécessaires, parce qu'elles servent à faire connaître mes acteurs, et parce qu'aucun fabuliste n'a encore songé à en donner de semblables. Le goût que tous les hommes ont pour l'Histoire naturelle, les progrès que cette science a faits de nos jours, la lumière qu'elle répand sur tous les objets qui nous environnent, la haute idée qu'elle donne de l'Être tout-puissant qui a su mettre tant d'ordre et tant de variété dans les ouvrages sortis de sa main, me font espérer que ces notes [...] intéresseront le plus grand nombre de mes lecteurs. [22]

À première vue, ses fables, qui s'appuient fréquemment sur les travaux de Buffon et de Valmont de Bomare, présentent pourtant moins une conciliation de la science et de la fiction qu'une juxtaposition des discours : la narration de facture traditionnelle est accompagnée d'un appareil critique qui semble n'apporter qu'un vernis savant au récit. Les notes entrent même parfois en dissonance avec la fable : ainsi, *La colère du lion* (II, 6) dénonce la susceptibilité et l'injustice du roi des animaux qui, après s'être emporté contre un « injurieux couplet », fait pendre abusivement l'un de ses sujets. La note renvoie à Buffon qui multiplie au contraire les termes valorisants pour qualifier le fauve : il est « le plus

[21] Abr Trembley, *Mémoires pour servir à l'histoire d'un polype d'eau douce à bras en forme de cornes* (Leyde: Verbeek, 1744), 246.

[22] Abbé Aubert, « Avertissement », *Fables nouvelles* (Paris :Moutard, 1773), XXVI.

noble et le plus courageux des animaux », « sa taille est si bien prise et si bien proportionnée que son corps paraît être le modèle de la force, jointe à l'agilité ». L'écart entre la fable et le discours du naturaliste permet en revanche d'éclairer le parti-pris idéologique de Buffon qui établit « une échelle graduée de valeurs[23]» entre les êtres vivants et distingue des espèces nobles, comme le lion ou le cheval, et des animaux insignifiants ou pervers, comme le loup et le chat. L'homme est évidemment au sommet de cette hiérarchie. L'anthropomorphisme caractéristique de l'apologue attire en quelque sorte l'attention sur la démarche contradictoire du savant cherchant à resituer l'espèce humaine au sein du vivant tout en lui conférant un statut à part.

Il arrive aussi qu'Aubert reprenne les couples auxquels recourt Buffon pour tirer parti des contrastes : celui-ci met par exemple en parallèle le dévouement, la docilité du chien et « la malice innée », « le caractère faux[24] » du chat. « Les deux Chiens et le Chat » (III, 1) oppose de même à l'amitié fidèle des premiers qui s'entraident l'hypocrisie du second cherchant à les séparer. C'est bien l'association des deux animaux qui fait signe[25] et la façon dont Aubert attire l'attention sur leurs rapports à l'homme : le « dévot chat » donne la réplique à « l'honnête homme de chien » tandis qu'une note, empruntée au *Dictionnaire portatif des mots français* de l'abbé Prévost[26], précise : « le chat est hypocrite. Il tient un peu du tigre par sa forme et sa légèreté. On prétend que sa cervelle est une sorte de poison. » Cette idée d'une nocivité de la chair du chat était répandue au Moyen Age et si Buffon ne la reprend pas à son compte, il la

---

[23] Thierry Hoquet, « Logique de la comparaison et physique de la génération chez Buffon », *Dix-huitième siècle*, 2007/1 n° 39, 598.

[24] Buffon, *Histoire naturelle, générale et particulière, avec la description du cabinet du roi*, 1756, t. VI, 3. Il situe ensuite les deux animaux par rapport à l'homme : « ils n'ont que l'apparence de l'attachement ; on le voit à leurs mouvements obliques, à leurs yeux équivoques ; ils ne regardent jamais en face la personne aimée ; soit défiance ou fausseté, ils prennent des détours pour en approcher, pour chercher des caresses auxquelles ils ne sont sensibles que pour le plaisir qu'elles leur font. Bien différent de cet animal fidèle, dont tous les sentiments se rapportent à la personne de son maître, le chat paraît ne sentir que pour soi, n'aimer que sous condition, ne se prêter au commerce que pour en abuser; et par cette convenance de naturel, il est moins incompatible avec l'homme, qu'avec le chien dans lequel tout est sincère. » (4).

[25] L'hypocrisie du chat est en effet traditionnelle dans les fables et, si le chien est souvent présenté comme un animal stupide ou glouton, il sait aussi faire preuve de dévouement. Voir par exemple Le chien fidèle de Phèdre.

[26] *Manuel lexique ou Dictionnaire portatif des mots français,* par l'abbé Prévost et Thomas Dyche (Paris : Didot, 1767), article « chat ».

transpose dans le domaine moral. Jean-Luc Guichet se demande alors quel crédit accorder à cette imagerie « qui penche finalement plus du côté d'une fable que d'une science[27]». Même s'il est probable que le naturaliste n'accorde pas une valeur scientifique à ces caractérisations morales, il fait de l'animal le support métaphorique d'une idéologie. Cette orientation anthropomorphique de l'histoire naturelle ne pouvait qu'inspirer les fabulistes en leur proposant à la fois un imaginaire familier et un nouveau soubassement épistémologique. Ainsi, la fable « Le Pongo et les Sapajous » (II, 8) participe humblement à la réflexion des contemporains sur les similitudes troublantes de l'homme et des grands singes :

> « Un Singe que Pongo l'on nomme,
> Dont l'intelligence surprend,
> Et qui n'est pourtant seulement
> Qu'un peu plus que Singe et moins qu'homme,
> Fit grand bruit autrefois dans notre continent [...]
> Or voici le portrait de ce singe étonnant:
> De longs bras, un grand corps, figure presque humaine,
> Et comme nous se soutenant
> Sur deux pieds sans aucune peine ;
> À table, comme nous, mangeant, et découpant
> Dans de l'argent ou de la porcelaine,
> De la serviette s'essuyant,
> Remplissant lui-même son verre,
> Et les convives saluant
> Sans répandre une goutte à terre.
> Chacun criait miracle en le considérant.

Aubert ajoute : « Tout ce que je dis ici, M. de Buffon le raconte, dans son *Histoire naturelle*, d'un orang-outang qu'il a vu. » La fable quitte l'univers de la fiction pour se réclamer explicitement d'un témoignage que le naturaliste rapporte en effet pour montrer à quel point l'orang-outang, ou plutôt le chimpanzé, diffère des autres singes[28]. Aubert développe à son

[27] Jean-Luc Guichet, *Rousseau, l'animal et l'homme : l'animalité dans l'horizon anthropologique des Lumières* (Paris: éditions du Cerf, 2006), 154.

[28] Buffon, *Histoire naturelle* (Paris : Imprimerie royale, 1766) t. XIV, 53 : « J'ai vu cet animal présenter sa main pour reconduire les gens qui venaient le visiter, se promener gravement avec eux et comme de compagnie ; je l'ai vu s'asseoir à table, déployer sa serviette, s'en essuyer les lèvres, se servir de la cuiller et de la fourchette pour porter à sa bouche, verser lui-même sa boisson dans un verre, le choquer, lorsqu'il y était invité, aller prendre une tasse et une soucoupe, l'apporter sur la table, y mettre du sucre, y verser du thé, le laisser refroidir pour le boire, et

tour la description pour mettre l'accent sur les caractéristiques anthropoïdes de l'animal et, s'il aboutit à une moralité on ne peut plus banale sur l'imitation, c'est ce jeu de miroir brouillant les frontières des espèces et interrogeant les fondements de l'humanité qui ancre le texte dans son époque. Buffon pointait lui-même à partir de l'étymologie toute l'ambiguïté de ce face-à-face :

> Les habitants de son pays, les Indiens policés n'ont pas hésité à l'associer à l'espèce humaine par le nom d'Orang-Outang, Homme sauvage ; tandis que les Nègres presque aussi sauvages, aussi laids que ces singes, et qui n'imaginent pas que pour être plus ou moins policé l'on soit plus ou moins homme, leur ont donné leur nom propre (Pongo), un nom de bête et non pas d'homme ; et cet orang-outang ou ce pongo, n'est en effet qu'un animal, mais un animal très singulier, que l'homme ne peut voir sans rentrer en lui-même, sans se reconnaître, sans se convaincre que son corps n'est pas la partie la plus essentielle de sa nature.[29]

Si Buffon est cartésien et n'admet pas l'hypothèse d'êtres intermédiaires entre l'homme doué de raison et la bête brute, un tel passage fait du singe le support d'une démarche introspective et d'un questionnement de la notion de civilisation.

La Fontaine, qui tisse une véritable continuité entre les règnes, en aurait tiré un tout autre profit. Patrick Dandrey a montré que ses animaux parlants tenaient à la fois de l'emblématique, des traités de physiognomonie, des théories gassendistes et peut-être aussi du *Traité sur la connaissance des animaux* (1647) de Cureau de La Chambre. Le médecin philosophe refuse en effet de réduire le comportement des bêtes à des réflexes mécaniques et propose un compromis : il opère une distinction entre une « âme sensitive » et une « âme intellectuelle », ce qui lui permet de prêter aux animaux une pensée n'impliquant ni la conscience, ni l'immortalité mais fondée sur la puissance inventive de l'imagination[30]. Une telle philosophie est très proche de la démarche qu'adopte La Fontaine dans son « Discours à Madame de La Sablière ».

---

tout cela sans autre instigation que les signes ou la parole de son maître, et souvent de lui-même. »

[29] Ibid., 3-4.

[30]Cureau de La Chambre insiste sur le fait que ces opérations mentales n'impliquent aucun recours à l'abstraction : « […] quand l'âme reprend une image qu'elle a déjà formée pour la joindre avec une autre, elle ne se replie pas sur elle-même ni sur la connaissance, mais seulement sur l'effet de son action. Aussi, il n'y a point d'abstraction, ni par conséquent, de véritable réflexion » (*Traité De la connaissance des animaux*, cité par Albert Darmon, *Les corps immatériels. Esprits et images dans l'œuvre de Marin Cureau de La Chambre*. Paris :Vrin, 1985,. 56).

Le fabuliste commence, on le sait, par exposer la théorie cartésienne de l'animal-machine qui relève d'une métaphysique dualiste plus que des sciences naturelles. Toute sa poétique étant fondée sur la complémentarité de « l'âme » et du « corps » de l'apologue, La Fontaine entreprend de montrer au contraire qu'il n'y a pas de rupture entre l'homme et l'animal, entre la raison et l'imagination. À la métaphore de l'horloge, il oppose alors une « logique de la phantasia »[31] empruntée à Gassendi qu'il associe à une « mémoire corporelle[32] ». Loin de voir son action réduite à un enchaînement de réflexes instinctifs, la bête partagerait avec l'homme une faculté imaginative. La Fontaine appuie son argumentation sur des anecdotes animalières dont le statut est délibérément ambigu. Les exemples du vieux cerf qui déroute les chasseurs en brouillant ses traces ou en donnant le change avec un animal plus jeune, de la perdrix qui traîne son aile pour éloigner le chasseur de ses petits, ont pu être empruntés à Plutarque (*De l'amour et charité naturelle des pères et mères envers leurs enfants*) ou au *De corpore animato* de Jean-Baptiste Du Hamel, secrétaire de l'Académie des Sciences[33]. Mais ces sources érudites ne sont pas les seules. De telles ruses sont aussi décrites dans les traités de vénerie de l'époque et correspondent à une expérience empirique. En même temps, ces saynètes oscillent entre la narration fictionnelle et la zoopsychologie : elles sont portées par un anthropomorphisme qui permet au fabuliste de jouer sur des similitudes et des oppositions. Ainsi, le cerf multiplie les « stratagèmes » à l'instar des « grands chefs » militaires et la perdrix cherche à protéger sa famille mais tous deux contribuent à dénoncer la cruauté des chasseurs. La Fontaine va plus loin avec les castors. Il puise dans les témoignages récents des voyageurs et insiste sur les talents de bâtisseur de l'animal qui contraste avec la rudesse des habitants du Nord, « vivant dans une ignorance profonde » :

---

[31] Voir sur ce point Jean-Charles Darmon, *Philosophies de la Fable. Poésie et pensée dans l'œuvre de La Fontaine* (Paris : Hermann, 201) 225 et suivantes.

[32] « Discours à Madame de La Sablière », vers 142.
« L'animal n'a besoin que d'elle.
L'objet, lorsqu'il revient, va dans son magasin
Chercher par le même chemin
L'image auparavant tracée,
Qui sur les mêmes pas revient pareillement,
Sans le recours de la pensée,
Causer un même événement. » (vers 145-151)

[33] *De corpore animato* (1673). Jean-Pierre Collinet signale que l'exemple du cerf est mentionné dans cet ouvrage qui fait la synthèse de l'aristotélisme et du gassendisme (*La Fontaine. Œuvres complètes*, Paris :Gallimard, 1991), 1240.

Je parle des humains ; car quant aux animaux,
Ils y construisent des travaux
Qui des torrents grossis arrêtent le ravage,
Et font communiquer l'un et l'autre rivage.
L'édifice résiste, et dure en son entier ;
Après un lit de bois, est un lit de mortier.
Chaque castor agit ; commune en est la tâche ;
Le vieux y fait marcher le jeune sans relâche.
Maint maître d'œuvre y court, et tient haut le bâton.
La république de Platon
Ne serait rien que l'apprentie
De cette famille amphibie.
Ils savent en hiver élever leurs maisons,
Passent les étangs sur des ponts,
Fruit de leur art, savant ouvrage ;
Et nos pareils ont beau le voir,
Jusqu'à présent tout leur savoir
Est de passer l'onde à la nage.

Ni Hérodote, ni Pline, ni Aristote n'évoquaient cette habileté à construire des barrages. L'explorateur français Nicolas Denys le qualifie à l'inverse d'« architecte » et s'émerveille devant la discipline de cet animal qui a ses lois, son gouvernement, ses maîtres d'œuvre. Le renversement ironique des principes cartésiens n'est pas simplement appelé par un souci moral – valoriser l'animal pour inciter l'homme à faire valoir sa singularité –. Il tend à relativiser cette prétendue supériorité[34.] Parallèlement, La Fontaine plaide en faveur des pouvoirs pédagogique et esthétique de l'imagination. Ces exemples argumentatifs, mêlant étroitement la thèse et leur illustration au point de les confondre, témoignent des liens consubstantiels unissant raison et séduction narrative.

Un pas de plus est franchi avec la fable enchâssée « Les deux Rats, le Renard et l'Œuf » où La Fontaine glisse du savoir-faire à « l'invention ». Il se plaît à rapporter la réflexion des rats s'interrogeant sur la meilleure façon de transporter un œuf avant d'adopter la technique du traineau[35]. Certes, Pline prêtait le même comportement aux marmottes et, plus récemment, Guillaume le Vasseur assurait avoir vu ce manège de ses

[34] Houdar de La Motte va dans le même sens en faisant du castor le symbole d'un esprit indépendant, gardant ses distances avec les hommes pour ne pas devenir leurs « valets d'ateliers » (« Le Castor et le Bœuf », III, 6).

[35] Vers 186-189 :« Car comment sauver l'œuf ? Le bien empaqueter,
Puis des pieds de devant ensemble le porter,
Ou le rouler, ou le traîner,
C'était chose impossible autant que hasardeuse. »

propres yeux chez les Bobaques[36]. Il n'en reste pas moins que l'on touche ici aux limites de la vraisemblance, La Fontaine suggérant par ce cas extrême qu'il convient d'appréhender la réalité animale d'une autre façon. Sa fable esquisse une théorie de la connaissance fondée sur les notions de sympathie universelle, sur l'alliance de la réflexion et de l'imagination, de la science et de la littérature. Et si sans doute l'auteur masque les failles de son raisonnement derrière des analogies approximatives et un langage polysémique, la poésie lui offre le moyen de penser avec des images, d'approcher des réalités insaisissables et difficiles à conceptualiser.

Car avec cette querelle sur l'âme des bêtes, il s'agit bien au bout du compte pour le fabuliste d'appeler à un décentrement du regard. Il utilise le merveilleux pour susciter une interprétation distanciée, pour interroger la force de nos préjugés ou pour démonter les opinions aberrantes, « les pensers du vulgaires […] mettant de faux milieux entre la chose et lui,/Et mesurant par soi ce qu'il voit en autrui[37]». On songe à la colère de l'ours dans *Les Compagnons d'Ulysse* lorsque son ancien roi lui représente sa laideur :

> « Qui t'a dit qu'une forme est plus belle qu'une autre?
> Est-ce à la tienne à juger de la nôtre? » (vers 70-71)

On pense aussi au « premier qui vit un chameau » et « S'enfuit à cet objet nouveau »[38]. Tantôt l'homme raisonne abusivement par analogie, tantôt il manque d'imagination devant des faits qui dépassent l'entendement ou le cadre de l'expérience familière. Dans un monde où « tout homme ment » et où « les fous vont l'emportant[39] », le fabuliste

---

[36] Guillaume le Vasseur, *Description des contrées du royaume de Pologne* (Rouen : Jacques Cailloüe, 1660), 80-81 : « […] l'on dirait qu'ils ont des esclaves parmi eux, car ceux qui sont paresseux ils les font coucher sur le dos, et leurs chargent sur le ventre une grande poignée d'herbe sèche que le Bobaque tient embrassée de ses pattes, et pour plus proprement parler de ses mains, car ces animaux s'en aident presque comme les singes des leurs, puis les autres la trainent par la queue jusques à l'entrée de leur tanière, et ainsi cet animal leur sert de traineau, et de là lui font porter l'herbe dans leurs cachettes: je les ai vus plusieurs fois faire ce ménage , et me fuis arrêté par curiosité à les contempler des journées entières ».

[37] « Démocrite et les Abdéritains », VIII, 26 ; vers 1 et 3-4.

[38] « Le Chameau et les Bâtons flottants », IV, 10 ; vers 1-2.

[39] « Le Dépositaire infidèle », IX, 1 ; vers 20 et 11. Dans cette fable, un marchand, à qui son voisin a dérobé un quintal de fer en accusant les rats de l'avoir mangé, feint d'admirer « un tel prodige » et réplique par une histoire explicitement invraisemblable : un chat-huant a enlevé l'enfant du voleur.

mise sur des fictions extravagantes et truquées qui sont comme autant de loupes de nos dérives et qui incitent à revenir à la vérité par des voies obliques. Ce réajustement opéré par la raison suppose que l'on sache mettre à distance les chimères ou, pour reprendre les termes de La Fontaine, « développe[r] le vrai caché sous l'apparence[40]». Mais, dans le même temps, la fable s'adresse à l'esprit d'enfance du lecteur ; elle fonde son emprise sur sa capacité à émerveiller l'auditoire en enchantant le monde. Or les découvertes scientifiques peuvent nourrir ce désir de fable. Réaumur compare l'étonnement de ses contemporains devant les prodiges de la nature à celui du sauvage découvrant les merveilles d'une civilisation avancée :

> Les merveilles prodiguées dans la construction intérieure des insectes nous échappent. Nous ne laissons pourtant pas d'y voir bien des mécanismes surprenantes, et qui doivent exciter ceux qui étudient les insectes à pousser plus loin leurs recherches. On a découvert que les chenilles ont un cœur et une suite de cœurs, qui règne d'un bout à l'autre de leur dos ; on a découvert que la plupart des anneaux dont leur corps est composé, ont deux ouvertures ou deux bouches destinées à respirer l'air. [41]

La découverte de réalités inouïes invite bien à changer de regard. Comme le note fort justement Nathalie Vuillemin,

> Le merveilleux surgit non pas du phénomène lui-même, mais de la lecture qu'on en produit. Il faut donc peu à peu abandonner l'idée de vraisemblance, se dépouiller de tout jugement que l'expérience ne peut soutenir, ouvrir sur la nature un regard pur de toute attente pour parvenir à en appréhender efficacement les virtualités.[42]

Il faut admettre les limites de nos connaissances pour accepter de voir ce qui déroge à toutes les certitudes antérieures. Il faut aussi avoir le goût du merveilleux pour savoir l'apprécier dans la nature comme dans la fable. Tous les savants et tous les fabulistes n'en sont pas capables : à La Fontaine qui se prend à rêver à *Peau d'Âne* au moment d'écrire sa moralité et qui construit volontiers des châteaux en Espagne s'oppose l'obsession pédagogique de La Motte qui accepte d'amuser le lecteur en enfant à

---

[40] « Un animal dans la lune », VII, 17 ; vers 26.

[41] Réaumur, *Mémoires pour servir à l'histoire des insectes*, Paris, Imprimerie nationale, 1734, 15.

[42] Nicole Vuillemin, *Les beautés de la nature à l'épreuve de l'analyse. Programmes scientifiques et tentations esthétiques dans l'histoire naturelle du XVIIIe siècle (1744-1805)* (Paris : Presses Sorbonne Nouvelle, 2009), 274.

condition d'en faire un homme[43]. Et Réaumur qui s'extasiait devant les gâteaux de cire de son « petit peuple » fut raillé par Buffon qui préféra « désenchanter la ruche » en comparant les abeilles à « dix mille automates animés d'une force vive et tous déterminés »[44]. Gageons que tout réside dans cet élargissement du regard, dans cette conviction que le savant peut partager avec le fabuliste l'intuition de beautés énigmatiques se dérobant à l'esprit de géométrie.

Œuvres citées

Abbé Aubert. *Fables nouvelles*. Paris: Moutard, 1773.

Darmon, Albert. *Les corps immatériels. Esprits et images dans l'œuvre de Marin Cureau de La Chambre*. Paris: Vrin, 1985.

Darmon, Jean-Charles. *Philosophies de la Fable. Poésie et pensée dans l'œuvre de La Fontaine*. Paris: Hermann, 2011.

Du Hamel, Jean-Baptiste. *De corpore animato.*Paris: 1673.

Fénelon. *Œuvres.* Edité par Jacques Le Brun. Paris: Gallimard, « Bibliothèque de la Pléiade », 1983.

Furetière, Antoine. *Dictionnaire universel contenant generalement tous les mots françois, tant vieux que modernes, & les termes de toutes les sciences et des arts*. (1690) édition de 1725, revue et augmentée par Henri Basnage de Beauval & Jean-Baptiste Brutel de La Rivière, consultable en ligne : http://gallica.bnf.fr/ark:/12148/bpt6k50614b

Guichet, Jean-Luc. *Rousseau, l'animal et l'homme : l'animalité dans l'horizon anthropologique des Lumières*. Paris: éditions du Cerf, 2006

Hoquet, Thierry. « Logique de la comparaison et physique de la génération chez Buffon », *Dix-huitième siècle,* 2007/1 n° 39.

Hooke, Robert. *Micrographia: or, Some physiological descriptions of minute bodies made by magnifying glasses.* Londres: J. Martyn and J. Allestry, 1665.

La Fontaine, Jean de. *Œuvres complètes.* Edité par Jean-Pierre Collinet. Paris: Gallimard, « Bibliothèque de la Pléiade », 1991.

Le Brun. *Des fables et opuscules pédagogiques*.

Le Menthéour, Rudy. « De la ruche au polype : figures de l'organisation sociale », *Dix-huitième siècle*, 2009/1 n° 41, 208.

---

[43] « Le Renard prédicateur », V, 3 ; vers 12. « Que le conte soit fait pour la moralité ; Prenons si juste nos mesures, Que nous allions tout droit à notre vérité ».

[44] Buffon, *Histoire naturelle*, tome IV, « Discours sur la nature des animaux », 98. L'expression précédente est empruntée à Rudy Le Menthéour, « De la ruche au polype : figures de l'organisation sociale », *Dix-huitième siècle*, 2009/1 n° 41, 208.

Le Vasseur, Guillaume. *Description des contrées du royaume de Pologne.* Rouen:Jacques Cailloüe, 1660.

Malebranche. *Œuvres.* Paris: Gallimard, « Bibliothèque de la Pléiade », 1979.

Malpighi, Marcello. *Dissertation epistolica de bombyce.* Londini: Martyn, 1669.

Manuel lexique ou Dictionnaire portatif des mots français. Edité par l'abbé Prévost et Thomas Dyche. Paris: Didot, 1767.

Perrot, Jean. « Contes et fables pour les enfants du XVIIe siècle à nos jours », *Diogène*, 2002/2 n° 198.

Réaumur. *Histoire des abeilles.* Paris: Imprimerie nationale, 1734.

—. *Mémoires pour servir à l'histoire des insectes.* Paris: Imprimerie nationale, 1734.

Richer, Henri. *Fables nouvelles mises en vers.* Paris: Etienne Ganeau, 1729.

Swammerdam, Jan. *Histoire générale des insectes* (1669).

Trembley, Abr. *Mémoires pour servir à l'histoire d'un polype d'eau douce à bras en forme de cornes.* Leyde : Verbeek, 1744.

Vuillemin, Nicole. *Les beautés de la nature à l'épreuve de l'analyse. Programmes scientifiques et tentations esthétiques dans l'histoire naturelle du XVIIIe siècle (1744-1805).* Paris: Presses Sorbonne Nouvelle 2009.

PATRICIA PICCININI, UNE ŒUVRE FAITE DE CHIMERES QUI SE NOURRISSENT DE SCIENCE ET D'ANGOISSE

CELINE CADAUREILLE

Les illusions sont parfois le moteur des plus beaux projets, qu'ils soient scientifiques ou artistiques. Les chimères nous dépassent car elles représentent ces rêves incroyables, des projets qui semblent irréalisables. L'origine du terme « chimère » nous vient de la littérature grecque où de nombreuses créatures extraordinaires sont nées d'unions contre nature. On sait qu'Homère présente la chimère avec une tête de lion, un corps de chèvre et un arrière-train de serpent, Hésiode dit seulement qu'elle est immense, forte et que sa bouche crache toujours du feu. La chimère est généralement vue comme un monstre composite qui se re-construit, se ré-invente selon les époques et l'imaginaire des artistes ou des écrivains. Nous verrons ici comment le travail de Patricia Piccinini s'inspire de la science pour construire d'étranges fables où des créatures chimériques apparaissent en tant que limites.

Repères et Influences

Patricia Piccinini est une jeune artiste australienne, née à Freetown en Sierra Leone en 1965. Elle grandit et fait ses études en Australie au Victorian College of the Arts. L'œuvre de Patricia Piccinini s'exprime de manière pluridisciplinaire, elle fait des dessins, des photographies qu'elle retravaille numériquement, des sculptures aux finitions hyperréalistes et également des vidéos. Si elle ne s'attache pas à un médium précis ; la science, les recherches en biologie et les manipulations génétiques sont toujours là pour sous tendre ses projets.

Patricia Piccinini est fascinée par les avancées scientifiques mais elle reconnait qu'elle suit ces découvertes avec une certaine distance, comme une non spécialiste… Elle a longtemps observé les malformations et fait des dessins dans les musées de médecine, elle a eu aussi l'occasion de voir des cultures de cellules et de cellules souches en laboratoire…

En effet, en Australie, les rapprochements sont de plus en plus fréquents entre les sciences et les arts, il y a par exemple à l'école d'anatomie et de biologie de l'Université d'Australie-Occidentale, à Nedlands, un Laboratoire nommé SYMBIOTICA qui reçoit des artistes pour travailler avec des scientifiques. Ils peuvent avoir à leur disposition des cultures de tissus, travailler avec des organismes vivants, des outils chirurgicaux… Mais contrairement au projet de SYMBIOTICA, Patricia Piccinini n'utilise pas du vivant, elle observe ces études scientifiques pour mieux comprendre comment ces disciplines façonnent nos idées à propos de ce qui est naturel ou comment l'artificiel finit par se mêler intimement à ce que l'on croit être naturel.

Des références scientifiques

Les greffes de *Proteine Lattice*

Dès ses débuts, l'œuvre de Patricia Piccinini était pétrie de références scientifiques. En 1997 une expérience de greffe de tissu et de cartilage a particulièrement attiré son attention. Elle a vu au journal du soir cette expérience réussie d'une greffe d'oreille sur le dos d'une souris. En réalité, il ne s'agissait pas d'une vraie oreille mais d'une armature ensemencée par des cellules cartilagineuses qui permettaient de reproduire la forme de l'oreille[1]. Cette souris va l'intriguer et l'inspirer car elle porte en elle cette confusion possible entre l'artificiel et le naturel. Avec cette greffe, elle laissait entrevoir une figure tragique et héroïque que Piccinini cherche à représenter dans sa série photographique *Proteine Lattice* (1997).

[1] Armature qui s'avérait être très utile pour les greffes sur les accidentés et les grands brulés pour remplacer les cartilages endommagés.

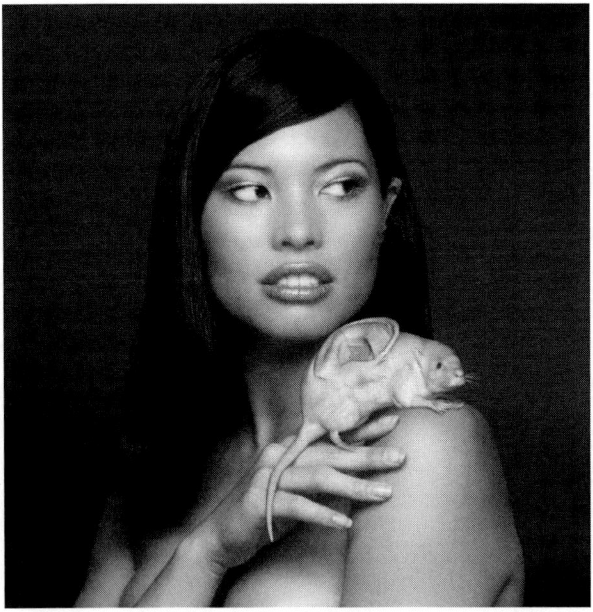

ILL. C-2 : Patricia Piccinini, *Protein Lattice* – Subset Red, Portrait, 1997. Digital C-Type Photograph, 80cm x 80cm. Courtesy of the artist.

Dans cette série intitulée *Proteine Lattice*, l'artiste croise et confronte l'image d'un mannequin à celle d'une souris qui a une oreille greffée sur le dos. La beauté lisse du mannequin se mesure à l'étrangeté de la souris de laboratoire qui supporte cette greffe d'oreille. Dans cette série, on repère aisément les codes esthétiques utilisés généralement pour des photographies de studio, des images publicitaires qui cherchent à vanter les bienfaits d'une crème de beauté ou d'un produit sophistiqué. Rien ne semble naturel dans cette série photographique, la femme comme la souris peuvent nous évoquer des formes chimériques, des êtres de synthèse. La beauté artificielle et les modes de séduction consuméristes sont ici au service de la science. Or cette présentation peut provoquer un certain malaise car elle induit l'idée que la science serait ainsi devenue un produit consommable comme les autres…

Biologie synthétique de *S02*

En 2000, Patricia Piccinini s'inspire à nouveau de la réalité avec son projet intitulé *S02* : une sorte de petit animal domestique à la peau rose. Ce titre S02 se réfère à des recherches en biologie synthétique qui touchent à la synthétisation de l'ADN, avec S01 de la nucléase S1. Même s'il n'est pas encore question de créer de la vie à partir d'éléments de synthèse, Patricia Piccinini distingue dans ses projets les désirs prométhéens de la science… L'artiste va chercher à révéler cette confusion en mêlant la fiction à la réalité et en présentant l'image de son faux S02 à des vrais zoologues pour le faire "expertiser".

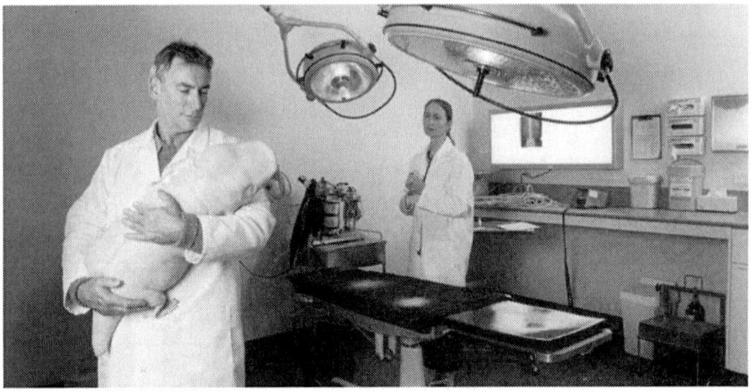

ILL. C-3 : Patricia Piccinini, s*o2 Laboratory Procedures*, 2001. Digital C-Type photograph, 100cm x 200cm. Courtesy of the artist.

Ainsi Paul Andrews, un taxinomiste du Zoo de Taronga lui donna le nom et la classification suivante : *Exallocephalla parthenopa*[2]. En se rapportant au mythe grec de Parthenope, l'artiste cherche à donner vie à quelque chose qui n'existe pas et elle retourne ainsi la science vers le merveilleux.

Comme pour renforcer ce semblant de vie, ses photographies de 2001 vont créer un contexte réel pour soutenir ce trouble scientifique, le mettre en scène et ouvrir un espace de narration à des fables contemporaines… On remarque par exemple dans *Laboratory Procedures* de la série *Science Story*, sa créature dans les bras d'un homme en blouse blanche, qui de part

---

[2] D'après le dictionnaire latin-français Gaffiot : *exallocephalla* indique une tête étrange et étroite et *Parthenopa* désigne une sirène.

son statut d'homme de science va comme valider l'idée que cet animal existe réellement dans nos laboratoires.

Elle réalise aussi une autre série de photographies qui propose un contexte tout autre, un contexte plus familial pour présenter S02. Pour sa série *Restless,* elle présente cette créature installée dans une voiture, sur la place du passager mais aussi, dans une autre photographie, on l'aperçoit entourée d'enfants. Patricia Piccinini s'amuse des idées que l'on se fait sur la faune australienne, de cette promiscuité qu'il peut y avoir entre les hommes et des animaux plutôt impressionnants. Les enfants semblent accueillir cet animal comme s'il s'agissait de n'importe quel autre animal sauvage que l'on croise parfois aux abords des villes australiennes.

Or, la présence de ces enfants va annoncer un tournant dans son œuvre car leur innocence et leur vulnérabilité va lui permettre de mettre en évidence l'aspect à la fois tendre et inquiétant des fables dans lesquelles ils prennent place.

Croisements chimériques

The young family

En 2003, Patricia Piccinini occupe le pavillon australien à la Biennale de Venise avec l'exposition *We are Family* qui nous a permis de mieux la connaître en Europe. Pour cette exposition, elle réussit à donner réellement forme à ses créatures imaginaires en utilisant les techniques du moulage. Il ne s'agit plus de photographies mais de sculptures qui prennent pied dans la réalité et qui se dressent face aux spectateurs. Pour ces sculptures, Patricia Piccinini travaille avec Sam Jinks, un artiste qui s'est spécialisé dans les techniques du moulage. Il parvient à reproduire l'aspect de la peau humaine avec une grande vraissemblance : les pores, la brillance, les pliures représentées paraissent très réalistes.

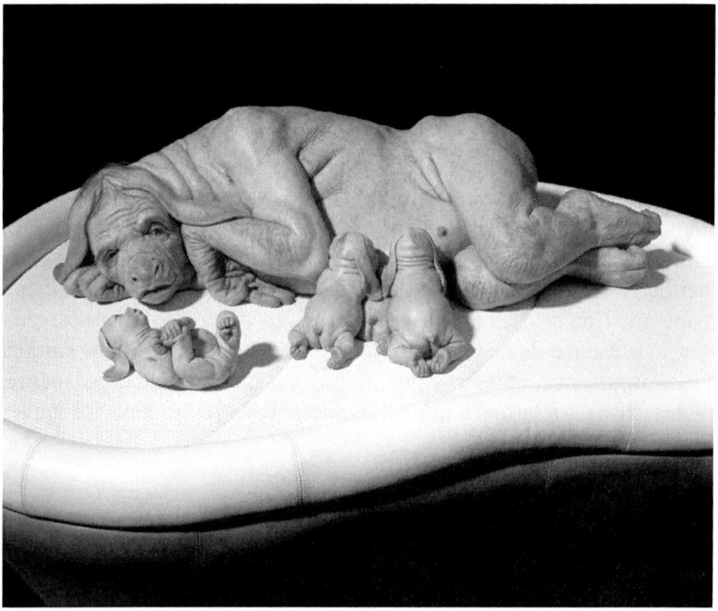

ILL. C-4 : Patricia Piccinini *The Young Family*, 2002-2003. Silicone, fibreglass, leather, human hair, plywood, 85cm high x 150cm long x 120cm wide approx. Courtesy of the artist.

Dans ces sculptures, on remarque aussi que le regard des hybrides paraît humain, ces yeux semblent être ceux d'une femme enfermée dans un corps animal. Ces détails font illusion et créent un véritable malaise car avec cette enveloppe peau presque humaine, ces animaux nous ressemblent et laissent entendre qu'un croisement serait possible.[3]

En effet, le titre *We are Family* indique ouvertement l'idée d'une animalité commune, d'un génome partagé. La chimère n'apparaît pas ici comme un monstre composite mais semble vouloir présenter des croisements interdits ou des manipulations génétiques entre l'homme et l'animal. On remarque d'ailleurs qu'en génétique on utilise le terme chimère pour désigner un organisme composé de deux variétés de cellules ayant des origines génétiques différentes, des génotypes différents.

[3] On note aussi que dans l'installation *The Young Family*, Patricia Piccinini se réfère à un animal qui semble s'apparenter au cochon alors que biologiquement le cochon paraît être assez proche de l'homme.

Ainsi, cette exposition de Patricia Piccinini nous interpelle en questionnant la manipulation génétique et ce qui relèverait d'une évolution naturelle de nos animaux domestiques.

Un certain malaise peut s'envisager car, comme l'explique Pierre Ancet dès les premières pages de son livre *Phénoménologie des corps monstrueux* : « *Le pire monstre est celui qui nous ressemble.*[4] »

Mais nous devons constater que l'impression étrange qui s'empare du spectateur face à l'œuvre de Patricia Piccinini ne relève pas seulement des traitements hyperréalistes de ses sculptures mais aussi des ensembles qu'elle crée pour ses installations.

Mise en scène et limites du contexte *Leather Lanscape*

La mise en espace de cette exposition semble relever d'une véritable mise en scène car en mêlant le monstre au familier, elle crée des tableaux d'une grande quiétude. Les formes animales prennent place dans des espaces qui peuvent rappeler nos intérieurs : ils reposent sur des structures molles en cuir blanc, sur des meubles hybrides entre le sofa et la panière pour chien, des paysages de cuir comme nous l'indique le titre *Leather Landscape*. Si le contexte semble ici futuriste, il est également familial et tranquille.

Avec cette exposition, Piccinini laisse entendre que les monstres, après avoir été des phénomènes de foires, pourraient s'infiltrer dans nos appartements et ce titre *We are family* nous indiquerait alors que, dans le futur, nous adopterions au sein de notre famille des bêtes aussi étranges. Nous formerions ainsi une seule et même famille, une génération monstrueuse. Cette impression d'inquiétante familiarité nous renvoie au texte de Sigmund Freud à propos du terme *unheimlich* qu'on a traduit en français par *Inquiétante étrangeté*. Dans cet essai il présente une analyse des réactions de malaise qu'on observe quand on discerne des éléments familiers dans ce qui nous dérange, quand on perçoit une ressemblance à travers ce que l'on croit être l'altérité.

En mêlant si intimement l'artificiel au naturel, Patricia Piccinini déplace insidieusement les limites et les catégories… Car on ne sait plus bien qui est qui dans cette promiscuité des genres et on ne différencie plus l'animal de l'humain, le réel de la fiction.

Dans l'installation *Leather Landscape* on remarque avec surprise la présence d'une sculpture hyperréaliste représentant un bébé bien humain qui se tend au bord de cette structure en cuir pour voir un peu mieux la

[4] Pierre Ancet, *Phénoménologie des corps monstrueux* (Paris: Puf, 2006), 1.

scène… Patricia Piccinini intègre dans son exposition des sculptures hyperréalistes représentants des enfants, or ces sculptures paraissent tellement réelles qu'elles peuvent se confondre aux vrais spectateurs car on les découvre généralement dans un deuxième temps, après avoir observé les créatures les plus étranges. On remarque alors que la limite entre l'œuvre et le spectateur n'est pas nette et cette mise en scène déploie au final un dispositif englobant le spectateur. L'aspect familier et paisible du contexte qu'elle crée fonctionne alors comme un miroir dont les reflets finissent par piéger celui qui évolue dans ces espaces, entre ces sculptures.

Les enfants qui sont ainsi représentés dans l'œuvre de Patricia Piccinini se révèlent être des leviers qui soulèvent et animent nos craintes face à une innocence qui peut paraître en danger dans la promiscuité entretenue avec ces étranges créatures.

Nature morte et cellules souches

Toujours dans l'exposition de la biennale de Venise, *We are family*, cette installation *Still Life With Stem Cells* représente une enfant jouant à même le sol avec des volumes difficiles à désigner et qui s'apparentent à l'informe monstrueux. Si nous reprenons le tableau de tératologie que Gilbert Lascaux a réalisé dans son livre *Le monstre dans l'art occidental* [5] nous pourrions nous reporter à la référence « 1-5 Cinquième ordre du monstre par indétermination ou détermination insuffisante, 1-5-4 quatrièmes familles formes molles et fluides ».

Ces formes étranges peuvent aussi nous évoquer un artefact de la science fiction qui émane du cinéma de David Cronenberg quand il imagine pour son film *Existenz* (1999), des « pods game ». Ces éléments aux couleurs chair qui ressemblent à des consoles de jeux vidéo mais qui se fixent un peu comme un cathéter en bas de la colonne vertébrale…

Dans cette installation, l'enfant ne semble pas s'inquiéter de ces formes, elle joue tranquillement avec elles comme si cette situation lui était familière, comme si elle avait l'habitude de s'amuser avec ce que le titre nomme des cellules souches. L'extraordinaire apparaît ici comme quelque chose d'ordinaire.

Le titre de cette pièce, *Still Life With Stem Cells,* est assez ambigu puisque *Still life* signifie en anglais une nature morte et se traduit mot à mot comme une vie immobile et silencieuse. *Still life with stem cells* présente une nature morte avec des cellules souches dont le potentiel ne semble pas encore s'être exprimé complètement, comme si le vivant était

[5] Gilbert Lascaux, *Le monstre dans l'art occidental* (Paris: Klincksieck, 2004).

encore silencieux, immobile, comme si une menace était contenue dans cette scène. Avec ce titre, l'artiste semble vouloir présenter l'enfant au même niveau que ces formes informes et laisse entendre que cette petite fille fait partie de la composition et de la nature morte. Ainsi, on peut croire que Piccinini fait glisser le vivant vers ce qui est inanimé et qu'elle souhaite poursuivre son œuvre de confusion en proposant de nouveaux renversements sémiotiques.

Depuis la Biennale de Venise, Patricia Piccinini a réalisé de nombreuses installations et expositions qui continuent à confronter ou confondre l'enfance et la monstruosité…

Dans son installation *Undivided* (2005) elle fait rentrer la chimère dans l'intimité de ce petit lit où un enfant dort avec une étrange peluche… On remarque que le nounours a été écarté de la scène et qu'il git au pied du lit comme pour laisser toute la place à cette créature qui semble moins rassurante qu'un *teddy bear*. La forme monstrueuse embrasse complètement l'enfant, elle invoque une menace latente qui, dans ce contexte nocturne, mêle la tendresse du rêve à l'horreur du cauchemar.

Dernièrement, l'œuvre de Piccinini semble s'être déplacée de la science à la fable. Lors d'une interview menée par Laura Fernandez Orgaz, Patricia Piccinini s'explique à ce sujet :

> Je pense que mes créatures sont en fait plus mythologiques que scientifiques. Ce sont des chimères que je conçois afin de raconter des histoires qui expliquent le monde dans lequel je vis sans le comprendre ni le contrôler totalement.[6]

Nous pouvons donc comprendre que l'œuvre de Patricia Piccinini ne délivre pas de message engagé par rapport à la science mais pose des questions aux spectateurs. Elle nous fait évoluer dans des installations où nous pouvons nous sentir mal à l'aise, car ces espaces semblent tellement calmes et familiers qu'ils inquiètent la réalité. Les illusions s'envolent et les limites qui enferment le monstrueux semblent se déplacer pour laisser place à des angoisses presque enfantines. Dans ces contextes familiaux, elles ne sont plus le monstre que l'on montre, celui qui nous renvoie à la racine latine de *monstro-monstranum,* se référant aux monstres de foires;

---

[6] "I think my creatures are actually more mythological than scientific. They are chimeras that I construct in order to tell stories that explain the word that I live in but cannot totally understand or control". Extrait de l'interview menée par Laura Fernandez Orgaz et publiée dans le catalogue de l'exposition *Tender Creatures* à l'Artium (2007, Espagne). Cette interview est lisible également sur le site de l'artiste Patricia Piccinini, www.patriciapiccinini.net/, sous le titre *The Naturally Artifial World.*

les créatures de Piccinini nous indiquent le sens d'une autre racine étymologique, celle qui provient de *monere* avec *monstrum*, le monstre qui montre, celui qui indique le mauvais présage…

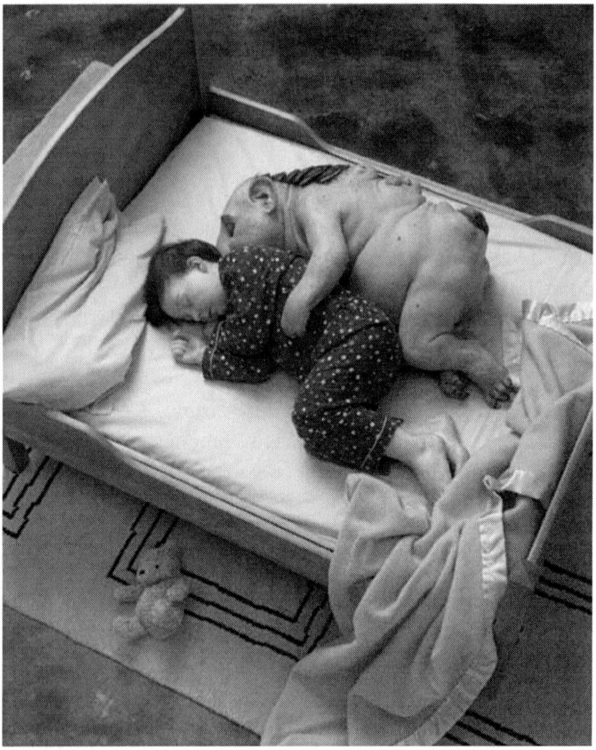

ILL. C-5 : Patricia Piccinini, *Undivided*, 2004,
silicone, fibreglass, human hair, flannelette, mixed medium,
101 x 127 x 74 cm high. Courtesy of the artist.

Le monstre nous fait craindre ce que nous ne connaissons pas et ne maitrisons pas, d'une certaine manière il révèle ainsi nos propres angoisses. D'après Sigmund Freud l'angoisse est cet état inquiété par un danger qui nous est inconnu[7], un état que l'enfant peut appréhender

[7] Nous pouvons nous référer à la définition que Freud fait des termes : « peur (Furcht), angoisse (Angst) sont des termes qu'on a tort d'utiliser comme synonymes, leur rapport au danger permet de bien les différencier. Le terme d'angoisse désigne un état caractérisé par l'attente du danger et la préparation à

lorsqu'il est, par exemple, inquiété par les ombres d'un espace, d'un monde qu'il ne connait pas encore très bien. On peut donc penser que la chimère est un monstre particulièrement angoissant puisqu'elle n'est pas clairement définie et évolue dans des espaces qui nous sembleraient inconnus. Dans l'ouvrage intitulé *Chimères*, Didier Ottinger situe ainsi ces créatures :

> Ces créatures ont en commun de sévir dans des zones de friches, là où les frontières s'estompent ; entre le monde sacré et le profane, entre la terre des hommes et le royaume des dieux, là où le domaine des vivants se mêle à celui des morts. Le monde de Chimère est celui de la confusion, de l'indétermination. A son nom est associé un principe de passage, une loi de continuité.[8]

Le fait de s'approcher d'une chimère nous indique donc que nous nous rapprochons un peu plus de ces espaces dont on ne connaît rien du tout… ces espaces qui nous semblent interdits. Les manipulations génétiques et la biologie synthétique nous permettraient de glisser vers ces zones de friche, ces zones de limites, où l'on se mesure à la fois aux monstres et aux dieux ; où l'on doit trembler d'angoisse car on ne sait pas exactement où cela peut nous conduire et c'est bien le propre de la recherche.

L'œuvre de Patricia Piccinini semble activer certaines angoisses car elle nous montre ce qui pourrait advenir si la science-fiction n'était plus fiction, si les sciences étaient accessibles aux profanes, aux non-spécialistes et si ces sculptures inanimées prenaient vie. L'hyperréalisme des installations de Patricia Piccinini cherche à offrir un semblant de vie à ce qui est artificiel et ainsi on peut croire qu'elle joue au biologiste qui va tenter de reproduire la vie avec des matériaux de synthèse (dans ce cas, des résines et des silicones). Ici, les projets de l'artiste et du scientifique se rejoignent comme ceux de Prométhées[9] contemporains, ils ne cherchent pas seulement à imiter la vie, ils la provoquent.

---

celui-ci, même s'il est inconnu ; le terme de peur suppose un objet défini dont on a peur, […] ». Sigmund Freud : « Au-delà du principe de plaisir » in Essais de psychanalyse (Paris: PBP, 1981), 50.

[8] Didier Ottinger, in *Chimères* (Arles: Actes Sud, 2003), 9.

[9] Prométhée qui créa l'homme à partir d'une motte d'argile d'après la théogonie d'Hésiode.

Œuvres citées

Ancet, Pierre. *Phénoménologie du corps monstrueux*. Paris: PUF, 2006.

Cronenberg David, *Existenz*, 1999, production canadienne et britannique, 1h36 min.

Freud, Sigmund. *L'inquiétante étrangeté et autres essais*. Paris: Gallimard, 1991.

—. « Au-delà du principe de plaisir » in *Essais de psychanalyse*. Paris: PBP, 1981.

Lascaux, Gilbert. *Le monstre dans l'art occidental*. Paris: Klincksieck, 2004.

Ottinger, Didier. *Chimères*. Arles: Actes Sud, Arles, 2003.

Pastoureau, Michel. *Le cochon : histoire d'un cousin mal aimé*. Paris: Gallimard, 2009.

Piccinini, Patricia. *We are Family Australia*, catalogue biennale de Venise, 2003.

SECTION II

DES LOIS BIOLOGIQUES AUX SCIENCES HUMAINES DANS KILLING MISTER WATSON DE PETER MATTHIESSEN

MARIE-CHRISTINE NOAILLES

Explorateur, naturaliste, anthropologue, voire ethnologue, Peter Matthiessen est doté d'une immense culture qui lui permet, dans son œuvre romanesque, des rapprochements insolites mais pourtant conformes à ses convictions de bouddhiste zen. Dans la philosophie zen, en effet, non seulement le monde et la vie sous toutes leurs formes ne font qu'un, mais le corps est indissociable de l'esprit, et surtout des domaines mentaux aussi traditionnellement écartelés que la science et la religion – toutes disciplines et confessions confondues – peuvent s'intégrer harmonieusement dans une vision unique mais globalisante à laquelle semble parvenir la quête littéraire de ce scientifique devenu écrivain.

Des lois génétiques à une forme de genèse, tous les thèmes s'imbriquent dans *Killing Mister Watson*. Le principe de vie s'avère, en fait, identique pour toute chose, et les contradictions d'un monde, d'un individu, ne sont que les fausses apparences d'une identité sous-jacente fondamentale. Comme par hasard, il n'y a dans le milieu barbare de *Killing Mister Watson* que deux personnages ayant reçu éducation et culture, mais aux antipodes l'une de l'autre. Ce sont Jean Chevelier et Jane Watson, lesquels représentent respectivement la science et la religion, soit les connaissances techniques qui permettent de comprendre la vie terrestre et de la rendre plus confortable, et l'ouverture spirituelle susceptible d'élargir le champ de la conscience et de la communion des hommes à l'ensemble de la création, tout en les élevant au-delà du contingent et du rationnel pour les rapprocher du divin. Malheureusement, sur le plan diégétique, la rencontre de ces deux tendances qui distinguent l'homme de l'animal, sera sans lendemain. Mais ces personnages peuvent exercer une influence dans leur entourage respectif, et véhiculent des idées qui sous-tendent les thèses de l'auteur. Celles-ci sont, à l'origine, scientifiques et posent pour première condition de la survie des individus l'hybridation.

Jean Chevelier est un homme de science français, égaré par sa passion d'ornithologue et par sa quête d'un trésor indien, dans cette région sans foi ni loi du sud de la Floride dont il connaît bien les origines topographiques, humaines et toponymiques. C'est lui qui établit ouvertement le lien entre les sciences de la vie et la ou les religions. Il prétend, en particulier, que les hommes de cette contrée ne peuvent s'organiser en une communauté soudée (au sens littéral de religion) tant que des différences de couleur, donc de race les opposeront. Il se fait l'écho de Darwin lorsqu'il affirme que les hommes étaient à l'origine tous identiques, car issus d'une seule et même espèce, laquelle se serait ensuite, à cause de la dispersion propice à des différenciations individuelles, subdivisée en races – l'équivalent des « variétés » chez les autres êtres vivants. Il induit de ce postulat que, conformément aux lois de la nature, le seul moyen de reconstituer l'espèce dans son uniformité originelle serait de croiser ces races entre elles jusqu'à ce qu'elles perdent toutes leurs différences et fusionnent dans la spécificité de leur origine commune.[1]

> This whole darn foolishness of blood will be the ruin of this country. As Old Chevalier told Daddy Richard, human beings was all one shade when they first appeared on earth, and only turned into different-colored races when they scattered out across the continents. The way they breed around these days, the Frenchman said, they were sure to wind up all one color again, and the sooner the better, too, he'd say, because life was terrible enough without this useless misery of color. (108)

Dans le premier chapitre, « Variation under domestication », de son traité sur l'origine des espèces, Charles Darwin pose une hypothèse semblable en prenant pour exemple le chou :

> Nevertheless, as our varieties certainly do occasionally revert in some of their characters to ancestral forms, it seems to me not improbable, that if we could succeed in naturalising, or were to cultivate, during many generations, the several races, for instance, of the cabbage, in very poor soil (in which case, however, some effect would have to be attributed to the direct action of the poor soil), that they would to a large extent, or even wholly, revert to the wild aboriginal stock.[2]

Darwin explique que ce cas de "réversion parfaite", aux caractéristiques de l'espèce d'origine par hybridation répétée, dépend de la préservation de conditions de vie tout à fait identiques et du maintien des variétés en

[1] Peter Matthiessen, *Killing Mister Watson* (New York: Random House, 1990).

[2] Charles Darwin, *On the Origin of Species by Means of Natural Selection or the Preservation of Favoured Races in the Struggle for Life,* 1859 (Oxford: Oxford University Press, 1919), 14.

nombre considérable ; ce qui permettrait l'élimination des moindres déviations ou différences.

Ce postulat, commun à Darwin et au personnage savant de Chevelier, s'appuie sur une appréhension biologique du principe même de la vie organique. Selon ce principe, il existe un enchaînement de subdivisions qui, à partir de la cellule de base, font proliférer d'autres cellules, jusqu'à formation d'un être vivant complet et complexe, gardant une identité fondamentale avec l'organisme d'origine – en accusant toutefois certaines différences –. L'organisme ainsi obtenu va ensuite amorcer une lente érosion, voire désintégration, par l'usure et la perte progressive de toutes ses cellules, jusqu'au retour à la matière organique une et indivisible.

Les processus de création – avec des variations conditionnées par l'adaptation –, de décomposition et de régénération sont des constantes, car dans sa dynamique cyclique d'évolution et de révolution, le mouvement de la vie est perpétuel et universel. Il concerne en premier lieu la nature, plantes et animaux inclus, mais aussi l'homme, et pourquoi pas, tout son champ d'action, et le cosmos dans ses moindres composantes. Darwin, lui-même, envisageait une application de sa loi de l'évolution à l'étude de la psychologie et à celle de l'origine et de l'histoire de l'humanité. Selon toute apparence, c'est la démonstration ambitieuse qu'entreprend Peter Matthiessen dans son roman floridien.

L'origine, la souche ancestrale, qu'évoquent Darwin et le personnage de Jean Chevelier, à propos de la "reversion" serait-elle Dieu ? Il est bien évident que, dans la logique d'un scientifique pur et dur comme Chevelier, cette idée d'un être spirituel, désincarné, est irrecevable. A plus forte raison si les hommes, prétendument "créés à son image", sont de couleurs différentes, ce Dieu perd le principe d'unicité qui le définit. Et, dans cette optique, il ne reste plus que l'homme au centre d'un univers qu'il s'efforce d'expliquer :

> The Frenchman come right out and said he didn't hold with no Father Who art in Heaven. Man is made in God's ee-mage? Who say so? Black man? Red man? Which man you talk about? White man? Yellow man? God is all these color? Say tabsurde ! *Homo sapiens,* he got to shit, same like any foking animal. You telling to me your God got to shit too? (38-39)

Pourtant, sans avoir du tout conscience que, sur ce point, il accorde la thèse scientifique avec celle de certaines spiritualités – primitives ou orientales, principalement –, il lui arrive de désigner comme seul « Grand Mystère » ou « Grand Maître » – selon ce que l'on veut bien entendre – le soleil, source de vie universelle : « Or that old man might point quick at the sun, point at a silver ripple in the water, saying, "Look quick ! See there? *That* is God ! *That is le Grand Mees-taire"* He meant "Big Mister", case you don't speak French. (39) » Mais il ne s'aventure guère plus loin dans la voie spirituelle, Il ne s'intéresse en fait qu'à la réversion qui prouverait l'origine une et commune à tous les hommes, et il se contente de prôner avec Darwin l'hybridation ou, en termes humains, le métissage.

Sans métissage, en effet, deux dangers menacent les hommes dans ces zones écartées de la Floride : une dégénérescence progressive, pouvant aller jusqu'à l'extinction de la race, résulterait du nombre réduit d'individus et d'une consanguinité importante – c'est le cas des Indiens Mikasukis (35), derniers "sauvages" authentiques ayant refusé toute compromission avec les Blancs ou autres tribus. Les parias de la société blanche réfugiés dans ces parages risquent, quant à eux, de régresser à l'état bestial de pure nature – comme le montrent les épisodes violents du roman ; cela corroborerait l'argumentation de Darwin sur la résistance des caractères génériques, par rapport aux caractères spécifiques – soit en termes moraux ou religieux, la survivance d'une certaine animalité en l'homme : "Species inheriting nearly the same constitution from a common parent and exposed to similar influences will naturally tend to present analogous variations, and these same species may occasionally revert to some of the characters of their ancient progenitors."[3]

Toutefois, à plus ou moins long terme, le métissage des races humaines s'avère inévitable dans ces marais des *Everglades*, dernier refuge des bannis d'une civilisation blanche conquérante, des Indiens déracinés, des Noirs en fuite, et creuset originel, peut-être, d'une nouvelle "race arc-en-ciel. "

Le métissage serait alors le fondement même de cette vision totalisante et unifiante de l'univers dans sa triple dimension : physique, philosophique et métaphysique. La famille Hamilton s'affirme comme la preuve vivante de cette hybridation extensive, résultant de mariages multiraciaux répétés au fil des générations. Richard Hamilton et son épouse Mary Weeks sont d'ascendance mixte très confuse : pères de pure souche britannique et mères indiennes dans les deux cas. Mais le grand-père maternel de Richard appartenait à la tribu des Calusas. Ces Indiens transférés à Cuba par les Espagnols et très vite réimportés en Floride ont su, en s'alliant aux esclaves fugitifs, semer le trouble et freiner la conquête anglo-saxone (32). Puis, à la suite du massacre des Calusas, la mère de ce Richard Hamilton s'est ralliée aux Choctaws, les « bons Indiens » qui ont aidé les Blancs à éliminer leurs congénères (29). Enfin, la mère de Mary est d'origine séminole, autre tribu avec laquelle les esclaves en fuite ont partagé tous les

[3] *On the Origin of Species*, 153.

combats et la même vie (41). Ainsi, les aléas de la mixité se compliquent, avec les sympathies des divers groupes indigènes, pour les colonisateurs blancs ou les opprimés noirs, – suivant les coalitions de l'histoire nationale, tribale et familiale (40).

La traduction visible de ce métissage complexe se trouve dans les nuances de couleur – blanc, noir, métissé – toutes plus ou moins soutenues, qui différencient physiquement les cinq enfants du couple Hamilton, et ce, dans une proportion conforme aux lois de Mendel (trois enfants plutôt blancs et deux de couleur). Ces lois biologiques affectent tous les gènes. Et, à en juger par la description de Carrie, le même mélange nuancé de l'hérédité physique, mais aussi psychologique, transmise par les deux parents caractérise les trois enfants que l'impétueux Watson eut de sa douce et pieuse épouse (68-69).

Il est à noter, à ce stade, que chaque être, bien que partiellement identique à l'un ou l'autre de ses géniteurs ou aux deux confondus, possède, dans sa complexité biologique et mentale, une intégrité, une identité propres. Et, au-delà de l'individu, se confirme une identité beaucoup plus vaste, pouvant couvrir dans les deux sens toute la chaine héréditaire du dernier survivant jusqu'au premier homme ; inversement le métissage prend ainsi force de loi naturelle dans les rapports de l'un avec le multiple ; il donne à la diversité l'identité d'une origine commune.

C'est par le processus de l'hybridation, soit le croisement répété de caractéristiques individuelles différentes, que parvient à se constituer un tout dont les éléments sont interdépendants, l'identité totalisante et compacte d'une pluralité d'individus, fondus dans une unité de plus vaste envergure : l'individu est la souche de la famille, la famille, la cellule de base de la société, et la société un microcosme du monde. Dans le cas des Hamilton, l'individu étant métissé l'identité du groupe familial s'élargit à la nation américaine puisqu'elle inclut toutes les races d'homme qui peuplent ce continent : "We had all colors in the Hamilton clan, and that's for sure. Jean Chevelier called Hamiltons the true New World family", because Richard Hamilton never thought about your color."108)

Sur le plan scientifique, il n'est pas question de pureté originelle, mais d'homogénéité, au sens littéral d'unité de base, "génétique", car l'hybridation permet, inversement, par différenciation, donc subdivision répétée de l'identité complexe, un retour à l'homogénéité constitutive :

> According to Jean Chevelier's way of thinking, there ought to be a law where *any* man who don't marry a different color would get castrated. That way *Homo sapiens* would stop his misery and plain damn stupidhood about his races and go on back to the color of First Man, which in Chevelier's opinion would work out pretty close to Richard Hamilton. (41)

Mais, en érigeant en loi le métissage obligatoire, sous peine de castration, afin d'améliorer l'espèce humaine par la restitution progressive de ses qualités premières, ce même Chevelier semble s'appuyer sur les théories de l'eugénisme scientifique pour mieux en détourner les méthodes.

Francis Galton (1822-1911), cousin de Darwin et fondateur de cette science, la définit comme suit: "the scientific study of the biological and social factors which improve or impair the inborn qualities of human beings."[4] L'idée de Galton était effectivement d'améliorer l'espèce humaine, en appliquant à l'homme la doctrine évolutionniste récemment décrite par son cousin. Ce concept n'a rien de répréhensible en soi. Mais, malheureusement, comme l'a démontré l'histoire, on a eu trop facilement recours à cette science pour tenter de justifier les pires excès des régimes totalitaires.

En outre, cette théorie présente, au départ, un défaut majeur: si, à la fin de l'ère victorienne, les qualités humaines particulièrement souhaitables ne font aucun doute, un siècle plus tard on comprend que la plupart d'entre elles sont fonction de déterminismes culturels, et que la résistance biologique dépend du maintien de la diversité, plutôt que de manipulations tendant vers un idéal universel. Toutefois, on ne peut reprocher à Galton de ne pas avoir tenu compte des mécanismes de l'évolution, ni de son étalement dans le temps, car il écrivait à une époque où l'on n'avait pas la moindre notion des fonctionnements de l'hérédité.

Pour en revenir à l'eugénisme dans le roman naturaliste et sociologique de Matthiessen, il s'agit bien pour Jean Chevelier de purifier la race des hommes, en revenant à l'homo-sapiens d'origine, mais par la multiplication des hybrides, et non des "purs" ; la stérilisation demeure, mais les "inaptes", les "dégénérés" qu'elle menace, sont justement ces "purs" qui s'opposent au métissage. Et ils sont relativement nombreux dans cette zone pourtant multiraciale mais désormais conquise par les Blancs.

Le personnage d'Eugène Hamilton, dont le prénom révélateur prend une tournure encore plus significative sous sa forme diminutive de "Gene" en est un bon exemple. Il est celui qui rejette la négritude : il ne fréquente que des blancs, renie jusqu'à ses propres frères, mulâtres comme lui, mais à la pigmentation plus foncée, et, par voie de conséquence, se renie lui-même: "Once in a while one of his Weeks or Daniels cousins would get

---

[4] Sir Francis Galton, *F.R.S.: The Legacy of his Ideas*, ed. Milo Keynes (London: Macmillan, 1993). Cité par John A. Lee, maitre-assistant en pathologie, à l'université de Sheffield, "Darwin's cousins and other animals," *New Scientist*, December 25, 1993, January 1, 1994, 61.

drunk and tease Eugene -Your brother Leon, now, he looks almost like a white man, don't he, Gene? – and all Gene's rage would get roiled up, he'd fight to prove that he was white till he was black-and-blue. And in the end, it was Walter got the blame." (107-108)

L'eugénisme est exactement le contraire de la "réversion" prônée par le personnage incarné par Chevelier sous l'influence de Darwin. Tous les critères répertoriés par le célèbre naturaliste pour soutenir sa démonstration étaient réunis dans la Floride de ce siècle. Cette région avait conservé de nombreuses espèces indigènes, enrichies d'éléments ou groupes allogènes, dans le règne végétal, animal et, bien sûr, humain ; et ce, d'autant mieux que la répartition de chacune était large et non isolée. L'arrivée massive des blancs menaçant de supplanter, ne serait-ce que par le nombre les Indiens qui, constitueraient l'espèce parentale (selon les termes darwiniens) a amené certaines tribus à pratiquer l'épuration ethnique, et ce, bien avant que l'eugénisme ne soit scientifiquement reconnu: ces autochtones ont systématiquement tué tous les nourrissons nés de relations avec les Blancs et, si possible, leurs géniteurs. Mais, ce faisant, ils se sont irrévocablement coupés des Blancs, à qui ils voudraient maintenant ressembler, pour mieux se démarquer des parias noirs ; ils ont donc porté préjudice à la survie de leur propre race et les pratiques eugéniques se sont avérées pour eux, comme pour Eugène Hamilton, autodestructrices (108).

Enfin, certains Blancs revendiquent eux aussi l'élimination de leurs dépravés dès le berceau, afin de purger la région de ses futurs criminels (294). Et beaucoup verront dans la seule victime de l'ouragan cataclysmique d'octobre 1910, un enfant de cinq mois, (rejeton illégitime du criminel Watson), un signe de purification divine, les encourageant dans la voie de l'eugénisme :

> Josie hollered that her boy was taken by the hand of Gad and the Chokoloskee men said that was sure right, because the boy was the accursed infant of that bloody-handed sinner who had brought God's punishment down upon them in the first place, that's why his life was the only one lost in the Islands. (351-352)

Sur un plan strictement biologique, l'inéluctabilité du sort des Indiens s'explique assez bien par la théorie darwinienne de la sélection naturelle dans la lutte pour la vie. En fait, ils n'ont guère le choix. L'auto-exclusion dans leur habitat d'origine, de même que la relocalisation dans les réserves de l'Ouest américain les vouent, à plus ou moins long terme, à l'extinction définitive. Dans le territoire des *Everglades*, déjà limité en zones habitables à cause de son exposition aux cyclones, de l'intrusion profonde de la mer et de ses marécages permanents, tout comme dans les zones arides et d'implantation restreintes de l'Oklahoma, par exemple, leur race peu nombreuse n'a plus ni les conditions biologiques ni le temps de se diversifier dans sa structure et dans ses habitudes ; elle ne peut s'améliorer, pour s'adapter et continuer à se propager, car l'hybridation ou sélection sexuelle n'affecte que les individus nomades. En outre, étant donné leur nombre insuffisant, leur dispersion et une occupation imparfaite de l'espace malgré le regroupement de quelques tribus parentes, toujours en vertu des critères de Darwin, un affrontement sporadique avec les envahisseurs blancs bien supérieurs en nombre ne peut que leur être fatal. Le texte de Matthiessen en donne une illustration par la voix de Richard Hamilton :

> Speaking of blood, my grandfather was real pure-blood Spanish Indin, didn't want a thing to do with the Muskogee and Mikasuki Creeks - the Seminoles - that was taking over his Calusa country. But finally he understood what Chief Tecumseh warned us, that if Indin people didn't put away our feuds and fight the whites all in a bunch, there wouldn't be no land left to fight over. Sure enough, the white men lied and broke ail their agreements. Here in Florida, they aimed to pen up any Indin they hadn't killed, ship that redskin sonofabitch to Oklahoma. (32)

Il ne reste donc que le métissage systématique avec l'espèce dominante, les Blancs, comme seule chance de gagner la bataille de la vie. Darwin soutient qu'un croisement entre deux variétés différentes, ou deux individus de même variété mais de lignées différentes, renforce la vigueur et la fertilité de la descendance, assurant ainsi la perpétuation de l'espèce, alors que des croisements d'individus très proches les diminuent.[5] Malheureusement, cette autre forme de lutte pour la survie chez les uns est contrecarrée par l'instinct de domination chez les autres.

Le roman floridien de Peter Matthiessen dresse toute une liste des méthodes utilisées par les Blancs pour venir à bout de la résistance indienne: depuis les guerres intestines, en passant par la corruption et les duperies, jusqu'à l'intégration déguisée de l'un d'eux afin de faciliter une déportation massive, dont le seul avantage est d'éviter, pour un temps, l'extermination totale :

---

[5] *On the Origin of* Species, 88.

> The Florida Wars was the only Indin wars the U.S. Army never won, had to trick and bribe and steal to get the job done, get one Indin fighting with another. Finally they got to Billy Bowlegs, who had started out with Chekaika on the Calusa Hatchee. Took Qld Billy over there to Washington, D.C., give him the name Mr. William B. Legs, snuck him into sorne upstanding hotel where no durn greasy redskins was allowed. A few years after that, they made him rich, and Billy took his people west, out to the Territory. (33-34)

Mais la généalogie et l'histoire de Richard Hamilton et de sa famille confirment la validité de la théorie darwinienne de l'hybridation comme seule voie de salut dans la lutte pour la survie et la propagation de l'espèce. Pour subvenir à leurs besoins sa mère, d'abord, puis lui-même, vont tirer parti du déracinement, et n'hésiteront pas à se compromettre en collaborant étroitement avec les Blancs: l'un se convertit très tôt au catholicisme et l'autre s'engage dans l'armée de l'Union où, déjà, les hybrides se distinguent par leur vigueur et leur bravoure :

> My mam went west to Wewoka, Oklahoma, with Billy Bowlegs's people from Deep Lake, signed right up with the Catholic mission so's to get her kids a bite to eat. I was out there in the Indin Territory all through my youngerhood. Later I went for a soldier in the Union Army, whole cavalry regiment of breeds and nigras slapping leather and raping and carousing all over the Territory and beyond. Some of them men was half-red and half-black, come down from strong slaves that run off across the wilderness and were taken in by Indins who prized their bravery, and they was the biggest. strongest men I ever saw. Indins called us buffalo soldiers cause the darker ones was buffalo color, with the same dark woolly nap. (34)

Plus tard, Richard Hamilton aura le courage de quitter les Blancs, pour retrouver ses racines indiennes dans sa terre natale de Floride (35), où il contractera mariage avec la fille d'un Blanc, Mary Weeks, et s'affranchira définitivement de toute dépendance et exploitation raciales :

> Old Man John Weeks, he passed for white, he had his honor there to think about, but it took me a while to figure out white people's attitudes. If I was a Choctaw, like I said, then I was a "good Indin", And if I was mulatta. like they claimed, then I was a free man, a free citizen. (35-36)

Le résultat est, non pas l'assimilation, mais une intégration réussie: "[…] live and let live […] Let the Blessed Lord Above take care of His own sinners, cast 'em straight down to perdition on the Day of Judgment" (36), telle est la devise de la famille Hamilton ; autrement dit, il faut affirmer sa différence, tout en respectant celle des autres. Grâce à la progéniture nombreuse et largement métissée du couple Hamilton, la race indienne se perpétuera encore sur plusieurs générations: “As for Mary Weeks, *her* mother Elizabeth was full-blood Seminole, supposed to be, the granddaughter of Chief Osceola. They can call us mulattas all they want, but we are Indin. Why heck, if this Hamilton bunch ain't Indin, then they ain't no Indins left in the U.S.A.” (37)

En outre, tout laisse à penser qu'avec le temps, et selon les lois de l'évolution, ces mulâtres seront de mieux en mieux intégrés dans la société cosmopolite et multiraciale qui se dessine. En effet, conformément aux lois biologiques, l'hybridation des hommes ne peut que s'intensifier, à cause de la mobilité des populations, jusqu'à ce qu'il n'y ait plus aucune différence de couleur, de race et de culture : “People move around these days, get all mixed up. Don't matter what our color is, we all going to be brown boys when the smoke clears” (42). Richard Hamilton – ou l'auteur, derrière lui – se rit, ici, des préjugés raciaux, mais il a de la suite dans les idées: il n'y a aucune raison pour que les peuples oubliés désespèrent d'être, un jour, réhabilités.

Toujours à propos de la diversification d'une espèce imposée par la nécessité de s'adapter, on peut aussi s'intéresser aux phénomènes de mutations transitoires. Darwin signale que, si une population a atteint son nombre maximum dans un milieu donné, elle ne peut plus s'accroître qu'en modifiant sa descendance, pour que celle-ci empiète sur le territoire d'autres populations. Ces modifications sont souvent dues à un simple changement de nourriture ou d'habitudes.[6] Il explique comment les variétés, sous l'effet de la sélection naturelle, creusent leurs différences au fil des générations, jusqu'à former des sous-espèces, puis de nouvelles espèces distinctes qui, dans leurs caractères, divergent l'une de l’autre et de leur origine commune.

On pourrait donc voir dans le gigantisme et dans la force sans égale de Hannah Smith et de ses “petites sœurs” une phase précoce et incomplète d'une diversification possible de l'espèce humaine, surtout si on les compare à l'érudite mais fragile Jane Watson, tellement inadaptée à ce milieu et à ce mode de vie qu'elle en mourra. D'un point de vue physique, Hannah et Sadie Smith ont quelque chose de monstrueux, mi-animal, mi-humain, mi-homme, mi-femme; elles seraient plutôt emblématiques d'une mutation précipitée et excessive, imposée par des conditions de vie particulièrement surhumaines dans un vaste territoire vierge et morcelé, dans lequel leur famille, originaire de Georgie, s'est très tôt transplantée :

---

[6] *On the Origin of Species*, 102.

> Now Sadie Smith was well knowed as the Ox-Woman, and when she found out that Everglade was clean over to the other side of Florida, maybe two-three months around to there by land and sea, she bided her time until the dry season, then hitched up two young oxen and cut and burned and hacked her way across the Everglades by cart. First time *that* was ever done, might been the last. Went north along Shark River Siough and followed the Injun dugout trails west through the Big Cypress, dug out wheels and chopped whole forests down to get that cart through. She come out near the head of Turner River and come on down to Chokoloskee Bay. Just showed up one day in her black sunbonnet, smelled like a she-bear (233).

Toutefois, avec le peuplement massif et l'exploitation très rapide du milieu, cette mutation n'aura pas le temps de s'installer définitivement pour s'étendre à l'ensemble de l'espèce humaine. “That was the old style of pioneer womenfolk, come down out of the Appalachian mountains. Don't make females like that no more, or we'd have one running the whole country”(233), dit avec soulagement Bill House, l'un des personnages les plus virils.

Mutations, hybridation ou eugénisme, les détails, bien que dispersés, sont trop nombreux pour être gratuits. Ils montrent à quel point Peter Matthiessen adhère aux thèses des grands spécialistes des sciences naturelles dans son étude ethnique. Il va jusqu'à étendre l'application des lois biologiques à tous les aspects de la vie humaine. L'émergence et le développement de la société des hommes, telle qu'il la représente dans ces marécages floridiens, semblent, en bien des points, fonctionner comme ceux des espèces animales et végétales.

Plusieurs narrateurs font valoir l'origine commune de leur microsociété. Comme aux premiers temps, les humains se trouvent d'abord isolés et en nombre réduit dans ce secteur de la Floride ; ils s'accouplent les uns aux autres de manière animale, en changeant s'il le faut de partenaire, si bien qu'ils finissent par être tous apparentés de proche ou de loin, pour ne former qu'une seule et grande famille, dont les membres présentent moins de singularité individuelle que de légères variations de l'un à l'autre.

A long terme, le métissage de ces variétés peut effectivement conduire à une uniformisation de la communauté toute entière. Les enfants adultérins de Watson et de ses diverses compagnes sont déjà nombreux et, dès la seconde génération, la population des *Everglades* est en grande partie composée d'individus ayant entre eux des liens de sang, à en croire Sammie Hamilton :

> Now my uncle Lewis Hamilton was married a little to Jennie Roe, who claimed that she got raped by Mister Watson. Nobody put too much stock in that one. Jennie Roe was a beautiful young woman, but she wasn't so particular. Her mother might been Josephine Parks, unless it was Henrietta Daniels - had to been one of them two sisters, what I heard. Them Caxambas families was all tangled up, and both half sisters had little girls by Mister Watson. (244)

En fait, si l'on s'en tient au seul cas de Watson, il semble avoir instinctivement appliqué la loi darwinienne du premier occupant, et, par la multiplication de ses épouses, maîtresses et enfants légitimes ou non, il s'est apparenté à de nombreuses familles de la région, ce qui est aussi une manière d'élargir son occupation territoriale avant d'en venir à l'impérialisme délibéré de la future société mercantile.

Du point de vue de l'organisation socio-économique, Darwin a montré que la sélection naturelle a donné une variété de fourmis produisant des femelles stériles qui sont les seules à travailler pour la communauté de ces insectes dits "esclavagistes" ; et il en a induit que dans les groupements humains, la division du travail procède de la même loi naturelle et utilitaire.

Watson a un peu ce comportement esclavagiste: à mesure que s'étend son territoire, Il emploie de plus en plus de gens avec qui il agit en bon maître – peut-être pour mieux assurer son intégration ; alors qu'en embauchant Smith, alias le dangereux Cox, son but – on le comprendra plus tard – est de mieux contenir le comportement asocial de l'individu et la menace que celui-ci représente pour lui et pour ses proches.

L'adaptation indispensable à la survie n'est pas qu'une exigence de la nature. Elle entre également dans le contexte socio-économique où, comme dans règne naturel observé par Darwin, toute variation profitable à l'individu d'une espèce – en l'occurrence d'une couche sociale – dans ses rapports complexes avec d'autres organismes – ou classes – et l'espace naturel externe, tend à préserver cet individu, qui le transmet généralement à sa progéniture ; d'où le mariage d'intérêt de Carrie Watson avec un riche éleveur proche des notables, organisé précocement par ce père accusé de tous les crimes :

> His son-in-law, Walter G. Langford, was a friend of sheriff Frank B. Tippins from their cowboy days, which doubtless encouraged a certain lenience towards Langford's father-in-law. In addition, Mr. Watson was protected by Langford's powerful friends, including the aforementioned Jim Cole. (113)

Enfin, sur le plan moral, une épuration radicale s'avèrerait, d'après la démonstration de Matthiessen, tout aussi suicidaire que l'eugénisme scientifique dans le domaine biologique. Parmi tous les Blancs écumant ces contrées marécageuses, il n'en est pas un seul, en effet, qui ne soit un gibier de potence: " But pure-in-hearts was never plentiful around the Bay, and once the sinners went to Hell, it might of got pretty lonesome around here, crying in the wilderness and all like that." (280)

Mais certains narrateurs ne désespèrent pas de l'homme, même du plus dépravé. Ted Smallwood crédite Watson de qualités indéniables, et voit en lui une possibilité de s'amender sous l'effet d'influences bénéfiques :

> Ed Watson's a very good farmer", he reminded me [Mamie Smallwood), starting in on the little speech that all the women got to hear that night in every shack on our scared little island. "He's a hard worker with a good head for business, and a generous neighbor, too, always ready to help - they ain't a family in the Islands won't say the same".
>
> This time he heard himself, the echo. "All right", he said. "But maybe a new young wife and family will steady him down. Ed opened an account this evening, paid out two hundred dollars just for credit. So I got no choice but to give that man a chance, cause he's the only customer we have that ain't behind. (215)

Il semble, en effet, que les vertus chrétiennes des épouses légitimes de Watson aient, à la longue, quelque peu tempéré sa dangereuse impulsivité; et, de toute façon, la probité de l'ensemble de sa progéniture finit par le racheter moralement et spirituellement, montrant ainsi à quel point cette autre forme d'hybridation entre les forces du bien et celles du mal est bien plus salutaire que toute purge.

De proche en proche, les lois biologiques de l'hybridation se répercutent dans toutes les manifestations de la vie humaine, et, à l'eugénisme scientifique de Galton, Peter Matthiessen préfère, de toute évidence, le transformisme d'une évolution régénératrice, exposé par Darwin et défendu par le personnage de Jean Chevelier.

Il va même plus loin et démontre que, du brassage des races à celui des cultures rapprochées par les hasards de l'histoire, il n'y a qu'un pas : le métissage conduit inexorablement à un syncrétisme culturel, grâce auquel chaque individu peut préserver une identité distincte tout en partageant des sentiments devenus universels. On a vu d'abord les conflits contre l'usurpateur blanc rapprocher, dans cette région de Floride, Indiens séminoles et esclaves noirs en fuite, lesquels ont adopté les mœurs de leurs compagnons d'infortune : "Back in the First Seminole War, the runaway slaves fought side by side with Seminoles, and lived as Indins, a lot of' em." (41)

Puis, par souci d'intégration à la race blanche dominante, ces mêmes Indiens peuvent avoir assimilé une éducation catholique sans renier pour autant le respect de la Terre Mère inhérent à leur culture d'origine; c'est l'expérience de Richard Hamilton et des siens :

> If you live Indin way, then you are Indin, color don't matter. It's how you respect the earth, not where you came from. Mary Weeks, she's kind of Catholic, and our kids is Catholic, and I go along with it somewhat, and read my Bible, because I was raised up in a Catholic mission back in Oklahoma. But in my heart I am still Indin, which is why I kept on drifting south to Lost Man's River, as far from those mean-mouthed cracker folks as I could get. (41)

Pour d'autres, comme Billie Conapatchie, un Creek Mikasuki, le métissage cuturel est visible dans la tenue vestimentaire: couleurs vives de la chemise indienne, qui jurent avec la sobriété du chapeau melon et du pantalon des notables blancs. Leur vie publique – fréquentation de l'église, des assemblées politiques, des théâtres d'amateurs – atteste un effort d'adaptation à la race dominante, mais c'est dans les écoles indiennes qu'ils tentent de sauvegarder les racines culturelles de leurs enfants. (196)

Plus tard, à cause des désastres écologiques du prétendu “progrès” de la civilisation, les blancs reviennent aux valeurs premières des Indiens et se soucient de protéger l'environnement naturel et vital de l'homme – sans toutefois le sacraliser ; la société Audubon est fondée au tout début du siècle et dénonce des excès remontant aux années 1880. (45-46)

Autre forme de syncrétisme culturel, la créolisation des langues participe de cette vision totalisante et unifiante de la vie. Darwin étaye, par l'exemple des langues, sa théorie d'un système naturel, se présentant comme un immense arbre généalogique dans lequel, à partir de l'origine commune, toutes les branches se ressemblent mais se modifient en accroissant leurs différences au fil des générations – il précise d'ailleurs que c'est cette ressemblance, ou affinité cachée, attestant l'origine commune, que les naturalistes devraient toujours rechercher :

> It may be worth while to illustrate this view of classification, by taking the case of languages. If we possessed a perfect pedigree of mankind, a genealogical arrangement of the races of man would afford the best classification of the various languages now spoken throughout the world; and if all extinct languages, and all intermediate and slowly changing dialects, had to be included, such an arrangement would, I think, be the only possible one. Yet it might be that some very ancient language had altered little, and had given rise to few new languages, whilst others (owing to the spreading and subsequent isolation and states of civilisation

> of the several races, descended from a common race) had altered much and given rise to many new languages and dialects, The various degrees of difference in the languages from the same stock, would have to be expressed by groups subordinate to groups; but the proper or even only possible arrangement would still be genealogical; and this would be strictly natural, as it would connect together all languages, extinct and modern, by the closest affinities, and would give the filiation and origin of each tongue.[7]

Il est bien évidemment difficile de reconstituer dans le roman de Matthiessen la généalogie du vernaculaire floridien et, à plus forte raison, de remonter à l'origine linguistique commune. Cependant, l'auteur n'hésite pas à montrer les correspondances et dérivations de différents idiomes.

D'après le personnage de Jean Chevelier, il y aurait peu de survivance du dialecte calusa, si ce n'est quelques noms d'oiseaux et de lieux, dans ces contrées dont les indigènes ont d'abord été déportés à Cuba par les Espagnols pour ensuite servir de force armée dans leur guerre contre les Américains (29). En revanche, le brassage des émigrants européens a des répercussions dans la langue souvent mixte de la population locale. L'emploi de termes espagnols ou français dans des phrases anglaises est très spontané ; ainsi Richard Hamilton: "His plan was to shoot this hombre quick, get the story later" (26) ; ou encore Jean Chevelier : "In this foking country even enfant are shooting pipples like I shooting birts !" (26)

En ce qui concerne l'ornithologue, son anglais à peine phonétique, émaillé de termes français, donne lieu à des homophonies cocasses qui dévient le sens de phrases : le *"Commaung* ?" du français devient un "Come on !" pour les autochtones américains (31).

En d'autres occasions, cette homophonie abusive, ou confusion phonétique d'une langue avec l'autre, peut involontairement trahir une identité de pensée plus profonde, un de ces sentiments universels justement. Ainsi, lorsque dans une discussion sur la religion, Chevelier, désignant le soleil et son reflet d'argent sur l'onde le qualifie de "Grand Mees-taire", la traduction par "Big Mister" ne s'écarte guère de l'idée première d'une puissance dans l'au-delà ; elle lève en quelque sorte ce " mystère ", et identifie la source naturelle de la vie à un dieu anthropomorphe. (39)

Un même type de confusion phonétique, mais avec identité totale de sens cette fois, donne lieu à une assimilation cocasse entre les morphèmes suivants: "mavericks", "hairy dicks" et "heretics" (190); tout vient d'un double-entendre volontaire au départ, puisque le locuteur, évoquant la vie

[7] *On the Origin of Species*, 380.

des vachers, le premier terme peut être pris dans son sens propre de “veaux non marqués” et non dans son sens figuré de “dissidents”.

Mais ce sont surtout les noms propres qui subissent des déviations de sens d’une langue à l'autre. L'américanisation phonétique du nom de “Monsieur Le Chevallier” ou plus familièrement “Jeen Chevelier” prononcé “Shovel-leer ”, présente une distorsion qui correspond peut-être mieux à ce que ce personnage est effectivement devenu (45). La noblesse de la racine, (cheval) dans le nom paronymique d'origine, s'est pervertie en symbole de rapacité concupiscente dans l’appellation d'arrivée plus conforme au réel, puisque cet homme a toute sa vie fouillé la région en quête de plumes d'oiseaux rares, mais surtout d'un trésor indien. Il y a, dans la transformation de son nom, comme une sorte de palimpseste héraldique.

En outre, la toponymie subit le même sort. Selon les fouilles historiques de l’auteur, elle se modifie en fonction des vicissitudes de l'histoire : le site des Everglades eut d'abord pour nom Haïti Potato Creek parce que le sol y était propice à la culture du manioc (nourriture de prédilection des Indiens) ; ce fut ensuite Allen’s River, du nom du premier pionnier blanc en 1875, et enfin, avec la première famille de colons en 1890, fut attribuée l'appellation définitive faisant référence à un havre de verdure et de fraîcheur.

D'autres noms de lieu sont franchement hybrides : c'est le cas de “Chevelier Bay” en l'honneur de l'ornithologue (45), ou de “Pavilion Key” théâtre d'une scène de barbarie sordide perpétrée par un pirate espagnol (27). Le caractère topographique est le plus souvent en anglais, tandis que le substantif placé en position d'adjectif peut être d'origine latine, comme dans les exemples précédents, où indienne: “Chatham River” serait une légère modification de “Chitto Hachee” soit “Snake River” (28), ou encore “Chokoloskee Island” qui vient du calusa “Chukkoliskee”, signifiant “vieille maison”, et désignant probablement un temple protégé des ouragans par la surélévation de ce monticule de coquillages que constitue l'îlot (35). Enfin, les dérivations phonétiques toponymiques peuvent parfois détourner le sens initial : l'archipel de “Key West”, à l'extrême sud de la Floride, doit son nom actuel à une prononciation américanisée de l'appellation espagnole d'origine “Cayo Hueso”, (le Récif de l'Os) : la visualisation topographique a donc été gommée au profit d'une localisation géographique qui peut prêter à confusion – ces îlots étant plutôt au sud-ouest. (58)

Pour Peter Mallhiessen, de toute évidence, l'hybridation ne concerne pas que les seules sciences de la nature, elle affecte aussi considérablement les sciences humaines. Comme les animaux et les plantes, des hommes de races et de cultures différentes ont été amenés à vivre ensemble, ce qui a très vite créé entre eux des liens de sang. Ils ont dû, dans un premier temps, mêler leurs diverses langues pour communiquer, et se forger une culture commune, par simple instinct de survie individuelle, puis familiale, du fait des alliances mixtes et, enfin, sociale, étant donné les pratiques polygames; d'où une créolisation des idiomes pour arriver à une langue commune, et un syncrétisme culturel à valeur de religion dans la mesure où il relie ces hommes, spirituellement au moins, en une communauté viable.

Mais on pourrait pousser l'analyse plus loin, et voir dans le roman floridien de Matthiessen une correspondance entre les sciences de la vie et celles de la matière et, de là, accéder à la métaphysique. Il faut, pour cela, repartir de l'hypothèse de Chevelier selon laquelle le mulâtre Richard Hamilton, défini comme le pilier de “l'authentique famille du Nouveau Monde” et, en tant que tel, dernier maillon en date de la chaîne évolutionniste, serait par sa pigmentation très claire l'image la plus approchante de l'homo-sapiens des origines (41).

Œuvres citées

Darwin, Charles. *On the Origin of Species by Means of Natural Selection or the Preservation of Favoured Races in the Struggle for Life*. 1859, Oxford, Oxford University Press, 1919.

Galton, Sir Francis. *F.R.S.: The Legacy of his Ideas*, edited by Milo Keynes (London : Macmillan, 1993).

Lee, John A. "Darwin's cousins and other animals," *New Scientist*, December 25, 1993, January 1, 1994, 61.

Matthiessen, Peter. *Killing Mister Watson*, New York, Random House, 1990. New York, Vintage, 1991.

L'ESTHETIQUE QUANTIQUE : UN REGARD CROISE ARTS ET SCIENCES

MICHEL CAFFAREL ET MONIQUE MARTINEZ

Du côté de l'art

En 1994 naît une association internationale à Grenade connue sous le nom de Salon des Indépendants, qui revendique le quantique comme fer de lance d'une nouvelle esthétique « révolutionnaire ». Selon ses fondateurs, la physique quantique aurait joué un rôle fondamental dans le « désordre » de la postmodernité et aurait servi de terreau fertile à l'élaboration des grandes théories (philosophiques, historiques, sociologiques, esthétiques) du siècle dernier et du XXIème siècle. Dans le domaine de l'art, la physique quantique, en remettant en cause les principes de la physique classique, en opposant le discontinu à la continuité, le hasard à la causalité, l'interdépendance des atomes à la séparabilité et à l'objectivité pourrait permettre de mieux comprendre et de mieux produire les œuvres artistiques de demain.

Dans le *Manifeste de Valence* issu de ces rencontres, les signataires (une soixantaine d'artistes dont les écrivains Juan Goytisolo et Javier Tomeo) revendiquent clairement l'esthétique quantique, nécessaire à la rénovation formelle et éthique dans une Espagne encore trop engluée, selon eux, dans le réalisme du XIXème siècle. Gregorio Morales, l'un des pionniers du mouvement, a approfondi sa réflexion sur l'esthétique quantique dans plusieurs ouvrages : *le cadavre de Balzac. Une vision quantique de la littérature et de l'art (1998) ; Le principe d'incertitude* (2003) ; *L'île du fou* (2005). Il a également participé à l'ouvrage collectif du Salón de los Independientes.[1]

Selon Gregorio Morales[2], l'art a eu quelques décennies d'avance sur les découvertes de la physique quantique et a montré, sous des formes

[1] Manuel J. Caro et John W. Murphy, *The World of Quantum Culture* (Westport, CT: Praeger, 2002).

[2] Nous reprenons ici les grandes lignes d'un séminaire donné par Gregorio Morales à l'Université de Toulouse en 2008 dans le cadre du programme de recherche de LLA/CREATIS, le théâtre quantique, en janvier 2009.

plastiques, ce qui devait ensuite se convertir en principes et en équations. Il affirme que l'esthétique quantique a été l'un des premiers mouvements qui, consciemment et délibérément, a mené à bien ces retrouvailles entre sciences et lettres. Il ajoute que c'est une tendance désormais inévitable. Même des scientifiques comme Edward O. Wilson, militants du plus fervent positivisme, plaident en faveur de cette union. C'est ainsi que Wilson a élaboré, pour s'y référer, le concept de « consilience ». « L'entreprise majeure de l'esprit, affirme-t-il, a toujours été et sera toujours la tentative de relier sciences et humanités ». Et d'ajouter : « ni les sciences ni les arts ne peuvent être complets s'ils ne combinent pas leurs points forts respectifs. La science a besoin de l'intuition et du pouvoir métaphorique des arts et les arts ont besoin du sang neuf qu'apporte la science »[3] La dimension révolutionnaire de l'esthétique quantique réside dans le fait qu'elle n'a pas besoin d'échapper à la science pour inaugurer un paradigme qui dynamite les vieux canons et sème le scandale et la liberté sur son passage.

En d'autres termes, si c'est la science qui, au XVIIIème siècle déjà, a, la première, imposé à l'art la voie du réalisme, c'est cette même science qui vient, désormais, rendre à l'art la grandeur, les possibilités et le mystère qu'il avait perdus. Le créateur quantique a donc la gigantesque tâche de reformuler la réalité dans laquelle nous vivons depuis ces perspectives nouvelles et de nous expliquer, pour la première fois dans notre vie et de manière *réelle*, que le destin peut être moi-même ; que je ne suis pas seul et qu'en moi vivent les actions de mes ascendants et celles de mes descendants ; que je peux communiquer avec tout l'univers ; que mes intuitions peuvent être vraies ; que nombre de hasards ne sont rien d'autres que le destin ourdi par nos esprits ; que la complexité est à l'œuvre dans la vie de chacun de nous ; que nous sommes vivants et morts à la fois ; que nous sommes simultanément mortels et immortels ; qu'il y a un terrible et effrayant foisonnement de réalité cachée et que l'assimiler est peut-être l'une des principales fonctions de notre vie. Il faut tout réinterpréter depuis la nouvelle réalité qui s'est ouverte à nous, écrire une nouvelle *Comédie humaine* où les univers parallèles et leurs mystérieuses manifestations dans notre réalité quotidienne auront cours et entreront en relation les uns avec les autres.

Toujours selon Gregorio Morales, l'esthétique quantique définit les principes fondamentaux de son approche de façon détaillée et nous en citerons quelques-uns :

[3] Edward Wilson, *L'unité du savoir* (Paris: Laffont, 2000).

Le principe de complémentarité. En fonction des mesures que nous faisons, un corpuscule peut se comporter comme une onde ou comme une particule.

Le principe d'incertitude. Il est impossible de déterminer en une seule mesure la position et la quantité de mouvement d'une particule.

Le principe anthropique. L'observateur modifie l'expérience par sa simple observation.

La non-séparabilité. Nous avons l'habitude de considérer que l'univers est composé d'objets séparés mais, selon la physique quantique il ne pourrait s'agir que d'une illusion. L'univers semble être, en effet, un tout *interrelationné* et sans limites (non local) où, pour employer l'expression poétique de Francis Thompson, « on ne peut toucher une fleur sans déranger une étoile »·

L'a-causalité. Dans notre monde « normal », nous sommes habitués à ce que tout phénomène soit toujours et nécessairement causé par un autre mais, dans le monde quantique où règnent les probabilités, cet état de chose n'est pas nécessairement vrai et un fait peut être interprété comme étant a-causal.

La complexité. Bien que tout dans le cosmos obéisse à la loi de l'entropie, il existe, néanmoins, dans les organismes, de la particule à la cellule, en passant par le gaz interstellaire, une tendance à s'organiser en structures toujours plus complexes.

L'ubiquité. Quiconque croit que les particules peuvent être localisées dans un endroit concret se trompe ; en fait, celles-ci se comportent comme si elles pouvaient être en de nombreux endroits au même moment, c'est-à-dire comme si elles avaient le don d'ubiquité.

Les champs morphogénétiques. Vraisemblablement, la connexion existant entre tous les points du cosmos donne lieu à différents modèles, responsables du fait que, par exemple, il existe un esprit universel ou un « air du temps ». Mais on peut même aller plus loin et voir dans des connexions déterminées les instincts d'une espèce, ses règles de conduite ou d'apprentissage. Autrement dit, les *champs morphogénétiques* sont ce qui, en matière de structures, de règles, de comportements, d'idées, ou de tendances, donne sa forme à des aspects déterminés de la réalité.

L'univers comme hologramme. L'Univers peut être vu comme Information. Une Information qui, en outre, est holographique : chaque partie contient le tout.

Ordre impliqué et ordre manifeste. L'univers aurait deux types d'ordre. Le premier d'entre eux, l'ordre *manifeste*, est celui qui apparaît, évident, sous nos yeux ; le second, l'ordre *impliqué*, est celui qui se cache à notre vue et qui ne se montre qu'à des moments donnés, ponctuellement.

La non-distinction entre matière et énergie, entre corps et esprit. La matière est énergie ; Par analogie on peut émettre l'idée que ce que nous appelons « esprit » est aussi une autre manifestation de la matière.

Ce mouvement autour de l'esthétique quantique initié par l'Espagne semble être devenu « planétaire ». Un courant important d'artistes internationaux revendique cette filiation Arts et Sciences. On pourrait citer dans le désordre et sans prétention d'exhaustivité ;

- pour le théâtre le français Claude Régy: *Espaces perdus* (1991) et *L'état d'incertitude* (2002) ;

- pour le roman quantique : le français Michel Houellebecq : *Les particules élémentaires,* le japonais Haruki Murakami *: Kafka sur le rivage ;*

- pour la musique quantique : l'américain John Cage ;

- pour la science-fiction : l'américain Robert Anton Wilson : *Schrödinger's Cat Trilogy* (1980-1981).

D'autre part, dans certains milieux avertis et autorisés, l'adjectivation « quantique » semble devenu un label « de chic et de choc » pour la création contemporaine. Au-delà de ce phénomène de mode, danger toujours possible de la quête effrénée de nouveaux concepts épistémologiques, qu'en est-il vraiment de cette appropriation de la science par l'art ? Qu'est-ce que l'imaginaire artistique doit aux théories quantiques ? Ou plus précisément quelle perception de la science traduisent, par leurs discours ou par leurs œuvres, les créateurs quantiques ? Quelle est la perception, aigue ou fantasmée, des grands principes des découvertes du siècle dernier ?

Nous n'avons pas la prétention de traiter dans un court article d'une problématique aussi vaste et complexe. Nous nous contenterons de questionner quatre principes de base auxquels se réfèrent les créateurs : le principe anthropique, le principe d'incertitude, l'a-causalité, la non-séparabilité.

Le principe anthropique est fondamental pour comprendre les enjeux de la création contemporaine. L'œuvre « ouverte » devenue un espace co-construit par le récepteur justifie l'analogie avec l'idée d'un observateur qui modifierait l'expérience par sa simple observation, rendant impossible toute tentative objective de fixation du réel. Ainsi, pour donner un exemple concret, dans la pièce *Sang de lune* de Sinisterra (sur le même thème que *Parle avec elle* d'Almodóvar), l'esthétique du translucide qui est à l'œuvre permet au spectateur de construire son interprétation : des bribes d'informations, des indices polysémiques, des vides sémantiques font que chaque récepteur s'invente son histoire et individualise son rapport à la fiction.

Le principe d'incertitude, quant à lui, a été particulièrement fécond pour l'imaginaire des créateurs. Forts de l'idée qu'il est impossible de déterminer en une seule mesure la position et la quantité de mouvement d'une particule, les artistes contemporains ont revendiqué ce qui était le propre de toute poïétique : la génération de formes émancipées des référents « réels » qui permettent les transgressions, les fantaisies, les déviances par rapport aux normes. Ainsi, dans *Le quantique des quantiques* d'Alberto Velasco, Abel Capricorne est un sujet poétique quantique, capable d'être à plusieurs endroits en même temps : dans un monde où une lotion pour les cheveux fait repousser les poils de vieux mocassins, dans une réalité où on est transporté par la pensée dans des pays étrangers, tout devient possible ; surtout, la fantaisie se trouve enfin légitimée par la science.

De la même façon, le principe d'a-causalité a donné lieu à des expériences surprenantes : la prise de conscience que tout phénomène n'est pas toujours et nécessairement causé par un autre, comme dans le monde microscopique. John Cage a poussé cette pratique du facteur d'imprévisibilité à l'extrême jusqu'à faire du chaos le fondement même de son œuvre (musicale ?), 4'33. L'interprète joue en silence pendant quatre minutes et trente trois secondes une composition musicale destinée à accueillir n'importe quel son de manière imprévue.

Les créateurs se réfèrent souvent au concept de non-séparabilité. L'idée que l'univers est un tout interrelationné, sans limites, non local a permis d'aller contre l'idée de la rationalité du monde et de la cohésion du personnage. Marguerite Duras, dans son tryptique *Aurelia Steiner* construit trois personnages superposés, Aurelia Melbourne, Aurelia Vancouver, Aurelia Paris qui sont trois états d'une même enfance. Leurs voix avancent comme la succession d'une même onde.

Du côté de la physique

C'est l'essence même de l'art de ne se fixer aucune limite quant aux sources de son imaginaire et de sa créativité. La physique quantique, en tant que création/découverte (on laissera le soin à l'épistémologue de discuter ce point) ne doit évidemment pas y échapper, et c'est ce que fait avec enthousiasme et imagination l'esthétique quantique dont les principes fondateurs viennent d'être évoqués. Cette appropriation, quelque légitime qu'elle puisse être, peut cependant apparaître suspecte aux physiciens : les mots et les expressions sont proches, les concepts, patiemment bâtis sous la double contrainte sévère des mathématiques et de l'expérience, valsent au gré de l'imagination de l'artiste, et, un sentiment étrange de confusion peut finalement étreindre le physicien. Toutefois, le scientifique doit éviter de juger la création artistique à l'aune de ses propres procédés, tout comme l'artiste ne doit pas avoir la prétention de quantifier ses propres intuitions. Si rencontre féconde il y doit y avoir, elle ne peut avoir lieu que sur le terrain commun de l'imagination, sous ses formes artistiques et scientifiques, et non sur celui de la technicité propre à chaque démarche. Chacun des deux mondes peut alors se nourrir des intuitions, des concepts et des images de l'autre. Dans cet esprit-là revenons à deux des principes de l'esthétique quantique évoqués plus haut.

Principe anthropique : « L'observateur modifie l'expérience par sa simple observation »

Dans le cadre de l'esthétique quantique, ce principe est présenté comme ayant une nature éminemment quantique. Nous proposons ici de mettre cette idée à l'épreuve. Pour cela, considérons une expérience de pensée au caractère indubitablement « classique » (par opposition à « quantique »). Il s'agira du mouvement de la Lune autour de la Terre, le système Terre-Lune étant considéré comme totalement isolé du reste du monde. La Figure 1 illustre notre propos. En accord avec les principes de la physique newtonienne, ce problème dit « à 2-corps » (Terre-Lune) est totalement prédictible. Par prédictible, nous entendons ici qu'il est possible à chaque instant de calculer avec une précision arbitrairement grande la vitesse et la position de la Lune par rapport à la Terre.

Introduisons maintenant un observateur. Cette personne (Manu, sur la figure 2) n'est évidemment pas un être neutre extérieur au problème. En tant que sujet observant, il s'insère dans le système Terre-Lune, jusque-là totalement isolé de son environnement. Notre système à 2 corps devient alors un système à 3 corps (Terre-Lune-Manu, figure 2), qui peut, de nouveau, être considéré comme isolé du reste de l'Univers. Mais à cet instant s'opère une transformation essentielle. La Nature nous joue un tour inattendu: le nouveau système à trois corps, bien qu'obéissant toujours à la physique newtonienne, perd sa prédictibilité au sens défini précédemment: la position de la Lune ne peut plus être connue à tout instant avec une précision arbitraire.

De par son intervention l'observateur a changé la nature *qualitative* de la relation Terre-Lune. Surgit donc ici une forme inattendue mais incontournable du « Principe anthropique » mais sous une version *classique* cette fois. Afin de comprendre ce changement singulier, il nous faut nous référer à un phénomène découvert au vingtième siècle, à savoir l'existence de comportements « chaotiques » dans les systèmes décrits par les équations de Newton. Il faut insister sur le fait que ce phénomène est loin d'être anecdotique puisqu'il apparaît de fait dans la plupart des systèmes dynamiques. L'existence de ce chaos signifie que toute perturbation - aussi petite soit-elle - sur les conditions opératoires peut changer de manière radicale le destin du système à des temps suffisamment longs. Même sans intervention extérieure, il n'est donc pas possible de prédire ce que le « déterminisme universel » du XIXème siècle avait pourtant pris pour acquis. Dans le cas de notre système Terre-Lune-Manu la perturbation provoquée par notre observateur est évidemment extrêmement faible, la masse de Manu dans ce problème à 3-corps étant insignifiante par rapport à celle des planètes.

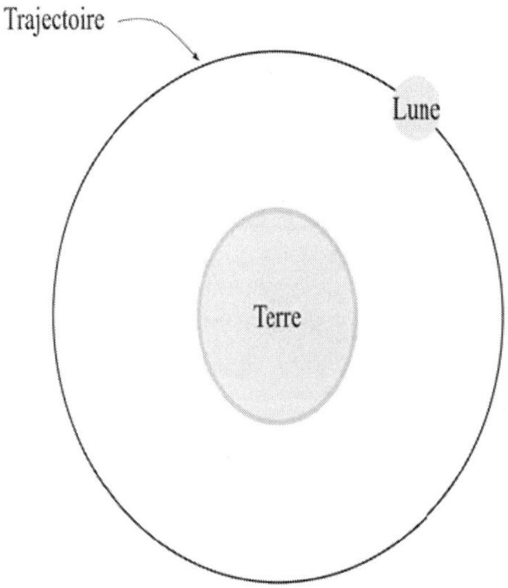

On peut ainsi garder l'impression pendant une durée assez longue (en fait, bien supérieure à celle de la vie de l'observateur) que Manu n'a pas changé grand' chose à la trajectoire Terre-Lune. Mais c'est une erreur puisque le système présente ce caractère imprédictible, le couple Terre-Lune et l'observateur doivent être considérés comme intimement et irrémédiablement imbriqués. Notons que du point de vue mathématique, le problème est redoutable ; pour en donner un exemple frappant, les calculs les plus précis ne permettent toujours pas de savoir si notre propre système solaire considéré comme un problème classique à N-corps, est stable ou non à très longs temps, c'est à dire au-delà de quelques dizaines de millions d'années, voir par exemple Laskar (1994).

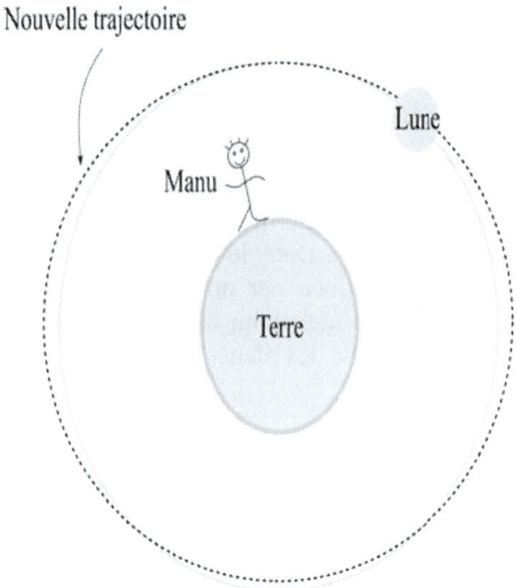

Qu'un observateur opère un changement essentiel à un système observé, quel que soit le schéma théorique utilisé pour le représenter (ici les lois de la physique classique ou quantique) ne doit cependant pas nous surprendre. Le caractère radicalement humain de toute connaissance est une caractéristique fondamentale de la pensée humaine, qu'elle soit philosophique (« L'homme est la mesure de toute chose ») ou scientifique.

Principe de non-séparabilité

Comme nous l'avons vu plus haut, le principe de non-séparabilité est au cœur de nombreuses œuvres quantiques. Il est à l'origine des concepts mentionnés précédemment de « mondes interrelationnés », de « champs morphogénétiques », d'ubiquité, et d'Univers considéré comme un hologramme. En physique, on fait généralement remonter la critique des conséquences paradoxales de ce principe à l'article d'Einstein, Podolsky et Rosen (1935), le fameux « paradoxe EPR » : en quelques mots, les principes fondateurs de la mécanique quantique impliquent que tout système sans interaction avec le reste de l'Univers doit être considéré comme un tout indissociable. Une action (= une mesure) sur une quelconque de ses parties est donc censé agir *instantanément* sur toutes les autres parties, et ceci indépendamment de leur éloignement géographique (accompagné d'une violation possible de la relativité). En laboratoire, des efforts considérables ont été déployés pour réaliser les conditions exceptionnelles permettant de détecter ce phénomène. Cela a été réalisé avec succès lors de célèbres expériences dues à Stuart J. Freedman et John F. Clauser (1972) aux Etats-Unis et à Alain Aspect et ses collaborateurs en France en 1981-1982.

Du point de vue fondamental, le principe de non-séparablité est directement relié à la nature ondulatoire - et donc spatialement délocalisée - des particules élémentaires. De ce principe découle un phénomène complémentaire à l'existence de corrélations EPR, connu sous le nom de « localisation quantique»[4]. Un des mystères de la physique classique - quelque peu oublié en ces temps de mystères quantiques - est qu'il s'agit également d'une théorie non-séparable: en mécanique newtonienne, deux corps interagissent à toute distance et instantanément. Ce fait, qui est évidemment apparu comme problématique aux fondateurs de la théorie classique, a donné lieu au cours des siècles à diverses échappatoires (pensons par exemple au concept d'éther[5]). Ici, on réalise que l'aspect ondulatoire de la matière proposé par la mécanique quantique est une solution très élégante nous permettant de sortir de cette impasse.

Grâce à un jeu subtil d'interférences constructives ou destructives, une particule quantique peut se re-matérialiser (« se localiser ») sous la forme d'une particule quasi-classique localisée spatialement et ayant perdu ces corrélations si étranges qui semblaient la lier pour l'éternité au destin de toutes les particules de l'Univers. On peut donc soutenir que, grâce à la physique quantique, notre monde quotidien est redevenu séparable et local comme notre sens commun nous le suggère. C'est donc la théorie quantique qui remet de l'ordre dans notre monde! Faudra-t-il alors opposer le côté « raisonnable » et « intuitif » de la physique quantique à la bizarrerie et à l'extravagance de la physique classique? Une fois de plus, la Nature se joue de nous, renversant nos points de vue et nos projections mentales.

De la même manière, les autres principes invoqués par l'esthétique quantique pourraient être examinés plus avant. Cependant, à partir de deux des principes fondamentaux, on réalise déjà que l'opposition entre un monde classique qui ne serait fait que de déterminisme implacable et de

---

[4] On parle également de « décohérence » dans le contexte de la dynamique des systèmes quantiques, voir par exemple Zeh (1970).

[5] L'éther est une substance hypothétique qui remplirait le vide de l'Univers et permettrait à la lumière de se propager comme le son dans l'air ou une vague sur la mer. Cette notion introduite en physique dès l'Antiquité et très discutée à l'époque moderne a été abandonné au début du 20ième siècle avec l'avènement de la relativité restreinte.

causalité triste et un monde quantique enchanteur où règnent l'indéterminisme, la liberté et le foisonnement, doit être questionnée. La Nature ne se laisse pas contraindre à nos catégories ; malicieuse, elle se plaît à nous jouer ses tours facétieux (chaos classique, décohérence, localisation, etc.) au sein même de nos constructions les plus élaborées, qu'elles soient classiques, quantiques ou autres. Le physicien est donc perplexe quant à la nécessité du label quantique pour « ...rendre à l'art la grandeur, les possibilités et le mystère qu'il avait perdus.... ». Cette grandeur et ce mystère existent, ils sont là, tapis dans les replis de nos visions du monde et de nos élaborations artistiques ou scientifiques les plus osées. Sachons peut-être mettre le monde au défi et ne pas catégoriser trop vite ce qui est révolutionnaire ou non.

Œuvres citées

Abuín González, Anxo. *Escenarios del caos. Entre hipertextualidad y la performance en la era electrónica*. Valencia : Tirant lo blanch, 2006.

Aspect, Alain, Philippe Grangier and Gérard Roger, *Phys. Rev*. Lett. Vol. 47 (1981):460.

Aspect, Alain, Philippe Grangier P. and Gérard Roger. *Phys. Rev*. Lett. Vol. 49 (1982):91.

Caro, Manuel J. and John W. Murphy. *The World of Quantum Culture*, Westport: GPG, 2002.

Corrons, Fabrice, Monique Martinez-Thomas and Agnès Surbezy. "Le théâtre quantique : ordre et désordre dans l'Espagne postmoderne." In *L'Annuaire théâtral*. Québec: Université Laval, n°43-44, 2009.

Freedman Stuart J. and John F. Clauser. "Experimental test of local hidden-variable theories", *Phys. Rev. Lett*. 28 (1972): 938.

Einstein, Albert, Boris Podolsky and Nathan Rosen. "Can Quantum-Mechanical Description of Physical Reality Be Considered Complete?" *Phys. Rev*. 47 (1935), 777-780.

Hébert, Chantal. "De la *mimesis* à la *mixis* ou les jeux analogiques du théâtre actuel." In *Multidisciplinarité et multiculturalisme,* edited by Chantal Hébert et Irène Perelli-Contos, 25-39. Théâtre, Québec: Nuit Blanche Editeur, 1997.

Laskar Jacques. "Large-scale chaos in the solar system." *Astronomy and Astrophysics*, vol. 287 (1994): L9–L12.

Martinez-Thomas, Monique and Agnès Surbezy, Agnès, "Le traitement quantique du thème de la mère empêchée." In *Nouvelles figures maternelles dans la littérature espagnole contemporaine. Les mères empêchées*, edited by Nadia Mékouar-Hertzberg, 137-154. Paris: L'Harmattan, 2009.

Morales, Gregorio. *El cadáver de Balzac. Una visión cuántica de la literatura y del arte*. Alicante: De Cervantes Ediciones, 1998.

Nicolescu, Basarab. "Stéphane Lupasco. Du monde quantique au monde de l'art". *Mélusine*, Lausanne, Suisse, n° XXVII (Le surréalisme et la science) (2007), 181-196.

Régy, Claude. *Espaces Perdus*. Besançon: Les Solitaires Intempestifs, 1998.

Régy, Claude. *Le principe d'incertitude*. Besançon: Les Solitaires Intempestifs, 2002.

Rykner, Arnaud. « L'univers quantique de Marguerite Duras et la critique des dispositifs. » In *Marguerite Duras et la pensée contemporaine*, edited by Eva Alhstedt and Catherine Bouthors-Paillart, 181-193. Göteborg : Université de Göteborg, 2008.

Shepherd-Barr, Kirsten. *Science on Stage: from Dr Faustus to Copenhagen*, Princeton, Princeton University Press, 2006.

Surbezy, Agnès. "Ondes et corpuscules : le personnage quantique dans la *Trilogía de mujeres medievales* d'Antonia Bueno." In *Théâtres dramatiques d'orient et d'occident ; 1968-2008*, edited by Carole Egger, Isabelle Reck and Edgard Werber, Edgard. Strasbourg : Presses Universitaires de Strasbourg, 2012.

Surbezy, Agnès, « L'ailleurs existe-t-il encore pour le chat de Schrödinger ? Là et non-là dans *El lado oeste del Golden Gate* de Pablo Iglesias Simón. » In *L'ailleurs dans les dramaturgies hispaniques*, edited by Carole Nabet Egger and Isabelle Rouane Soupault. Marseille: PUP, 2013.

Surbezy, Agnès, ed. *Le Théâtre quantique, une utopie ?*, Carnières: Lansmann, à paraître, 2013.

Valmer, Michel. *Théâtre de sciences*. Paris: éditions du CNRS, 2006.

Wilson, Edward. *L'unité du savoir*. Paris: Laffont, 2000.

Zeh, Hans Dieter "On the interpretation of measurement in quantum theory", *Foundations of Physics*, Volume 1, Issue 1 (1970): 69–76.

DES CHIMERES BIONANOTECHNOLOGIQUES : L'HUMAIN AUX PRISES AVEC LES IMAGINAIRES TECHNOSCIENTIFIQUES

MATHIEU QUET

Bionanos et visions du corps à venir[1]

Les débats autour des bionanotechnologies accordent souvent une importance particulière aux transformations à venir du corps humain. Ces visions du (corps) futur sont variables, plus ou moins ambitieuses et plus ou moins sombres – de la crainte de la pollution ou de la contamination du corps humain par les nanoparticules aux discours sur les mutations « radicales » du genre humain et à la notion de post-humanité. Par exemple, au cours de la série de débats publics consacrés aux nanotechnologies et organisés par la Commission Nationale du Débat Public fin 2009, début 2010, un certain nombre de questions ont été abordées par des acteurs très différents : des groupes écologistes ont insisté sur les problèmes de pollution que risque de soulever à l'avenir la production de nanoparticules, les industries pharmaceutiques ont insisté sur les bienfaits que pouvaient apporter les nanosciences avec l'émergence de nouveaux traitements, comme la thérapie génique, et un groupe dit « transhumaniste » a évoqué des nanorobots infiltrés dans le corps et la mutation du genre humain. Chacun de ces groupes produisait ainsi des prédictions particulières sur les conséquences à venir des nanotechnologies pour le corps humain.

[1] Ce texte s'inscrit dans le cadre d'un projet collectif (financé par une ANR sur les nanotechnologies) qui rassemble des chercheurs en sociologie des controverses scientifiques (le GSPR à l'EHESS), des chercheurs en sciences de l'argumentation (du Laboratoire Communication et Politique au CNRS) et des chercheurs en sciences du sport (le CERSM à l'Université Paris Ouest Nanterre). Il résulte donc d'un projet interdisciplinaire qui mêle approche sociologique, analyse de discours et anthropologie du corps. L'auteur adresse tous ses remerciements aux participants du projet pour leurs conseils sur des versions antérieures de ce texte, et en particulier à Assimakis Tseronis.

Dans ce contexte, on peut essayer d'appréhender les discours d'anticipation au sujet des effets des nanotechnologies sur le corps humain dans leur hétérogénéité, et tenter de comprendre comment ces discours s'articulent, par-delà leur variété et leurs différences. La production discursive de chimères, ou de transformations plus ou moins imaginaires du corps humain, dans le cadre des multiples débats et controverses sur les nanos, est l'un des modes d'inscription des nanosciences dans l'espace public. La production de prédictions concernant l'avenir du corps, voire de la nature humaine, est donc l'une des médiations par lesquelles les acteurs sociaux tentent de saisir l'objet « nano », de penser son encadrement social. Et il n'est pas l'apanage de quelques illuminés, mais bien un mode d'inscription partagé par tout un faisceau d'acteurs : des chercheurs, des ingénieurs, des militants, des journalistes, et bien d'autres. Dès lors, les chimères et autres imaginaires technologiques remplissent un rôle essentiel dans l'appréhension sociale des effets des nanosciences.

Nanos et imaginaire SF

Plusieurs observateurs ont souligné que le lien entre nanosciences et science-fiction avait une teneur particulière, et ont notamment insisté sur le rôle des imaginaires de science-fiction convoqués depuis le début du projet « nano »[2]. Certes, les nanos n'ont pas l'exclusivité des imaginaires de science-fiction : le nucléaire ou l'informatique ont aussi eu leur pendant romanesque, et plusieurs articles du dossier montrent que la relation féconde entre science et fiction ne date pas d'hier. Un aspect toutefois intéressant des nanosciences est la présence précoce de cet imaginaire, dès le lancement du projet, et le fait qu'il y joue un rôle central, à la fois fédérateur pour les recherches et catalyseur pour les débats. En schématisant, on retient généralement deux noms dans l'histoire des nanosciences. Le premier est celui de Richard Feynman, physicien qui, en 1959, tient une conférence célèbre dans laquelle il annonce « il y a plein de place en bas » (« *there's plenty of room at the bottom* ») et invite à poursuivre des recherches autour de l'infiniment petit. Le second est celui du principal promoteur des nanotechnologies à partir des années 1980, le

[2] Sylvie Catellin, « Le recours à la science-fiction dans le débat public sur les nanotechnologies : anticipation et prospective, » *Quaderni* 61, (2006) « La fabrique des nanotechnologies », Automne : 13-24. Sylvie Catellin, « Nanomonde : entre science et fiction. Quelles visions du futur, » *Alliage* 62, (2008) « micro & nano », avril : 67-78. Marina Maestrutti, « Prendre au sérieux la fiction. La mise en débat des nanotechnologies. » *Alliage* 62, « micro & nano » (avril 2008).

chercheur-ingénieur Kim E. Drexler, qui publie en 1986 un livre de prospective dans lequel il essaie d'imaginer les usages à venir des nanosciences[3]. Ce livre est lui-même nourri de science-fiction. Il imagine en particulier un instrument qui a fait l'objet de nombreuses discussions : l'assembleur (nano) moléculaire, qui pourrait décomposer la matière de n'importe quel objet afin de recomposer un autre objet (en poussant à l'extrême le principe « rien ne se perd, rien ne se crée, tout se transforme »). Il imagine aussi des nanomachines capables d'autoréplication (et donc aptes à s'autonomiser de l'humain). Drexler va être un ardent promoteur des politiques nanotechnologiques à partir des années 1980, créant une fondation de soutien et de prospective, constituant des réseaux humains, financiers et techniques autour des nanos, jusqu'à intervenir au Congrès états-unien pour défendre son projet. C'est un véritable entrepreneur scientifique, et de façon notable, il fait partie de ces chercheurs-ingénieurs pétris de culture technologique au sens le plus large, c'est à dire férus à la fois de technique et de science-fiction.

Issu du *Massachusetts Institute of Technology*, centre de formation à l'innovation technologique par excellence, il a appartenu à une société typique de la contre-culture ingénieure américaine dans les années 1970, la *L5 Society*. Ce groupe rassemble des ingénieurs, des auteurs de science-fiction comme Robert Heinlein ou Isaac Asimov, et son objectif central a été de lancer des vols spatiaux sans le contrôle de la NASA – imaginaire sf, individualisme, contre-culture y font très bon ménage. Bref, si l'on s'en tient au parcours de Drexler, l'histoire des nanos comme projet scientifique et politique est placée sous le signe de l'imaginaire science-fictionnel.

L'autre exemple célèbre et fréquemment cité de la marque de la science-fiction sur les nanotechnologies est celui d'un rapport remis à la *National Science Foundation* par Mihail Roco et William Bainbridge en 2002 et intitulé *Converging Technologies for Improving Human Performances*[4]. Le rapport est extrêmement optimiste et promet, grâce aux nanos et à la convergence entre nanotechnologies, biotechnologies, informatique et neurosciences, un bien-être universel, la paix dans le monde, et de nouvelles formes d'interaction entre hommes et machines « intelligentes ». Comme le rappelle Jean-Pierre Dupuy, le rapport

---

[3] Eric K. Drexler. *Engines of creation: The coming era of nanotechnology* (New York: Doubleday, 1986).

[4] Mihail Roco and William S. Bainbridge, *Converging technologies for improving human performance: Nanotechnology, biotechnology, information technology and cognitive science* (Arlington: U.S. National Science Foundation, 2002).

conjecture que « *l'humanité pourrait bien devenir comme un 'cerveau' unique, [dont les éléments seraient] distribués et interconnectés par des liens nouveaux parcourant la société* »[5]. Le ton du rapport est d'autant plus étonnant qu'il est supervisé par le principal organisme de financement de recherche américain. Mais il s'explique en partie par l'appartenance de l'un des auteurs, William Bainbridge, à l'Association Mondiale Transhumaniste, qui diffuse ardemment l'idée d'une mutation de l'humanité, et constitue l'horizon du néopositivisme technologique qui porte le développement des nanotechnologies.

L'économie des promesses

Ces deux exemples permettent d'illustrer de façon schématique les liens les plus saillants entre le projet nano et l'imaginaire de science-fiction. Reste alors à expliquer cette relation. La première remarque qui a été faite à ce sujet concerne les conditions économiques de développement d'un projet tel que les nanotechnologies. Si les promoteurs des nanos promettent monts et merveilles, c'est bien parce qu'ils ont quelque chose à vendre à des investisseurs. Dans cette perspective, un certain nombre d'analystes des sciences ont parlé du développement d'une « économie des promesses ». Cette économie des promesses prend une importance croissante dans les processus d'innovation. Elle force les chercheurs et les entrepreneurs scientifiques à mettre en place un *storytelling* du futur, et à proposer des récits prospectifs sur les conséquences d'une innovation, d'un projet ou d'un champ de recherche en développement. Selon le contexte, les publics auxquels il s'adresse et les ambitions de ses narrateurs, le récit peut porter sur du court, moyen ou très long terme, et envisager des transformations sociales plus ou moins radicales et plus ou moins globales. Au sein d'une telle économie, chercheurs, entrepreneurs scientifiques et pouvoirs publics se trouvent engagés dans des projets dont la viabilité repose en priorité sur l'apparence de robustesse d'une série de prédictions sur les bienfaits à venir de la recherche[6]. Si Drexler ou Bainbridge

[5] Cité par Jean-Pierre Dupuy, « Pour une évaluation normative du programme nanotechnologique », *Annales des Mines* (février 2004) : 28.

[6] Pierre-Benoît Joly (ed.), *Démocratie locale et maîtrise sociale des nanotechnologies. Les publics grenoblois peuvent-ils participer aux choix scientifiques et techniques ? Rapport de la Mission pour la Métro* (22 septembre 2005). Ulrike Felt, ed., *Taking European Knowledge Society Seriously* (*Report of the Expert Group on Science and Governance to the Science, Economy and Society Directorate, Directorate-General for Research, European Commission*, 2007).

produisent des imaginaires aussi ambitieux, c'est dans la perspective d'alimenter les espérances du marché et des investisseurs (et on peut dire aussi, afin de drainer des chercheurs et de fédérer un domaine qui, malgré tout, reste atomisé).

Mais cette remarque en entraîne une autre. Car si « économie de la promesse » il y a, c'est dès lors forcément en un double sens : l'économie (financière) fondée sur les promesses en implique une autre : une « économie de la promesse », discursive cette fois, indissoluble d'un marché de la promesse. Ce marché-là n'est régi par aucune monnaie. En revanche, il met en concurrence des projections de l'avenir et des visions du futur, dont la distribution devient un aspect essentiel de la régulation économique des sciences et des technologies.

Deux économies imbriquées, donc. La première, traditionnelle mais appuyée sur des enjeux nouveaux, est une économie de l'innovation, de la recherche et de la technologie qui implique la mise en jeu de promesses pour drainer des investissements ; le financement des projets technoscientifiques contemporains, la croissance de l'économie de l'innovation actuelle nécessitent la production permanente de promesses. La seconde, essentielle pour comprendre la première, est une économie discursive et métaphorique, qui met en circulation et en concurrence différentes projections de l'avenir, les unes le dépeignant radieux, les autres imaginant ses aspects les plus sombres.

En effet, les projections de Drexler ou Bainbridge provoquent inévitablement des réactions, et appellent des visions contradictoires, à travers lesquelles les nanotechnologies sont vues comme un danger. Le cas le plus célèbre est probablement l'ouvrage de science-fiction de Michael Crichton, *Prey*, dans lequel l'humanité est prise en chasse par un nuage gigantesque de nanoparticules[7]. Mais dans les discours des groupes militants, écologistes ou non, on retrouve des contre-imaginaires, comme par exemple l'apparition d'une humanité à deux vitesses, opposant les humains augmentés technologiquement et d'autres non-augmentés[8]. Et ces visions du futur (optimistes et pessimistes) s'articulent, se concurrencent, au sein d'une économie sociodiscursive des prédictions.

[7] Michael Crichton, *Prey*. (New York: Avon Books, 2003).

[8] BEDE (Biodiversité : Echange et Diffusion d'Expériences) (2009) *Bang ou la convergence des technologies. Nanotechnologies et artificialisation du vivant*, rapport.

Acteurs de l'économie des promesses

Dans le cas des nanosciences et des nanotechnologies, cette économie des promesses est difficile à caractériser, car on a affaire à un très grand nombre d'acteurs occupant des positions très différentes dans l'espace social et à des modes d'interventions variables. En effet, si l'on schématise les acteurs centraux des débats autour des nanos, on retrouve des groupes aussi variés que des entrepreneurs scientifiques, promoteurs du projet nanotechnologique, des chercheurs, des groupes militants souvent du côté de l'écologie politique, des universitaires, souvent philosophes et/ou éthiciens, des courants contre-culturels (tels que le transhumanisme), des auteurs de littérature, des institutions de promotion de la démocratie délibérative, et bien d'autres encore. Cet inventaire, loin d'être exhaustif, donne une idée de l'hétérogénéité des personnalités qui produisent simultanément des discours sur les nanotechnologies et leurs conséquences pour le corps humain. A priori, ces groupes partagent peu de choses. Mais l'observation permet de saisir des moments particuliers où cette multitude d'acteurs et de discours tout à fait hétérogènes entrent dans un même espace discursif. Et c'est justement là que la notion d'économie des promesses est intéressante, en tant qu'elle pointe les moments où tous ces discours et prédictions se trouvent confrontés. Une fois encore, on peut évoquer le cas des débats publics de 2009-2010 sur les nanotechnologies, au cours desquels des acteurs très variés ont produit des avis ou participé, d'une façon ou d'une autre, au débat. Lors de ces rencontres, les différentes prédictions se sont trouvées mises en présence de façon très concrète.

Mais alors, quels sont les enjeux de cette mise en concurrence de prédictions, de chimères ? Pour le montrer, je vais m'appuyer sur l'un de ces « moments » au cours desquels des discours hétérogènes sont mis en présence. Il s'agit du débat sur la question des nanotechnologies dans le sport, et en particulier autour du dopage génétique. Le corpus donne lieu à une prolifération de chimères, de monstres représentant le fruit du mariage des technologies et de l'humain. De façon critique parfois, de façon utopique souvent[9]. Et ce bestiaire d'un genre nouveau permet aux acteurs de représenter à la fois les formes de vie émergeant conjointement aux avancées de la recherche, et les institutions nécessaires pour les encadrer.

---

[9] Cette seconde partie s'appuie sur des analyses réalisées sur un corpus de 96 textes consacrés aux enjeux du dopage génétique. Le corpus rassemble des textes issus de revues médicales, de sciences sociales, de médias généralistes, de blogs et sites internet, privilégiant la diversité des types de sources. Analyse menée avec le logiciel Prospéro.

Monstres et chimères biotechnologiques

On a ainsi affaire à une prolifération de personnages. Les débats sur le dopage génétique convoquent ainsi des athlètes « génétiquement dopés »[10], « améliorés technologiquement » au moyen de prothèses, des sportifs sélectionnés « sur leurs promesses génétiques »[11], des « athlètes génétiquement modifiés » sur le modèle des OGM[12], des « superman » aux performances démultipliées[13], des cyborgs, ou des athlètes cyborg, des « super soldats », des clones de Michael Jordan[14], des hommemachines (en un mot)[15], des Robocop et des « créatures bioniques qui peuplent la science-fiction »[16], des Genrich (humains génétiquement « enrichis »)[17]. Les animaux ne sont pas en reste, puisqu'il est aussi question de « souris schwarzennegger »[18], ni les machines, puisque les ordinateurs à ADN sont évoqués[19], tout comme les robots dont la conscience s'autonomise de façon croissante. Ces remarques mettent en évidence un modèle d'humanité transformée et hybridée par les biotechnologies et les technologies de l'information principalement.

Le corpus met en évidence un phénomène intéressant. Le vocabulaire par lequel sont produites ces chimères est un vocabulaire de l'image et de l'imaginaire, mais pas seulement : les prédictions font aussi bien appel à des imaginaires de fiction (comme Superman) qu'à des faits réels identifiés comme symptômes de l'hybridation. Par exemple Tiger Woods et son opération des yeux pour obtenir une vue d'aigle, la greffe du visage d'Isabelle Dinoir, les exosquelettes motorisés HAL, les prothèses

---

[10] « Bientôt des athlètes génétiquement dopés », entretien avec G. Dine, *Libération*, 28 août 2004.

[11] « Des athlètes génétiquement modifiés sont-ils déjà sur le podium ? », *Science et Avenir*, juillet 2009.

[12] Ibid.

[13] « Athlètes génétiquement modifiés, à vos marques », *Libération*, 25 mars 2002.

[14] « Le champion du futur : Stock, le début de l'aventure », *L'Equipe*, 3 janvier 2001.

[15] Pièces et Main d'Oeuvre, « Nanotechnologies : et maintenant, le tsunami de la communication », 17 octobre 2006.

[16] « Le dopage génétique se profile à l'horizon 2012 », *Les Echos*, 21 août 2008.

[17] Dominique Babin, PH1. *Manuel d'usage et d'entretien du post humain* (Paris: Flammarion, 2004).

[18] « Les surhommes et nous », *Le Figaro*, 31 juillet 2006.

[19] Patrick Couvreur et Jean-Claude Mounolou, « Diagnostic et thérapie, la révolution des nanotechnologies, » *Vivantinfo*, juin 2005.

commandées par le cerveau de Claudia Mitchell, la puce RFID injectée dans le bras de l'ingénieur Kevin Warwick sont convoqués comme autant de signes de la transformation de l'humain, parfois au même titre que des phénomènes imaginaires. Les intervenants dans les débats sur le dopage génétique s'appuient sur et reconduisent un effet de continuité entre des innovations perçues comme relevant de la science-fiction et des images de science-fiction.

Par ailleurs, ce ne sont pas seulement des « êtres » chimériques qui sont discutés, mais aussi les types de caractéristiques augmentées. Il est ainsi question d'amélioration de l'endurance, de surproduction de globules rouges (qui favorise la régénération des muscles), d'*enhancement* de la taille et de la mémoire, d'amélioration des humeurs, d'accroissement des capacités physiques et cognitives, d'amélioration du transport de l'oxygène, d'augmentation des capacités cardiaques, d'optimisation du développement des fibres musculaires, de meilleure gestion de l'énergie métabolique, de mémoires plus développées et d'humeurs plus heureuses, de lutte contre les troubles de la mémoire liés à l'âge, de la capacité d'apprendre rapidement le mandarin pour un homme d'affaire qui souhaiterait travailler en Chine[20], de longévité de 120 ans, du vieillissement comme maladie susceptible d'être guérie, de la capacité de gravir plus facilement et sans oxygène les sommets les plus élevés grâce à une plus grande aisance respiratoire, de modifications corporelles récréatives (implants, tatouage, piercing).

On voit ainsi qu'au cours des débats sur le dopage génétique, de multiples chimères sont imaginées ou désignées, à la fois sous la forme de nouveaux êtres dotés d'autonomie et sous la forme de nouvelles caractéristiques attribuées à ces êtres à venir. Les discours catalysent ainsi les imaginaires et toute une série de prédictions sur ce que sera, devrait devenir ou risque de devenir le corps humain. A partir de là, je ferai deux remarques.

L'appareillage biopolitique à venir

La première porte sur le cadre d'inscription de ces chimères. Il serait tentant de retenir seulement la dimension spéculative et abstraite de ces discours. Or, ce qui apparaît dans le corpus, c'est que ces prédictions de chimères n'existent pas indépendamment d'un discours sur l'appareillage technologique et institutionnel (réel ou imaginé) sans lequel les chimères ne pourraient exister. Les acteurs sociaux qui discourent sur les athlètes génétiquement modifiés, les « super-soldats » ou les « souris

[20] « L'homme augmenté, » *Encyclopédie de l'Agora*, novembre 2008.

schwarzennegger » ne le font pas sans réfléchir à la fois aux techniques qui peuvent ou pourraient les rendre possibles et aux institutions ou aux actions qui permettraient d'encadrer la création de tels êtres (ou de s'y opposer). Par conséquent, le futur et la spéculation sur le futur est ici un moyen de se représenter des usages des technologies, et de penser leur contrôle ou leur inscription.

Les participants évoquent tout d'abord régulièrement des technologies ou des formes techniques : l'implémentation de bras artificiels ou de prothèses sur le corps humain, la greffe du coeur, les implants sur le visage, etc. Ces techniques peuvent être abordées de façon générale et imprécise, ou au contraire à travers des descriptions relativement développées, comme dans cette description du C-Leg :

> le C-Leg va peut-être changer les choses. Il s'agit d'une prothèse de genou électronique, qui offre bien plus de possibilités que les genoux mécaniques actuels. Mi-hydraulique, mi-cybernétique, il assure le verrouillage du genou et la proprioception. Des capteurs placés dans le tube jambier mesurent 60 fois par seconde les forces exercées pendant la marche ; ces mesures sont instantanément traitées par un logiciel fonctionnant sous Windows. Le moteur est alimenté par une batterie de type GSM, et malgré la batterie et l'huile, le C-Leg est moins lourd qu'un genou mécanique.[21]

Encore une fois, certaines des opérations techniques et médicales discutées sont réelles tandis que d'autres sont imaginaires, considérées de façon plus générale ou prospective. De cette manière, les manipulations génétiques considérées en général côtoient le contrôle des humeurs par l'usage de drogues, les molécules capables d'améliorer temporairement mémoire et concentration telles que la choline, le modafinil, la ritaline, ou d'agir sur le développement musculaire, comme la myostatine et la myogénine, ou encore la possibilité d'inculquer des caractéristiques génétiques à un embryon à partir d'un diagnostic pré-implantatoire.

Mais ces formes techniques sont elles-mêmes fréquemment considérées dans un cadre politique plus global. Les acteurs sociaux impliqués dans les débats sur le dopage génétique ne se cantonnent pas à imaginer des formes de vie : ils pensent en même temps les techniques et les institutions de ces formes de vie. Un appareillage politique complexe voit ainsi le jour à travers les discours, nécessaire pour permettre le développement de ces

---

[21] « Le futur, c'est maintenant. Les prothèses pour sportifs handicapés, » *Equilibre*, mai 2008, 30-31.

nouvelles performances et libertés : des instituts de triage permettant d'obtenir des garçons ou des filles, des banques de sperme, ou encore le « marché de gènes » proposé par Robert Nozick :

> Robert Nozick est allé jusqu'à proposer un "marché de gènes" où les parents seraient libres de commander le programme de leur enfant sans que leur soit imposé un design unique pour toute la société. Le grand avantage d'un tel supermarché, précise-t-il, c'est qu'il serait à l'abri de toute décision centralisatrice qui limiterait le choix à un seul (ou à plusieurs) type humain. [22]

De même, de nouvelles institutions apparaîtraient pour soutenir ce développement, par exemple dans le monde sportif :

> Peut-on imaginer qu'un jour il n'y aura plus un seul Championnat NBA, mais deux : l'un garanti cent pour cent humain, l'autre réservé aux athlètes génétiquement modifiés ? C'est bien possible, et il sera intéressant de voir lequel de ces deux Championnats intéresse le plus le public. Personnellement, je pencherais plutôt pour le Championnat génétiquement modifié, parce que c'est là que les performances les plus incroyables auront lieu. [23]

Une observation attentive des discours sur les chimères technologiques provoquées par les nanotechnologies, telles qu'elles sont imaginées dans le cadre du débat sur le dopage génétique, montre que les chimères ne sont donc qu'un élément de représentation du monde technologique tel qu'il est en train de se constituer, avec ses régulations, avec ses normes. Et au cours de ce travail d'appropriation discursive, les acteurs sociaux cherchent à cadrer le déroulement de l'innovation autant qu'à l'imaginer.

Science fiction et science facts

Ma seconde remarque porte sur la tension qui existe entre ces discours et la perception qu'ont certains acteurs de leur inadéquation à la « réalité » du travail des chercheurs. De nombreuses personnes, au cours de ces débats, insistent sur la nécessité de distinguer « *science fiction* » et « *science facts* ». Cette exigence est appuyée sur l'idée que ces constructions de chimères sont assez éloignées du « vrai » champ de la « vraie » recherche. Or, cette entreprise de distinction n'existe que comme

[22] « L'homme augmenté, » *Encyclopédie de l'Agora*, novembre 2008.
[23] « Le champion du futur : Stock, le début de l'aventure, » *L'Equipe*, 3 janvier 2001.

une tension qui ne peut jamais être résolue. Bien entendu, la production de discours sur les chimères et la recherche en nanotechnologies sont deux choses très différentes. Mais chose plus importante, ce sont aussi deux modes de construction de l'objet « nano » dans l'espace public. Des modes concurrents si l'on veut, et fréquemment (pas tout le temps : fréquemment) entremêlés. Il faut en finir avec l'idée réductrice qu'il y aurait d'un côté un espace mass-médiatique abritant les spéculations sur l'hybridation homme-technique, et de l'autre le travail quotidien des chercheurs. Ces deux espaces sont poreux. Et les chercheurs eux-mêmes recourent à un mode de construction « imaginaire » dans la présentation qu'ils font de leur travail, et dans l'appréhension qu'ils en ont. Ceci a été rappelé plus haut avec l'exemple de Drexler et de ses assembleurs moléculaires. On peut citer aussi le cas de Richard Smalley, chimiste et prix Nobel, grande figure de la recherche nanotechnologique, qui est à l'origine d'une controverse avec Drexler au cours de laquelle il l'accusait de réaliser des projections farfelues, tout en reconnaissant que ces projections sont en grande partie responsables de certaines des directions qu'il a donné à son travail de chercheur.

Enfin, on peut citer le cas de François Berger, médecin et neuroscientifique rattaché au projet de clinique expérimentale Clinatec (Grenoble). Au cours d'une intervention à Lille en 2010, Berger dénonçait les visions transhumanistes, tout en reprenant certains éléments du vocabulaire que les transhumanistes contribuent à diffuser dans le domaine. Il s'interrogeait par exemple sur les « cyborgs » que seraient des personnes au cerveau assisté de puces électroniques. Et les scientifiques qui sont amenés à parler de leur travail, par exemple sur la question de la thérapie génique, ont fréquemment recours à des prédictions, qui ne sont pas seulement des projections inscrites dans le temps de la recherche.

Dans cette perspective, il apparaît clairement que les chimères des débats nanos, d'une part, sont l'un des « modes d'existence » publics des nanos, et d'autre part, sont, au moins en partie, entremêlées avec la recherche.

Œuvres citées

Babin, Dominique. PH1 : *Manuel d'usage et d'entretien du post humain.* Paris: Flammarion, 2004.

BEDE (Biodiversité : Echange et Diffusion d'Expériences) (2009) *Bang ou la convergence des technologies. Nanotechnologies et artificialisation du vivant*, rapport. http://www.bede-asso.org.

Catellin, Sylvie. « Nanomonde : entre science et fiction. Quelles visions du futur », *Alliage* 62, (2008) « micro & nano », avril : 67-78.

—. « Le recours à la science-fiction dans le débat public sur les nanotechnologies : anticipation et prospective. » *Quaderni* 61, (2006) « La fabrique des nanotechnologies », Automne : 13-24.

Couvreur, Patrick et Jean-Claude Mounolou, « Diagnostic et thérapie, la révolution des nanotechnologies. » *Vivantinfo*, juin 2005.

Crichton, Michael. *Prey*. New York: Avon Books, 2003.

Drexler, Eric K. *Engines of creation: The coming era of nanotechnology*. New York: Doubleday, 1986.

Dupuy, Jean-Pierre. « Pour une évaluation normative du programme nanotechnologique. » *Annales des Mines* (février 2004).

Encyclopédie de l'Agora, « L'homme augmenté, » novembre 2008.

Equilibre, « Le futur, c'est maintenant. Les prothèses pour sportifs handicapés, » mai 2008, 30-31.

Felt, Ulrike, ed. *Taking European Knowledge Society Seriously*. Report of the Expert Group on Science and Governance to the Science, Economy and Society Directorate, Directorate-General for Research, European Commission, 2007.

Joly, Pierre-Benoît, ed. *Démocratie locale et maîtrise sociale des nanotechnologies. Les publics grenoblois peuvent-ils participer aux choix scientifiques et techniques ?* Rapport de la Mission pour la Métro (22 septembre 2005).

L'Equipe, « Le champion du futur : Stock, le début de l'aventure, » 3 janvier 2001.

Le Figaro, « Les surhommes et nous, » 31 juillet 2006.

Les Echos, Pièces et Main d'œuvre. « Nanotechnologies : et maintenant, le tsunami de la communication. » 17 octobre 2006. « Le dopage génétique se profile à l'horizon 2012. » 21 août 2008.

Libération, « Athlètes génétiquement modifiés, à vos marques, » 25 mars 2002.

—. « Bientôt des athlètes génétiquement dopés, » entretien avec G. Dine, 28 août 2004.

Maestrutti, Marina. « Prendre au sérieux la fiction. la mise en débat des nanotechnologies. » *Alliage* 62, « micro & nano » (avril 2008).

Roco, Mihail and William S. Bainbridge. *Converging technologies for improving human performance: Nanotechnology, biotechnology, information technology and cognitive science*. Arlington: U.S. National Science Foundation, 2002.

Science et Avenir, « Des athlètes génétiquement modifiés sont-ils déjà sur le podium ? » juillet 2009.

Selin, Cynthia. "Expectations and the Emergence of Nanotechnology." *Science Technology Human Values* 32, no. 2, (March 2007): 196-220.

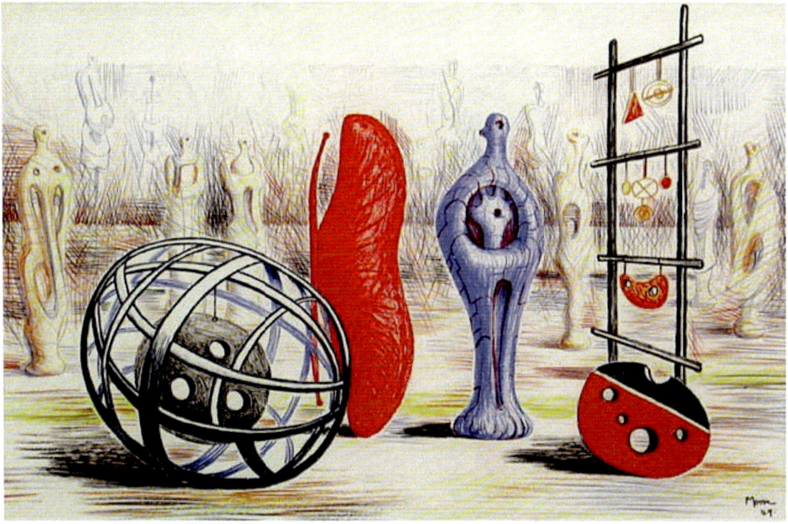

ILL. A 3 : Henry Moore, *Sculptural Objects*, lithograph, 1949 (University of Dundee Museum Services, D'Arcy Thompson Zoology Museum Art Collection. Presented by the Art Fund, 2012. Copyright Henry Moore Foundation.) Reproduced by permission of The Henry Moore Foundation.

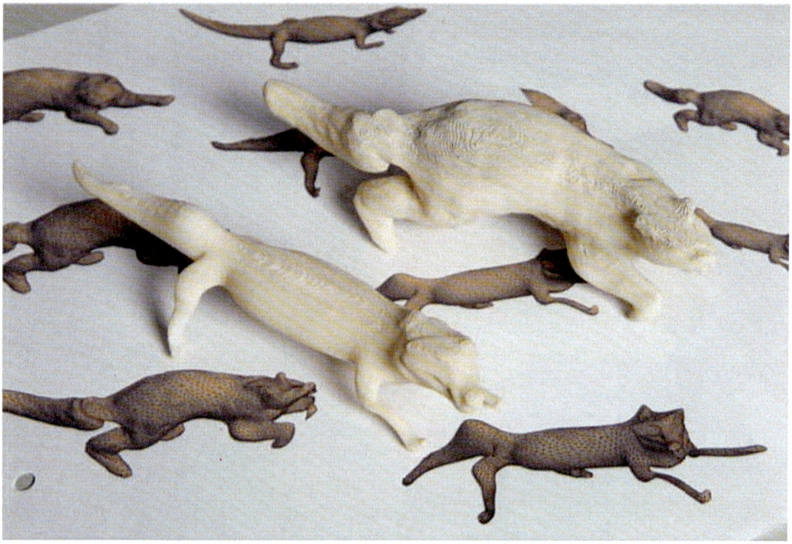

ILL. A 4 : Bruce Gernand, *Coded Chimera—Cat Crocodile Morphs*, rapid prototype models with digital prints, 2012 (University of Dundee Museum Services, D'Arcy Thompson Zoology Museum Art Collection. Presented by the Art Fund, 2012). Copyright the artist.

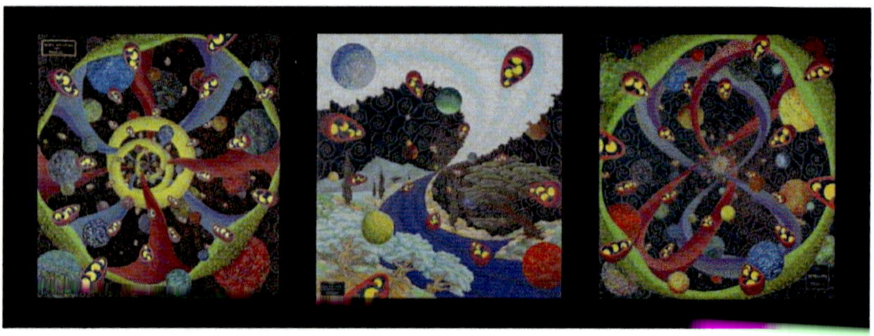

ILL. D 3 : François Miglio, *Triptyque EXPLORATION I*: *Holistic Vision* (H 85cm x L 80cm) – *Unions of Matrices* (H 85cm x L 80cm) – *Gemmation* (H 85cm x L 80cm)

PART D:

ARTISTIC VOICES

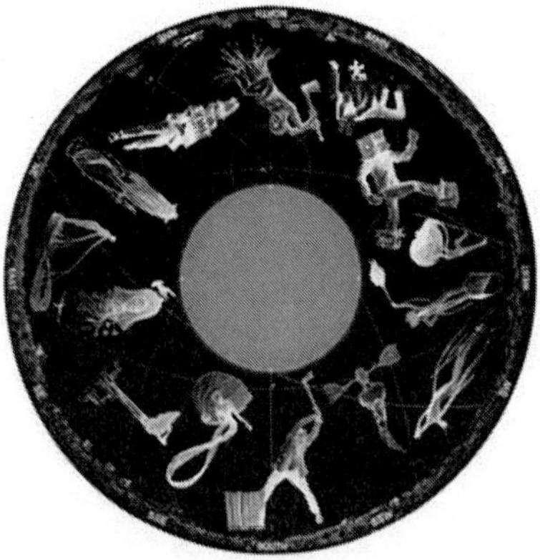

MAI BORÉAL
14 constellations en rêve et réalité

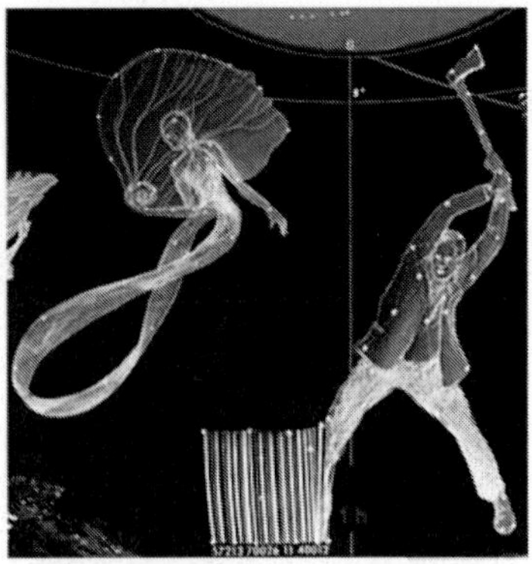

Détail sud, de gauche à droite :
Constellation de L'Homo-Möbius dite Homme solidaire.
Constellation de l'indignation.

ILL. D-1 : Hervé Delamare, Détail de « *MAI BORÉAL 14 constellations entre rêve et réalité*», mai 2011. Composition numérique, dimension variable.

SECTION I

FIGURES MOUVANTES : AUTOUR DE L'EXPOSITION « REVE DE CHIMERES »

LAURENCE ROUSSILLON-CONSTANTY

> Chacun d'eux portait sur son dos une énorme Chimère, aussi lourde qu'un sac de farine ou de charbon, ou le fourniment d'un fantassin romain. Mais la monstrueuse bête n'était pas un poids inerte ; au contraire, elle enveloppait et opprimait l'homme de ses muscles élastiques et puissants ; elle s'agrafait avec ses deux vastes griffes à la poitrine de sa monture ; et sa tête fabuleuse surmontait le front de l'homme, comme un de ces casques horribles par lesquels les anciens guerriers espéraient ajouter à la terreur de l'ennemi.
>
> —Baudelaire, *Chacun sa Chimère*[1]

De la fin du Moyen-âge à l'époque contemporaine, dans les bestiaires comme dans les livres de la nature, au fronton des cathédrales et jusque dans les espaces d'exposition contemporains, les chimères sont partout. Tantôt associées à l'univers fabuleux des légendes et fruits de l'imaginaire collectif ou rationalisées comme autant de preuves de l'existence de Dieu et des imperfections de la nature, elles traversent la littérature, la science et les arts de toutes les époques : dans les récits de fiction bien sûr, où les héros sans cesse doivent en triompher pour prouver leur bravoure, mais aussi dans les manuels de médecine, et jusque dans les cabinets de curiosités des amateurs d'art. A chacun sa chimère, dit l'adage, et chaque époque recèle de nombreux exemples qui illustrent la prolifération de ce motif à la croisée des arts et des savoirs.

---

[1] Charles Baudelaire, *Œuvres complètes, IV, Petits Poëmes en Prose.* Paris : Michel Lévy Frères, 1869, 18.

A défaut de pouvoir proposer une étude panoramique sur les occurrences et les diverses interprétations à donner à ces chimères de tous les temps – tentative éminemment chimérique s'il en fût – l'objectif du présent article est de prolonger la réflexion sur les chimères amorcée dans l'exposition « rêve de chimères », présentée à l'Université Paul Sabatier en juin 2011.[2] A travers quelques exemples d'images et de figures particulières empruntées à des époques très éloignées que nous avons eu la chance de pouvoir exposer, il s'agira ici de montrer la continuité entre quatre grands champs de savoirs que l'on mêlera volontairement : la littérature, la mythologie, la médecine et l'esthétique.

Chimère, qui es-tu ?

La Chimère (*khimaira* qui signifie « jeune chèvre » en grec) est une créature composite dont la légende se diffusa en Grèce dès le VIIème siècle avant notre ère. Née de l'union de la vipère Echidna et du monstre Typhon, elle est représentée comme un quadrupède à trois têtes (une de lion, une de chèvre et une de serpent) ou dotée d'une queue de serpent et d'un corps de chèvre. Dans tous les exemples classiques où elle apparaît, la Chimère est un être qui comporte dans son organisation des parties qui appartiennent à une espèce et d'autres qui appartiennent à une autre espèce.

Hésiode la décrit comme une créature terrifiante, immense, au pied rapide et dont l'haleine est une flamme ardente (*Théogonie,* v.319)[3]. On dit qu'elle habite en Lycie où elle est vite associée à un volcan. Longtemps réputée invincible, elle finit par être abattue par le héros Bellérophon, aidé de Pégase. La Chimère présente tous les attributs du Dragon : le feu, l'hybridité et la nature reptilienne. Dès l'Antiquité et dans toutes les civilisations, le combat contre la Chimère devient rapidement un mythe qui se retrouve dans toutes les mythologies. Chaque culture l'adapte à sa façon et chaque période le réinterprète en fonction de ses préoccupations. Dans ces récits, la chimère incarne la peur que l'on a du mal à se figurer autrement que comme monstrueuse, protéiforme et elle apparaît au héros mythique comme une illusion. Ainsi Virgile reprend-il par exemple cette idée dans l'épisode VI de l'Enéide dans lequel on voit Enée entrer dans le palais des Ombres :

---

[2] Un aperçu complet des livres exposés lors de l'exposition sont consultables sur le site d'Hervé Delamare, http://h.delamare.free.fr/Frameset.htm.

[3] Hésiode. *La théogonie*. Traduction de H.J.G. Patin (Paris: G. de Chamerot, 1872), 14.

> In medio ramos annosaque bracchia pandit
> ulmus opaca, ingens, quam sedem Somnia vulgo
> vana tenere ferunt foliisque sub omnibus haerent.
> Multaque praeterea variarum monstra ferarum,
> Centauri in foribus stabulant Scyllaeque biformes
> et Centumgeminus Briareus ac belua Lernae
> horrendum stridens, flammisque armata Chimaera,
> Gorgones Harpyiaeque et forma tricorporis umbrae.[4]

Confronté à ces formes monstrueuses hybrides que sont la Chimère, la Gorgone et le Centaure, le héros tente de faire preuve de force, comme autrefois Bellérophon. Cependant la Sybille de Cume avertit Enée et le met en garde : ces « images vaines et vides » ne sont que des vies sans corps, de pures projections sans âmes et rien ne sert de les affronter. Avertir (*monere*), mettre en regard tout autant que *porter au regard* : tous les pouvoirs dont la Chimère est dotée dans ces récits mythologiques classiques nous renvoient aux fonctions que désigne l'étymologie même du mot « monstre » auquel la créature se rapporte.

En effet, comme le soulignent de nombreux auteurs depuis Cicéron, le sens même du mot « monstre » justifie son nom, « car il annonce l'avenir et renseigne sur la volonté des dieux »[5]. Faisant partie de la famille des monstres, la Chimère a elle aussi comme fonction d'avertir et de mettre en garde, mais également de conserver la trace, la mémoire d'un passé douloureux.[6] Au Moyen-âge, la Chimère en vient à représenter le signe du paganisme ou du Mal pour la pensée chrétienne. En tant que signe, ou « hiéroglyphe » à déchiffrer, la Chimère fait alors l'objet d'une réflexion théologique et philosophique, à l'instar du monstre ou du prodige.

Dans la *Cité de Dieu*, Saint-Augustin s'interroge par exemple sur l'appartenance des « monstres et prodiges » à l'humanité et tente de rapporter leur existence à une causalité d'ordre transcendant. Dans son

---

[4] Virgile, *Enéide*, VI, 282-289.

[5] Cf. Festus, *De significatione* verborum et Cicéron, *De natura Deorum*, II, 3.

[6] Cf. le commentaire de Jean Clair dans son récent ouvrage, *Hubris* : « *monere*, avertir, prévenir, mettre en garde.Mais *monere*, c'est aussi, de manière tout aussi inattendue, et d'un sens apparemment contradictoire, conserver le souvenir, la trace, la mémoire. C'est l'injonction du tombeau, de l'inscription, de la statue. De *monere* dérive le *monumentum.* Le monstre et le monument ont donc partie liée dans l'évocation commune d'un passé lointain, mythique ou fabuleux, qu'il faut tout à la fois conserver mais dont il faut aussi se garder. » Jean Clair, *Hubris* (Paris:Gallimard, 2012), 9-10.

approche, la diversité ou la variété de la nature sont à interpréter soit comme un signe de la grandeur du Créateur, soit comme une preuve de l'impossibilité pour l'homme de comprendre les desseins divins et d'en interpréter les signes. Dans ses *Confessions*, Saint-Augustin réfute ainsi l'idée du contre-nature. S'inscrivant à cet égard dans la lignée d'Aristote ou de Cicéron pour qui les prodiges n'existent pas mais sont seulement des phénomènes naturels, il souligne que toutes les créatures, mêmes les plus inhabituelles, sont issues du premier homme façonné par Dieu.

En ce sens, et comme le remarque Anne-Isabelle Bouton-Touboulic, Saint-Augustin « prétend neutraliser [la] catégorie [des hommes monstrueux] au nom d'une définition de l'homme comme « animal raisonnable mortel » ; ces monstres participent à la beauté de l'univers que Dieu « tisse » par la ressemblance ou la diversité des parties »[7]. Cette justification théologique de l'existence des créatures hybrides ou chimériques se retrouve tout au long de l'histoire de la Chimère et nous permet d'effectuer une première distinction entre d'une part, les créatures hybrides que l'on trouve manifestement dans la nature, et celles qui ne se trouvent que dans les récits ou fables et sont des produits de l'imagination « des peintres et des historiographes » (Paré). Dans le même temps, et avec la Renaissance, on passe également du nom propre de Chimère au nom commun chimère qui en vient à désigner tout être composite imaginaire ou idée fausse. Cette évolution du sens du mot vers un terme générique et la confusion récurrente entre une interprétation littérale de figures mythologiques et une lecture symbolique de ces mêmes figures expliquent le croisement entre plusieurs discours, scientifique, théologique et littéraire.

L'exemple le plus célèbre d'une telle conflagration de sens se trouve dans les écrits de l'une des figures les plus éminentes de l'histoire de la médecine, auteur d'un très célèbre ouvrage sur les monstres et les prodiges parmi lesquelles figurent toutes sortes de chimères : le chirurgien et médecin Ambroise Paré.

Paré, un œil médical face aux chimères

Lorsqu'il publie en 1573 son ouvrage *Des monstres et prodiges*, Ambroise Paré est un chirurgien reconnu et l'auteur de plusieurs ouvrages

---

[7] Anne-Isabelle Bouton-Touboulic, « Monstres et merveilles dans les *Confessions* de saint Augustin », *La beauté du merveilleux*, ed. Aurélie Gaillard et Jean-René Valette (Bordeaux : Presses Universitaires de Bordeaux, 2011), 58.

destinés à « enseigner les jeunes chirurgiens »[8]. Le livre des prodiges fait d'ailleurs partie d'un ensemble de deux traités de chirurgie et se pose d'emblée comme le produit des observations d'un homme habitué à la pratique et qui a « eu le moyen de voir et connaître (…) tout ce qui peut être d'altération et de maladie au corps humain (…). »[9]

Par ailleurs, Paré est inévitablement un homme de son temps et l'on repère aisément dans son discours qu'il est l'héritier d'une tradition tératologique aux sources très diverses. Comme le souligne très justement Gisèle Mathieu-Castellani dans sa préface à la réédition de l'ouvrage, Paré entremêle dans sa description des monstres trois types de discours, allant du discours herméneutique le plus ancien au discours médical ou scientifique le plus récent, en passant par le discours « rationaliste » aristotélicien. L'ouvrage s'ouvre sur les treize causes de monstruosité repérées par Paré où l'on voit d'emblée que l'auteur n'opère aucune distinction claire entre les anomalies d'ordre naturel et les manifestations d'ordre surnaturel ou celles produites par une imagination excessive. De même, la frontière entre l'humain et l'animal est sans cesse franchie, révélant l'hésitation de Paré à trancher sur la nature de ces êtres chimériques tantôt du côté de l'humain, tantôt du côté de l'animal. Souvent, un attribut commun relie d'ailleurs les êtres décrits, comme si un organe ou une partie du corps servait à engendrer une véritable cosmogonie où règnent le composite, l'hybride. Contrairement aux auteurs dont il s'inspire (tels Belon ou Rondelet), Paré se soucie peu de taxinomie car pour lui, « […] ce qui constitue l'ordre du monde, c'est, par-delà les différences de leurs parties, le réseau serré d'analogies qui les rassemble dans l'unité.[10] »

De même, au niveau de la narration tout autant que de l'image, l'ouvrage est en grande partie fondé sur la reprise de légendes ou de motifs connus. Aussi, comme le remarque Jean Céard, le livre des *Monstres et des Prodiges* emprunte à tant de sources nombreuses et variées qu'il constitue un « remarquable essai de synthèse, qui, par bien des aspects, renouvelle la conception classique que les médecins se faisaient des

[8] Préface au discours de la licorne, 47. Pour une approche interdisciplinaire de la figure d'Ambroise Paré, cf. *Ambroise Paré, une vive mémoire,* ed. Evelyne Berriot-Salvadore. Paris: De Boccard Edition-Diffusion, 2012.

[9] Avis au lecteur de l'édition de 1575.

[10] Jean Céard, *La nature et les prodiges* : l'insolite au XVIe siècle (Genève: Droz, 1996), 309.

monstres.[11] » Dans l'énoncé, le discours de Paré est sans cesse émaillé de formules qui dénotent à la fois le scepticisme de son auteur et sa réticence à dénoncer en bloc la crédulité de ses contemporains.

Ainsi, lorsqu'il dépeint au chapitre III le monstre de Ravenne, fameusement décrit depuis sa naissance en 1511 ou 1512,[12] Paré construit son récit sur le modèle de ses prédécesseurs, dans un procédé que détaille Gisèle Mathieu-Castellani ainsi:

> Reprenant pour modèle de description le texte de Conrad Lycosthènes (l'éditeur du *Prodigiorum Liber* de Julius Obsequens), *Prodigiorum ac ostentorum Chronicon (anno domino 1511)*, publié en 1557, déjà allégué par Boaistuau au chapitre 40 de ses *Histoires Prodigieuses*, Paré donne le *Pourtraict d'un monstre merveilleux* au chapitre III (…), comme « autre preuve » témoignant de l'ire de Dieu.[13]

En parallèle, l'illustration qui figure dans l'ouvrage[14] reprend, à quelques variations près, les nombreuses gravures connues de l'hybride humain.[15] Cependant, comme l'attestent les différentes gravures du monstre figurant dans les ouvrages de Lycosthènes[16] (ou de Pierre Boaistuau[17]), Paré s'écarte ici d'une interprétation religieuse ou herméneutique en supprimant les signes potentiellement prémonitoires ou religieux que représentent la lettre et la croix présentes sur le thorax de cette chimère humaine. Même s'il retient les caractéristiques physiologiques de la créature (la bisexualité, l'animalité à travers la présence des ailes et d'une corne au front), Paré se garde de conclure à une

[11] Ibid. 293.

[12] Né en Italie en 1511, cet enfant se distinguait par « une corne », « deux ailes, un seul pied semblable à celui d'un oiseau de proie », « la présence d'un œil à la jointure du genou, et son hermaphrodisme ». Comme le rapporte Pierre Boaistuau, la naissance de l'enfant sera interprétée comme le signe tangible, incarné d'une malédiction divine : quelque temps après sa mise au monde, les troupes de la Sainte Ligue italienne succombent devant l'armée française et l'on impute leur défaite à l'hybride progéniture qui devient pour le parti victorieux le symbole des mœurs sodomites du peuple transalpin.

[13] Ambroise Paré, *Des monstres et prodiges* (Genève: Fleuron, 1996), 21.

[14] Pour un aperçu de cette image, cf. ILL. D-2, extrait du livre d'Ambroise Paré, *Les Oeuvres... avec les portraicts & figures de plusieurs monstres*. Paris: veuve Cl. Rigaud et Cl. Obert, 1633.

[15] Cf. ILL. D-2 où l'on aperçoit une illustration extraite du livre de Fortunio Liceti, *De Monstris… Amsterdam*, Andreas Frisius, 1655.

[16] Conrad Lycosthènes, *Prodigiorum ac ostentorum chronicon*, (Bâle: H. Petrus, 1557).

[17] Pierre Boaistuau, *Histoires prodigieuses* (Paris: Robert Le Mangnier, 1566).

corrélation étroite entre cette naissance insolite et les événements ayant eu lieu à Ravenne à l'époque de son apparition. Alors que *Monstres et Prodiges* reprend nombre de croyances et superstitions liées aux chimères, c'est une créature chimérique qui fournit à Ambroise Paré l'occasion de dénoncer les fausses idées véhiculées par les médecins et autres « charlatans » de son époque : en rédigeant son discours sur la licorne, Paré prend ainsi le risque de mécontenter ses contemporains.

Figure fabuleuse s'il en est, la licorne nous fait ainsi passer des « monstres biologiques » à un bestiaire exotique d'animaux chimériques qui occupent non seulement les traités médicaux mais envahissent les carnets des géographes et des voyageurs. Circulant entre texte et image, relayée par les récits des explorateurs, la licorne condense tous les récits et se trouve investie des pouvoirs les plus divers et les plus surprenants. Avec elle et à partir des nouvelles actions miraculeuses qu'on lui attribue, chaque époque semble réinventer une nouvelle généalogie, sans toutefois s'écarter d'un « fond commun » d'archétypes et d'images[18]. Oscillant entre plusieurs territoires, entre histoire naturelle et récit légendaire ou ésotérique, la licorne semble cristalliser tous les désirs et les espoirs de l'humanité.

Bête à cornes

Animal fantastique sans doute le plus ancien[19], la licorne (du latin *unicornis*, « qui n'a qu'une corne » équivalent du grec *monokéros*) possède un caractère ambivalent : solitaire et intrépide, elle est tantôt censée incarner la pureté, tantôt réputée posséder un esprit démoniaque et impitoyable. Comme pour la plupart des êtres chimériques, son aspect varie selon les récits et les époques : Ctésias[20] rapporte que c'est un âne

---

[18] Dans le chapitre sur l'unicorne du *livre des êtres imaginaires*, Borges note ainsi une grande constance dans la description de la licorne qu'il nomme « unicorne », si bien que « [l]a première description de l'Unicorne coïncide presque avec les dernières», Luis Borges, *Le livre des êtres imaginaires* (Paris: Gallimard, 1987), 215.

[19] Selon André Leroi-Gourhan, on trouve dans la très célèbre salle des taureaux de la grotte de Lascaux la première représentation iconographique connue d'une licorne.

[20] Ctésias de Cnide était le médecin (Ve/IVe siècle) d'Artaxerxès II *Mnémon* et un historien de l'Inde et de la Perse.

sauvage qui vit en Inde. Il aurait la tête pourpre, le corps blanc, les yeux bleus et une corne au front. Les auteurs classiques (Aristote, Pline l'Ancien et Elien) lui donnent des aspects divers : cheval, âne, cerf…ou porc ! Parfois, la licorne est assimilée au camphur, un autre animal fabuleux amphibie. Ses qualités morales varient également au gré des auteurs et selon les contextes dans lesquels elle apparaît.

Dans la Bible, la licorne est plutôt maléfique: trop grande, elle n'a pu trouver refuge dans l'Arche de Noé. Parfois, on la décrit dangereuse pour les autres animaux (y compris le lion, par exemple), qu'elle attaque avec sa corne. A partir du XIIème siècle, on la compare au Christ : on ne peut la prendre que lorsqu'elle s'endort sur les genoux d'une fille très pure. Sa mise à mort par les chasseurs la rapproche davantage du Christ.

En outre et au croisement entre médecine et magie, la corne de l'animal est considérée comme un remède contre les morsures et les poisons. Réalisée à partir de défense de narval, la potion magique est préconisée dans de nombreuses pharmacopées comme dans *Le marchand sincère, ou traité général des drogues* du botaniste et droguiste Pierre Pomet (1658-1699).[21] Au chapitre 2 de cet ouvrage de référence de l'époque, intitulé *De la Licorne*, Pomet ne nie pas l'existence de la fameuse licorne « que les Naturalistes nous dépeignent sous la figure d'un Cheval ayant au milieu du front une corne en spirale, de deux à trois pieds de long ». Bien plus, il agrémente son texte de plusieurs illustrations mettant en scène les créatures fabuleuses de façon à conforter les légendes que l'on véhicule depuis de nombreux siècles à leur sujet : au plus haut de la page, deux licornes se font face comme pour combattre, attestant le caractère belliqueux de l'animal. La légende de ces deux licornes indique qu'il s'agit du camphur que nous avons déjà évoqué et que mentionne Ambroise Paré dans son discours sur la licorne[22].

Plus bas, une licorne à deux cornes, également identifiée par Paré sous le nom de Pirassoipi[23] est représentée de profil, et dans la même posture que les licornes à deux cornes des bestiaires médiévaux anglais ou français. Contrairement à Paré qui met en doute l'efficacité de la « poudre

---

[21] Cf. ILL. D-2 pour un aperçu de la licorne dans le livre de Pierre Pomet, *Le marchand sincere, ou, Traite general des drogues.* Paris: chez l'auteur, 1695.

[22] Pour d'autres images représentant les licornes au combat, voir par exemple le manuscrit de l'Histoire du Grand Alexandre, f°257 v°, XVème siècle (Paris: Musée du Petit-Palais). Plusieurs images de ces scènes sont reproduites dans le remarquable ouvrage d'Ariane et Christian Delacampagne, *Animaux étranges et fabuleux ; Un bestiaire fantastique dans l'art.* Paris: Citadelles & Mazenod, 2003.

[23] Dans son livre, Paré décrit cette licorne à deux cornes comme vivant en Arabie (Livre XXI, Ch. LVI. Des venins et piqueures de bestes venimeuses).

de licorne » utilisée par de nombreux médecins, Pomet recommande l'usage de la poudre qu'il vend lui-même sous le nom de « corne de licorne » et décrit comme une poudre issue du « narval » islandais. Corne ou dent de narval, la corne de licorne se vend à Paris et dans toute l'Europe dès le XVIème siècle et jusqu'au XIXème siècle comme l'atteste l'estampe du tableau du peintre Le Nain, *Le Marchand de Corne*[24]. Au début du XXème siècle, on trouve encore trace de cette croyance et des efforts de la communauté scientifique pour lever définitivement le voile sur le commerce associé à ces pseudo-cornes de licorne, comme l'illustre par exemple un des comptes-rendus de la société d'histoire de la pharmacie datant de 1927 où l'on s'intéresse à la véritable nature d'un objet présenté comme une corne de licorne (et finalement identifié comme une dent de narval)[25].

Enfin, si la fortune iconographique de la licorne est abondante et amplement documentée, il faut également noter que la licorne apparaît en littérature dans de nombreux poèmes et récits où son existence se trouve confirmée ou raillée, et parfois présentée de façon ambivalente. Chez Rabelais, par exemple, elle est décrite à plusieurs reprises (dans le quart-livre et dans le cinquième livre) et de manière légèrement différente, ce qui reflète peut-être les hésitations de l'auteur à croire en son existence. Comme le souligne Paul Smith dans son article « Rabelais et la licorne », tantôt Rabelais s'appuie sur les récits des naturalistes pour décrire la créature fabuleuse, tantôt il présente une prolifération d'animaux extraordinaires, ce qui tend à mettre en doute ses pouvoirs et jusqu'à son existence même[26]. Dans tous les cas, et à toutes les époques, il est certain que la licorne suscite une incontestable fascination, symbolisant tour à tour la croyance aux êtres surnaturels, la recherche d'une forme d'idéal ou de pureté, une forme d'utopie.

---

[24] Le Nain, Bibliothèque nationale de France, département Estampes et photographie, RESERVEFOL-QB-201(152).
Notice de recueil : http://catalogue.bnf.fr/ark:/12148/cb41090066d.

[25] A.R.Pratt, « Corne de Licorne » : Communication de M.A –E Pratt, à la 19ème séance de la société, *Buletin de la société d'histoire de la pharmacie,* 15ème année, N.55, 1927, 1-2.

[26] Paul J. Smith, « Rabelais et la licorne », *Revue belge de philologie et d'histoire*, tome 63, fasc.3,1985, 477-503.

Epilogue

A travers les nombreuses représentations de la licorne dont nous n'avons pu offrir qu'un aperçu aussi partiel que partial, il semble que ce soit finalement cet aspect-là que l'on puisse retenir : sorte de chimère suprême, la licorne incarne en dernier lieu l'utopie, qu'elle désigne une contrée exotique ou un modèle de société idéale, un antidote à la peur de la prédation ou de l'extinction qu'incarnerait à l'inverse le dragon sous ses différentes formes. Comme le signale Michel Beaujour à propos de Rabelais[27], la licorne, par son aspect protéen, métamorphique, serait alors à considérer comme un symbole de l'illusion littéraire toujours renouvelée et de l'imagination créatrice à l'œuvre. C'est ainsi que les chimères de tout poil survivent aux avancées de la science et des connaissances, renaissent à chaque fin de siècle et ressurgissent des tréfonds de l'imagination fertile des créateurs, prêtes à adoucir nos angoisses les plus sombres et à enchanter nos rêves les plus fous.

Œuvres citées

Baudelaire, Charles. *Œuvres complètes, IV, Petits Poëmes en Prose.* Paris: Michel Lévy Frères, 1869.

Berriot-Salvadore, Evelyne. *Ambroise Paré, une vive mémoire.* Paris: De Boccard, Edition-Diffusion, 2012.

Beaujour, Michel. *Le jeu de Rabelai*s. Paris: Gallimard, 1969.

Borges, Jorge Luis. *Le livre des êtres imaginaires.* Paris: Gallimard, 2005 (1987).

Boaistuau, Pierre. *Histoires prodigieuses*, Paris: Robert Le Mangnier, 1566.

Céard, Jean. *La nature et les prodiges*. L'insolite au XVIe siècle. Genève: Droz, 1996 (1977).

Clair, Jean. Hubris. *La fabrique du monstre dans l'art moderne*. Paris: Gallimard, 2012.

Delacampagne, Ariane et Christian. *Animaux étranges et fabuleux ; Un bestiaire fantastique dans l'art.* Paris: Citadelles & Mazenod, 2003.

Festus, Pompeius S. *De significatione verborum*, 1471.

Gaillard, Aurélia et Jean-René Valette. *La beauté du merveilleux*. Bordeaux : Presses Universitaires de Bordeaux, coll. "Mirabilia", 2011.

Hésiode. *La théogonie.* Traduction de H.J.G. Patin. Paris: G. de Chamerot, 1872.

---

[27] Michel Beaujour, *Le jeu de Rabelais* (Paris : Gallimard, 1969).

Leroi-Gourhan, André. *Le Geste et la Parole, 1. : Technique et langage, 2: La Mémoire et les Rythmes*. Paris: Albin Michel, 1964-1965.

Lycosthènes, Conrad. *Prodigiorum ac ostentorum chronicon*, Bâle: H. Petrus, 1557.

Paré, Ambroise. *Des monstres et prodiges*, Genève: Fleuron, 1996.

—. *Discours de la Momie et de la Licorne*. Paris: Gallimard, 2011.

Pratt, A. R. « Corne de Licorne » : Communication de M.A –E Pratt, à la 19ème séance de la société, *Bulletin de la société d'histoire de la pharmacie,* 15ème année, N.55, 1927, 1-2.

Smith, Paul J. « Rabelais et la licorne », *Revue belge de philologie et d'histoire*, tome 63, fasc.3, 1985, 477-503.

Virgile, *Enéide*. Tome 1, édition bilingue de Jean-Pierre Chausserie-Laprée. Paris: La Différence, 1993.

ENTRETIEN AVEC HERVE DELAMARE, PLASTICIEN[1]

Dans l'exposition que vous avez présentée à Toulouse, les média utilisés étaient très différents, allant du dessin à l'impression numérique. Y a-t-il une grande différence entre ces deux approches ?

La technique, quelle qu'elle soit, est un moyen d'arriver à une fin. Il m'arrive souvent d'utiliser les outils numériques en amont d'un projet de peinture ou de sculpture. Il est vrai que pour *Mai boréal, 14 constellations entre rêve et réalité,* la chose s'est inversée. L'ensemble des 14 dessins se retrouve dans une composition organisée grâce à l'outil Photoshop. L'efficacité de cet outil s'exprime dans la version en négatif du même travail. Le résultat est toujours un ensemble de dessins, mais blanc sur fond noir, dégageant plus de tension et surtout plus proche d'un effet de constellation. Je suis en général plutôt attiré par la matière mais l'objectif prime, donc de nouveaux choix s'imposent.

Le 'disque' des constellations fait tantôt penser à un œil géant, à un CD ou à un kaléidoscope. Quelle était l'inspiration de ce travail ?

J'aime toujours quand le spectateur a sa part, quand il peut projeter son imaginaire. J'ai une affection pour le format italien dit *tondo*. De 2003 à 2005 cela a même été l'un des axes de travail dans ma peinture. Pour « *Mai boréal ...* » la forme du disque est pour l'essentiel inspirée des représentations classiques du ciel étoilé. Le fait d'avoir évidé l'anneau est né du nombre des constellations imaginaires que j'avais sur la table de l'atelier. Deux groupes de sept constellations que j'ai orienté nord et sud, induisant de ce fait cette composition en anneau.

Dans le prolongement de ce travail, je réalise actuellement une petite série de sept détails repris de la version en négatif de « *Mai boréal ...* » les encadrant dans des ipods, chacun d'eux étant reliés à un escargot.

---

[1] Les œuvres d'Hervé Delamare ont été exposées à la Bibliothèque universitaire de l'Université Paul Sabatier (Toulouse 3), lors de l'exposition *Rêve de chimères* (1er au 28 juin 2011).

L'Ipod-escargot, c'est la métaphore de la décroissance ?

Oui, mais pas seulement. Ce serait là la façon de souligner la nécessité d'adopter dans nos sociétés occidentales un rythme plus lent, plus organique. C'est une façon d'exprimer une utopie de la lenteur face à celle du capitalisme révolu. Il y a moyen de brancher de la lenteur sur la technologie. L'idéal serait justement de se détacher de l'immédiateté et de retrouver une forme plus organique.

D'ailleurs, dans votre travail, on retrouve souvent des antennes ou des racines – toutes formes qui suggèrent une ramification infinie, un ancrage souple dans la terre ou vers le ciel.

C'est vrai, cet imaginaire de la dichotomie végétale ou animale, du réseau, du filament est présent dans mes recherches depuis un moment déjà. Je pense à la série de peintures *Lagune*, aux sculptures de la série *Hybride*. Ce vocabulaire de formes est né je crois d'un ensemble de créations autour du thème de l'abeille dans les années 2000. La réflexion artistique autour de la figure humaine et de cette idée que l'avenir de l'homme peut se situer dans cette alliance avec le végétal m'intéresse beaucoup. La problématique de la chimère, que l'on associe souvent à l'hybride humain/animal est ainsi dépassée par l'idée d'une chimère végétale plus pacifique et positive. Une sculpture illustre cela : *Sur ses racines*. Il s'agit d'une grande silhouette très élancée de quatre mètres de haut, constituée de fils de fer rouillé entremêlés, marchant d'un pas franc sur ses racines. Ses quatre membres se prolongent en arborescences jusqu'au sol.

Lorsque l'on voit votre travail de sculpture, on ne peut s'empêcher quand même de penser à deux maîtres du genre, Giacometti et Brancusi qui proposent des formes plutôt raides mais également allongées. Est-ce qu'ils ont joué un rôle dans votre imaginaire ?

Tout à fait, ce sont deux mentors. *Sur ses racines* illustre bien l'influence de Giacometti. Brancusi est une influence plus récente, notamment dans mes séries plus abstraites intitulées *Germe* ou *Élixirs*. Les sculptures de Giacometti font quant à elle depuis longtemps partie de mon imaginaire et lorsque j'étais aux Beaux-Arts, je me souviens d'une série de petites sculptures qui mêlaient le langage de Giacometti et un travail graphique que nous réalisions sur les cours de nus.

A travers la légèreté ou la danse ?

Probablement, car la danse a également joué un rôle majeur dans ma façon d'envisager le mouvement. Je vis avec une chorégraphe depuis plus de 15 ans.

Est-ce que la danse renvoie pour vous au féminin ?

Oui, sans aucun doute. Lorsque j'étais étudiant, l'école d'art avait initié avec la maison de la culture et le Centre Chorégraphique National du Havre un trimestre entier ayant pour thème la danse contemporaine. Nous étions entre autres immergés dans l'univers de Bouvier/Obadia. Je me souviens avoir été très emballé par un groupe de danseuses assises sur des strapontins, adossées et dispersées sur un mur géant ! Au milieu de cette architecture géante, elles dansaient également sur un sol constitué d'une épaisse couche de terreau. Cette image m'a marqué et je l'associe à un certain éveil de la sensualité chez moi dans lequel la danse est forcément associée au féminin. Dans ma vie comme dans mon travail des mères de substitution m'ont souvent guidé et m'ont permis d'avoir une ouverture culturelle considérable. Je pense comme le poète que la femme est l'avenir de l'homme.

Comment le partenariat avec Total s'inscrit-il dans votre travail ? Est-ce que cela n'entre pas en contradiction avec la notion d'artiste engagé que vous défendez ?

Pour le moment j'essaye de m'engager dans la vie tout simplement, ou comme dirait Desproges : « je suis un artiste dégagé ». Forcément, cette collaboration m'interroge et me renvoie à la question de la place de l'artiste dans la société. Ce n'est pas le premier partenariat, cela m'intéresse toujours d'aller y voir de l'intérieur, ont y apprend toujours une quantité de choses sur le monde du travail, sur l'humain, sur les dogmes de telle ou telle entreprise ... C'est vrai celui-ci s'est construit en six mois, à un moment où parallèlement j'ai vécu cet 'éveil de la conscience', non pas que je n'en avais pas en écoutant souvent dans l'atelier France inter en début d'après midi, mais en découvrant dans le détail comment fonctionnent l'argent et le capitalisme aujourd'hui. Ce n'est pas le genre de truc qui attire en premier un artiste, mais le sujet est des plus présents. Et comme le chante Alain Souchon : « il faut voir comme on nous parle ». Malgré ses milliards de bénéfice Total est un grain de sable dérisoire dans un gigantesque casino qui ne connaît aucune limite, aucune morale, aucune peur, aucun camp.

Je préfère prendre de l'argent à Total que de lui en donner trop souvent à la pompe pour me rendre chaque samedi pour consommer dans une ZAC sans savoir pourquoi. Participer à un projet financé par la multinationale est également un bon moyen d'affronter nos interrogations : dans quel monde vit-on ? De découvrir et d'essayer d'analyser le discours interne. Dans ce partenariat, l'anneau mathématique de Möbius est un symbole fil-rouge de ma proposition. Pour moi il constitue un angle critique : l'anneau de Möbius est ainsi à la fois l'évocation des rubans de Total, la façon dont la pétrochimie parvient à optimiser et pérenniser les transformations pour obtenir le meilleur carburant et tout un ensemble d'autres produits et sous produits que nous avons tendance à oublier, et ils sont pléthore ! Mais l'anneau est également un terrible symbole de permanence, de quelque chose qui tourne en boucle, fermé sur lui même. J'ai toujours été attiré par le paysage industriel, qui est pour moi le théâtre de notre dramaturgie. C'est même un thème que j'ai commencé à interroger à partir de 1989, sous l'angle de la friche désertée.

Malgré tout, vous vous souhaitez donc du côté des indignés ?

Oui, et cela, depuis longtemps et justement avec ce travail sur le thème des friches urbaines et sub-urbaines. Tout était parti d'une série de peintures et de photographies sur les cabanes des jardins ouvriers, puis d'une friche immense qui était sur le point d'être rasée le lendemain où je l'ai découverte : *L'usine des Extraits du Havre.* Une ville dans la ville de 5 hectares. C'était la plus vieille usine d'importation de bois exotique du Havre. J'ai fait de cette friche un terrain d'exploration. Il s'agissait dans ce travail de montrer comment l'on était passé d'une activité florissante à une chute. Ce travail fut le point de départ d'une période que j'ai intitulée *Sabi*, un terme traditionnel japonais emprunté à l'essai l'*Eloge de l'ombre* du romancier Junichirô Tanizaki. Ce travail était à la fois une poétique mélancolique et un questionnement sur le monde industriel. J'y introduisais également en permanence les clins d'œil à une célèbre estampe qui m'interpelle toujours beaucoup : la *Mélancolia I,* d'Abrecht Dürer. L'ensemble de cette recherche déclinait ce que l'on nomme en peinture une nature morte de vanité.

Quelle a été votre évolution par rapport à cette esthétique romantique ?

Je me suis d'abord attaché à mon environnement immédiat, et d'autres sites m'ont inspiré des pistes de création. Notamment les jardins ouvriers de Quillebeuf sur Seine, l'usine des extraits tinctoriaux et tannants du Havre, l'histoire des destructions successives de cette ville, son quartier portuaire dit de L'Eure. Les paysages ferroviaires sur la ligne Le Havre Paris. L'usine à clous de Duclair. La friche en construction de la Faculté des sciences du Havre. La ville de Bolbec. L'usine Dresser du Havre ... Le travail s'est clôturé sombrement avec une série de peintures reprenant en l'actualisant le sujet de la nature morte, de la vanité. On retrouvait par exemple le crâne humain sur un tapis roulant d'usine posé sur un miroir à côtéd' un évier remplit d'une vaisselle en attente.

Comme *Le Penseur* de Rodin, que je vois comme une émanation de la gravure de Dürer, ce travail sur la friche est comme l'octaèdre mystérieux de la *Mélancolia*. Il souligne l'aspect sombre de l'humanité. Avec *Mai boréal, 14 constellations entre rêve et réalité*, j'aborde le côté sombre mais j'envisage également un pendant plus optimiste à ce projet.

Est-ce qu'il y aura un Mai austral ?

Sans doute, et j'espère qu'il y aura d'autres ciels étoilés. Dans les constellations que j'ai faites, il y a pour le moment plus de place laissée à la critique qu'à l'innovation positive. J'espère que les prochaines créations seront l'occasion d'offrir une vision harmonieuse de l'avenir, les mouvements de l'économie sociale, des villes en transitions ou des monnaies locales complémentaires sont porteurs de cela. En résumé je partage l'optimisme d'Albert Jacquard, tout n'en étant pas dupe de la vanité humaine.

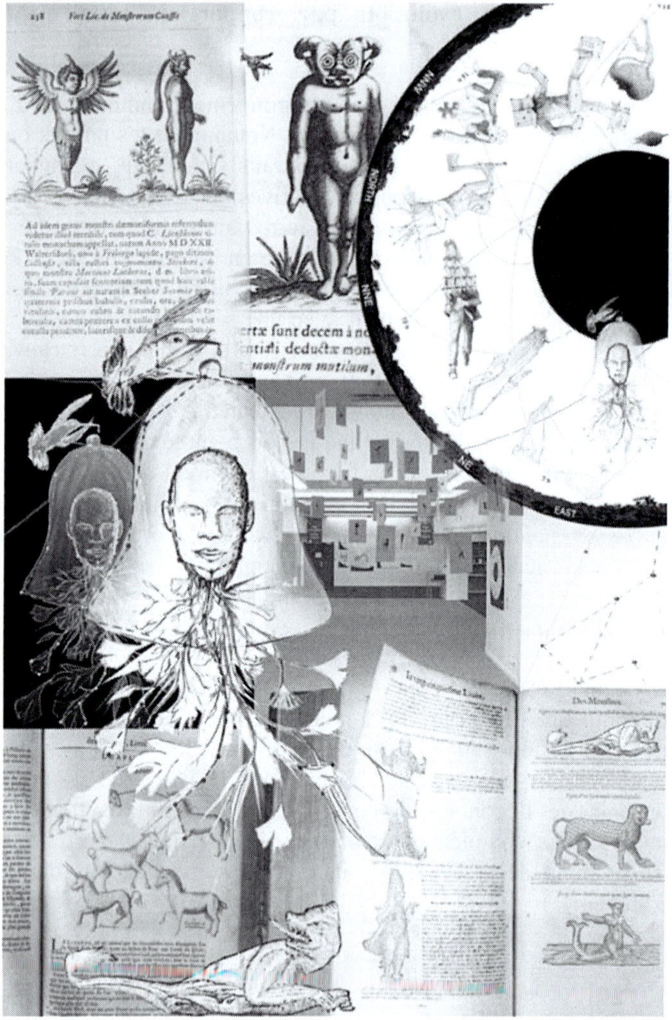

ILL. D-2: Detail from the "Rêve de Chimères" exhibition, B.U. Sciences, Université Paul Sabatier, June 2011. Reproduced by permission of the artist and the Université Paul Sabatier SCD.

FRACTALS IN MYTHS, LEGENDS AND FABLES

FRANÇOIS MIGLIO

Invisible
You reveal yourself, at last
unsuspecting world
Both from the same cradle
Restricted by the same boundary
By the fractal door
Men of Endless doubt
Will enlighten
Your somber light

Introduction

I am convinced that a better understanding of complexity depends upon our ability to have dialogues that are both horizontal and trans-disciplinary.

I consider myself a humble self-made artist. My testimony is based on intuition, and my artistic approach is founded in contemplation and poetry. Both my work and my approach are fueled by a deep questioning about the real and the sensitive world. My words will be those of the artist, not those of an erudite scholar. I'm also accustomed to saying that my paintings are doors that may open onto a threshold for others to view, but which remain forbidden to me. However, it is through my artistic experience, and mostly in my spontaneous and earliest pictorial representations, that my work has found an inseparable resonance with the fractal world. And this world appears to me as a rich territory to be explored.

Discussion

The fractal world has become more and more familiar to us. Who would not be puzzled when looking at the computerized representation of the Mandelbrot's Set? Although Mandelbrot was the first to name and define the mathematical structure of the fractal world, and to reveal its beauty as a purely mathematical construct, fractal representations by human kind are very ancient. We are surrounded, constituted and determined by the fractal world, which is visible all throughout the universe. These dynamic fractal structures explain a number of behaviors of the living and non-living worlds, at the border between order and chaos.

Could this omnipresent world have been left unseen to the first human eyes? Indeed, the Australian Aborigines first represented fractal patterns about 45,000 years ago, and images belonging to Parietal Art (about 35,000 years ago) all are inspired by fractal patterns.

These representations and fractal expressions have accompanied the history of mankind through its different cultural languages, at the crossroads of art and the sacred: the repetitive forms of music, songs, dances, paintings and architecture, and also in the repetitive structures of Mandalas, Zen gardens, Celtic knots, Hopi spirals, Arabesques and Rose windows.

In the nineteenth century, long after the intuitive and spiritual expressions of fractal patterns had begun, science revealed the interior structure of the fractal world with a new geometry. And today, fractal mathematics is used in various domains such as biology, economics, fluid mechanics, demography, and communications. How do we find meaning in chaos?Do myths, legends and fables go beyond reality? Are they simple interpretations of reality? Or, do they represent a part of reality itself? If they do, what is that part? How do we establish a link between the fractal world and complexity? And what do myths, legends and fables teach us? What has my artistic experience taught me?

The theory of emerging properties shows that the structure of nature is the same as the process of writing: it is based on simple elements that combine into new ones whose properties are different. The dynamic structures of fractals are one of the languages of nature and their chaotic and ordered expressions represent its different styles. Humankind does not stand apart from the fundamental laws of nature. Men also assemble simple things to create more complex and more sophisticated ones, whose properties are different. And when humans blow the dust of charcoal to put the outlines of their hands on the rock, they create a representation of their presence in the cosmos.

Does not the history of myths, legends and fables start as an imitation of this fundamental natural law? From the earliest times, we have known that these expressions are not still or static. Their very nature is dynamic, and thus, the oral tradition is constantly evolving as the result of iterative narration. Couldn't we consider these expressions as new elements with new properties and the result of additions of other elements? Isn't their structure the result of telling, which itself is the result of ideas and images that were built by countless stories that the story-teller remembers, that the audience enriches when they take part and thus, contribute to the story? When myths, legends and fables are created, it is through a complex process that also unfolds the complex adventure of the history of mankind entangled with the complex history of the universe.

Whatever we may call it – a play-within-a-play, images repeated to infinity as with mirrors, or a 'Chinese boxes' structure, - this process has, undoubtedly, always been perceived by humans. What remains interesting is the adventure of complexity in these forms of expression – that they represent the cosmos as the initiator of surprise, of creation, of chaos.

Man is surrounded by fractal expressions of nature that keep whispering to him in a language defining a passage from stability to chaos and from chaos back to stability. Man has naturally and spontaneously expressed or represented what he saw: spirals, waves, repetition, and self-similarity. Shamanic rituals, repetitive dances and musical rhythms, trances and other hallucinatory practices may be interpreted as man's attempts at echoing his chaotic environment while trying to explore the realm of the gods, of the dead, and dreams…in other words, of chaos itself. Do not myths, legends and fables draw part of their imaginative world from these practices?

This is why linkages to fractal patterns may be observed in numerous tales, legends and myths, such as, for example: *The Arabian Nights*, the myth of Sisyphus, the myth of the Flood and Noah's Ark also describes the passage of order into chaos and chaos back to order. Like a demiurge, Noah disrupts the original order, choosing separate, disparate elements that he re-assembles on the ark, which becomes the matrix. In doing this, Noah creates order from the chaos of the flood.

Through the representations of a complex and chaotic world, do not myths and legends teach us how to survive?

The possible content of fractal symbols does not provide ready-made answers to everything but it does give us a better understanding of natural structures, principles and behaviors so that we may be able to live within the complex and chaotic harmony of our environment as well as of ourselves. Why, for thousands of years, has man had an intuitive understanding of the chaotic and complex world that scientists and mathematicians have only recently explained rationally?

We've just seen that some myths, legends and fables describe models of exponential growth leading to chaos. In the middle of the nineteenth century, and with a simple equation, Verhulst explained the growth of a model which, after a certain number of forks, or " bifurcations ", reaches a point where it starts behaving in a chaotic manner, constantly "hesitating" between destruction and infinite growth.

What has my artistic career taught me?...A fractal intuition.

Indeed, as I proceeded to a sort of retrospective of my childhood drawings, I realized that right from the beginning they contained spontaneous representations of fractal patterns. Of course, the question of the presence of Man in the universe has always been one of my major concerns. And the sensations, the emotions, the data that were refracted by the prism of a dreamy contemplative spirit constituted my imaginary world very early on. But it is worth noticing that fractals – spirals, labyrinths, swirls, and arabesques – spontaneously emerged into my drawings.

Now, the freedom of expression I've been able to maintain from academic constraints has revealed a link with many other pictorial forms of expression. The fact that fractals have always been represented throughout times and cultures, the fact that they have been spontaneously expressed by children, or by people having mental issues, or by "art brut", that they've been enriched by sacred symbolism and aesthetic virtues, make them the most widespread and universal form of expression.

On the one hand, we can draw parallels between our conscious thoughts behaviors, subconscious mind, and forms of expression. On the other hand, the basic structures of nature constitute not only the natural environment, but also all of our anatomic systems - circulatory, skeletal, respiratory, digestive and nervous systems. Our lungs, arteries, veins are perfectly fractal as of course are our brains. This may suggest that Man has adopted an unconscious fractal behavior that may well be compared with Nature's, and that can also be replicated mathematically.

Conclusion

Such a fractal resonance demonstrates that Man's intuitive understanding of Nature is much deeper and more profound than has generally been admitted or understood: "Truth sometimes lies beneath the surface of things" . Over the years, I've tried to find meaning in the fact that my drawings and paintings seem to emerge spontaneously from my pencils, pastels, and tools, and finally show an interior world that is mine, and yet, at the same time, totally goes beyond my own being to be part of something universal. Through my understanding the fractal world, I have established a link between my imaginary world and the real world around me. I also think that, while using different languages, artists, mystics, and scientists all share a common heritage and obey the same laws of nature when it comes to the creative and expressive processes. Natural laws, including the fractal world, are the links that we can see between art, science and the sacred.

SECTION II

MOTHS AND FISHES, NARRATIVES OF ADAPTATION

MARGARET DRABBLE

Bad science can inspire good books, and impossible and scientifically naive or illiterate speculations may create or enforce profound myths. Mary Shelley's *Frankenstein* and H.G Wells's *War of the Worlds* and *First Men in the Moon* are memorably predictive although the scientific framework of all these tales is naïve. Wells had a sound scientific training in zoology and biology under the the celebrated evolutionist T. H. Huxley, the champion of Charles Darwin, but it in no way inhibited his wilder imaginative leaps into evolutionary end-of-the-universe distances. It gave him a launch pad, and he soared. Jules Verne's fantasies were scientifically more plausible than those of Wells, although he was not as good a scientist, but the power of his fables comes from somewhere else.[1] Zola's views on heredity are notoriously fanciful, but as a narrative device they work powerfully within the framework of his fiction. A writer doesn't need scientific credentials or plausibility to make good use of science.

We assume a natural hostility, a failure of communication, a latent aggression between what we label 'the sciences' and 'the humanities', but this is largely a symptom of an evolutionary struggle for survival, of a struggle for status and funding. The real interrelationship between the disciplines is of course much more complex and interesting than this.

Before I embark on a personal account of my own experiences of using scientific concepts both as inspiration and framework- as I hope I may- I feel I must pay tribute to what seemed an important moment in the history of the science-humanities debate. In 1959 an English novelist called C. P. Snow delivered in Cambridge a well-remembered lecture called *The Two Cultures and the Scientific Revolution*, in which he contrasted the culture of what he called 'literary intellectuals' with that of scientists, and described the increasing gulf between them. Literary people, he claimed, were scientifically illiterate, and he chose as his test case their ignorance

[1] I recently re-read his novel *The Mysterious Island*, which is weirdly predictive about volcanic eruptions and ash clouds- a strange mix of Robinson Crusoe and an ecological dystopia.

of the second law of thermodynamics- an illustration he came to regret, as it aroused much mockery in the savage attacks he received from the literary establishment, particularly in Cambridge.[2]

Snow later said that he should have chosen as his illustration the discovery of DNA, a simpler concept which did not 'involve serious conceptual difficulties... it needs very little mathematics to understand... There are few parts of the hard sciences which one can understand so much without mathematical training. What one needs most of all is a visual and three-dimensional imagination, and it is a study where painters and sculptors could be instantaneously at home.'[3] Snow continued: "this branch of science is likely to affect the way in which men think of themselves more profoundly than any scientific advance since Darwin's."[4] And he was right. DNA created a new paradigm, a new mythology.

It's worth noting that Snow was a research scientist at the Cavendish Laboratory in Cambridge before becoming a successful civil servant and a serious writer in a solid realist middle brow mode- he wrote a multi-volume *roman fleuve*, entitled *Stranger and Brothers*- , and he was of course abused as a failed scientist by the offended literary critics whose domain he attacked.

One of his novels, *The New Men*, gives an authentic firsthand account of the ethical problems faced by scientists working in the nuclear industry, very much a novel of its period, and one of which few other fiction writers would have been able to conceive. It was not a great novel, and is long out of print, but it told an important story in its time. It was an old-fashioned book about unmanageably and to most of us unimaginably new material. Snow in fact predicted the possibility of a nuclear bomb, in 1939, a suggestion he regretted having made public when the Second World War broke out. Re-reading *The New Men* this month, I was surprised to find as well as warnings about cracked containers and radiation sickness, the very phrase 'Ground Zero'. I had thought this to be of much more recent date. There is also a critical appraisal of the atomic scientists' sense of 'self-congratulatory Awe' not , the 'Shock and Awe' to which George W Bush

---

[2] The second law of thermodynamics was formulated by Baron Kelvin and predicted the death of the world by entropy- it's said to have contributed greatly to late nineteenth century fin-de-siecle pessimism.

[3] And so, of course, it has proved to be. The double helix is one of the most recognisable scientific models of all time, and one of the most suggestive.

[4] C.P.Snow, *The Two Cultures: and a Second Look: An Expanded Version of the Two Cultures and the Scientific Revolution* (New York: Mentor Book, 1963), 69-70.

and Tony Blair introduced us, but Awe, already on its way to 'Shock and Awe.'

Since Snow's day, and perhaps prompted a little by his challenge, science writers have made much greater efforts to communicate with the general public, and writers of fiction- and not just of science fiction- have become more aware of the metaphorical and actual richness of our expanding knowledge of the cosmos and of the mysteries of DNA. We reach across the divide. Just as George Eliot, Elizabeth Gaskell, Zola and Proust were inspired by Humboldt's explorations, by Darwin- there is a great deal of Darwin in Proust- by evolutionary determinism and by entropy, we now look to the double helix and the Hadron collider at Cern for plots and images and interpretations.

In my novel about genetic inheritance, *The Peppered Moth* - which is, in its way, a family saga - I made use of the story of the peppered moth, biston bettularia, a humble moth which has been claimed by entomologists as the only visible proof of evolution in action, of the hypothesis of the survival of the fittest observed in a recorded span of time. In brief, there are two species of this moth, one of which breeds successfully in pale woodland, the other on trees and walls blacked by industrial pollution, and studies of their contrasted populations is said to prove Darwin's theories of natural selection. This story has fascinated me for many years, partly because it is closely connected to the history of industrialisation, which, as a daughter of the steel and coal belt of the North of England, I feel is part of my own biography. I doubled this recent scientific theory with yet more recent research into the matrilineal nature of mitochondrial DNA, a form of DNA which, unlike most genetic material, can be inherited only through the female line. These two strands, twisted together, gave me the thread by which I traced my way back to the lives of my great-grandparents and grandparents, and to the spoiled landscape in which they lived. I read many books on moths and mitochondria, most of which I have forgotten, I interviewed geneticists, I had my own DNA tested in hopes of a Neanderthal connection, I talked with a history teacher whose own DNA can be traced directly back to a nine thousand year old skeleton preserved since the end of the last ice age in the limestone caves of the Cheddar Gorge. In fact it was the story of this teacher, as it reached the press, which first alerted me to the dramatic and narrative potential of mitochondria.

My novel was concerned with inertia, progress, social mobility, questions of genetic determinism, hereditary intelligence, and a host of other matters that make up a family saga, but the thread of mitochondria and the metaphor of the peppered moth gave me a story line, as well as a way of describing my mother's inability to settle and adapt to the new environments which her education, her upwardly mobile marriage, and her frustrated ambitions imposed upon her. The dark moth of the industrial landscape, which clung to the blackened bark of trees and the blackened brickwork of the northern cities, could not survive happily in the cleaner world of the south where its white counterpart thrived. Her accent, her habits, her background betrayed her and exposed her to predators. A scholarship girl at Cambridge, I believe she felt isolated amongst the richer, more middle-class, more sophisticated students. She was a great admirer of Virginia Woolf, but she belonged to a class that Woolf could not imaginatively encompass.

She passed her ambitions and her DNA on to her children, who were able to enter and adapt to a world in which she was never happy. Her children mutated, although at some cost.

My daughter Rebecca Swift published a very good poem about her grandmother, which I used as an epigraph to my novel, and I would like to read it to you.

It's amazing we got that far, loveless
As you were supposed to be, yet suddenly
I have a longing for your tripeish thigh;
Swallows, thronging to the eaves; a teasmade
Playing boring Sunday news and all sorts of
Rites and rituals that seemed notable but
Were really just trips in and out of the bathroom…
Loveless, the daughter said, years later when
The slow-lack peppered in their brains like a dust,
And life had grown as troublesome as thought.
Yet just tonight I am dreaming of your thigh,
And of the unconscious swallows thronging in the eaves.[5]

The Peppered Moth was a dark and difficult novel to write. *The Sea Lady*, which has a marine motif throughout, was much more fun. It took shape in my mind as I was chairing the judges of the Aventis Science Book Prize for the Royal Society in 2003. The submissions were books of popular science on subjects ranging from the cosmos, the environment, climate change and neurology to DNA and the intersexuality of fish. The prize winner in the adult section was a book on left-handedness, which

[5] Rebecca Swift. *Virago New Poets,* eds. Melanie Silgardo and Janet Beck (Virago: London, 1992), 111.

suggested that the universe itself is left-handed, but the children's winner was a beautifully illustrated book on ocean life. I found the books dealing with fish, the oceans and marine life particularly entrancing, for they reminded me of all the happy times we had spent as children on our summer holidays, gazing into rock pools, collecting little creatures in buckets, swimming, playing on the beach. If I had had any confidence in my abilities as a scientist, I would love to have been a marine biologist, and in *The Sea Lady* I had the enjoyment of creating an *alter ego* in the person of Professor Humphrey Clark, whose professional life has been devoted to the underwater world. A champion diver in his youth, as bold and adventurous as Jacques Cousteau, he becomes a distinguished academic, pondering the mysteries of evolution, yet retaining a vaguely mystical, pantheistic feeling for the sea. An ethnographer rather than a geneticist, he feels that seriousness is draining away from his own discipline- as many of us in the humanities have felt that power is flowing from the humanities towards the sciences- and he regrets this change.

> The computer and the DNA revolution were both in their infancy when Humphrey Clark embarked on postgraduate research. The vast fortunes and commercial wars of the patented genome lay in the future. It was still possible, at that time, to be a serious scientist and to adopt a quasi-mystical approach to the natural world. It was still possible to regard the sea as a sea of faith. Not perhaps as a sea of old-fashioned religious faith, but as a sea of faith in the glory and diversity of creation. The old traditions of respect and wonder flourished, in direct descent from those venerable Victorian collectors of snails and butterflies who had filled their cabinets and built the great museums. Those collectors of beetles and bones and the larval cases of caddis flies were his ancestors.[6]

In contrast, his counter part and antagonist, Ailsa Kelman, is still riding high in the buoyant tide of feminism, which she feels she has helped to create. It is a new paradigm- not a scientific paradigm, but a social paradigm, and it is creating a new mythology of monsters and mermaids. Gender, in Ailsa's time, has come under intense scrutiny, as we have come to question our assumptions about the male and the female and the social conditioning that makes us what we are. And yet it too has its scientific aspects, for we fill the environment with gender-affecting hormones, as the fish of our rivers and seas mutate and change sex, as many young men produce lower sperm counts and become less fertile. Humphrey was

---

[6] *The Sea Lady,* 144.

inspired to study marine life for its beauty, but Ailsa awards a Science Book Prize to a book called *Hermaphrodite: Sea Change and Sex Change*, which, she says, is "a survey of gender and sex in marine species... covering bold hypotheses about the evolutionary origins of sexual reproduction, followed by startling revelations about current female hormone level, current male infertility, and rising sexual instability caused by POPs and other chemical hazards..."[7]

I greatly enjoyed inventing this fictitious work of science, and in reading all the books and papers that enabled me to invent it.

I must confess to an appetite for scientific discoveries and hypotheses that I cannot really understand, and I go on reading and searching, perhaps because that is what human beings are programmed to do. Last week I rediscovered a paper titled 'Is Social Evolution daian or Darwinian?' by Geoffrey M Hodgson, which I must once have read with the deepest interest, and probably while I was writing *The Sea Lady*, because, re-reading it, I see have made notes and marks throughout. What was it that so gripped me?

I think by nature I am a hopeful Lamarckian, always hoping that by will and volition and good faith and good intentions we can change our destinies, but we have been told repeatedly that Lamarckian theories cannot possibly be true, and I fear that in my deepest self I believe in a pessimistic Universal Darwinism, as proposed by the ultra-Darwinian of our days, Richard Dawkins. Hodgson's paper attempts a reconciliation of the positions, and points out that 'all sciences embody metaphor...and these metaphors have a naturalistic ambience.' 'Metaphor permeates the most rigorous and realistic statements in science' 'Analogy is not an evasion of truth. On the contrary, we use analogy to get at the truth.' He proposes such fine compromises as this: "Biological evolution could be described as *metaphorically* Lamarckian as well as being *literally* Darwinian."[8] He really really wants to stick with Lamarck.

Reading this kind of material, even as ineptly as I do, provokes many questions about family inheritance and human behaviour. What is the unit of cultural evolution? Is it a gene, or an idea? Why do twins separated at birth find themselves liking the same patterns, the same songs, the same food, and having the same nervous habits?

We all know that we replicate our mothers, for good or bad, to a lesser or greater degree, and worry about whether we are able to pass on to our

---

[7] Ibid., 4.

[8] Geoffrey M. Hodgson, "Is Social Evolution Larmarckian or Darwinian?" in Laurent, John and John Nightingale, eds., 2001.

children or our readers our knowledge, our insights. And can we also free them from, alert them to, our faults?

The interplay of chance and destiny is one of the great themes of fiction. Science tells us that there is no chance, that all is genetically, neurologically determined, cosmically determined from the Big Bang more than six billion years ago. Yet each life story is unique. As a novelist, I have, or appear to have, the freedom to create accidents, incidents, chance encounters that produce love affairs, marriages, children, grandchildren. We track back for causation, but in each novel in progress, as in each life in progress, there is a future which is yet unknown. The story is not over yet and we do not know its plot.

A LA FRONTIERE DU REEL : LES LIMITES DU REALISME DANS LES ROMANS DE MARGARET DRABBLE

CLAUDINE PEYRE

L'itinéraire professionnel de Margaret Drabble est jalonné par des œuvres dans des genres aussi divers que le roman, la nouvelle, la biographie mais aussi, les nombreuses préfaces et éditions de livres, tels que *The Oxford Companion to English Literature* (1985), ouvrage de référence majeur. Plus récemment, elle a écrit un surprenant ouvrage qui mêle et entremêle autobiographie, biographie et histoire des puzzles, *The Pattern in the Carpet* (2009). Cette polyvalence dans l'écriture indique une curiosité, un intérêt pour tout ce qui peut enrichir le récit, ouvrir les perspectives du genre romanesque.

Le roman est une forme littéraire malléable, que l'auteur travaille et façonne, c'est un format perméable aux influences présentes et passées. Les romans de Margaret Drabble sont empreints de réalisme, dans la lignée d'une riche tradition victorienne dont elle se réclame, ils relatent l'histoire d'une narratrice, la plupart du temps et sont ancrés dans le quotidien qui s'intéresse progressivement à un monde de moins en moins confiné géographiquement. L'héroïne, qui n'est chaque fois, ni tout à fait la même, ni tout à fait une autre, devient moins narcissique, regarde ce monde, de sa fenêtre d'abord, prudemment, puis plonge dans la réalité, au fur et à mesure qu'elle mûrit. Et cette réalité s'inspire de la vie quotidienne : politique, faits divers, science s'immiscent, s'infiltrent dans l'écriture et influencent l'atmosphère des récits.

Une définition du réalisme de Margaret Drabble

Dans ses romans, Margaret Drabble utilise le contexte social et historique pour planter le décor. Depuis près d'un demi-siècle, la romancière a traversé, au fil de ses oeuvres, les différents courants de la littérature: elle démarre sa carrière à la fin de l'ère du modernisme et à l'aube du post-modernisme et cette place dans le paysage littéraire et artistique se reflète dans ses écrits par des stratégies d'écriture variées. Le réalisme dont son oeuvre est empreinte est un réalisme conscient qui ne se contente pas d'être un simple effet de miroir mais essaie d'harmoniser réalité et perception individuelle, comme elle l'avouait en 1989: "There is much more self-consciousness on the part of the narrator which dispels the illusion of realism."[1] Dans la lignée des auteurs modernes, la plupart des romans relatent plutôt le conflit de l'être humain avec son environnement que leur rapport harmonieux. La subjectivité du personnage, son portrait psychologique, ses aspirations et ses fantasmes tiennent une place importante, au même titre que la représentation plus neutre du contexte historique et la prise de conscience du rôle de l'écrivain.

Margaret Drabble n'a jamais nié son appartenance à une tradition littéraire et les références textuelles explicites abondent dans ses oeuvres. En revanche, certains de ses textes semblent échapper à une quelconque influence implicite: l'intertexte ne s'y manifeste pas clairement mais peut se dessiner parfois dans des bribes subtiles, dans une stratégie narrative car, selon Roland Barthes:

> Toute trace écrite se précipite comme un élément chimique d'abord transparent, innocent et neutre, dans lequel la simple durée fait peu à peu apparaître tout un passé en suspension, toute une cryptographie de plus en plus dense.[2]

Ce que Barthes appelle « cryptographie » apparaît comme un terme précurseur de « intertexte » et souligne l'importance des liens des textes entre eux. Ces échos ne sont, la plupart du temps, que des traces intertextuelles, mais ils contribuent néanmoins à révéler la structure de l'oeuvre et à souligner l'orientation littéraire de l'auteur. Margaret Drabble se détourne progressivement du réalisme classique par l'intégration dans la narration d'éléments qui marquent une rupture avec la peinture fidèle de la société britannique. En outre, au fil des romans, le lecteur devient partie intégrante du récit, il est privilégié et participe au récit. Interpellé fréquemment, il devient omnipotent: les nombreuses métalepses narratives cassent l'illusion réaliste et jouent sur l'ambiguïté entre auteur et narrateur :

> And then what, then what? Ah, do not ask. How can there be an ending for this most unsuitable of couples? (*The Radiant Way* 294)

---

[1] Claudine Peyre, "Three interviews with Margaret Drabble, 1989, 1992, 2011", *Cercles* 21, 112-130, http://www.cercles.com/n21/peyre.pdf

[2] Roland Barthes, *Le degré zéro de l'écriture* (Paris : Seuil, 1953), 16.

> Spot the one invented story if you can. No prize offered.[3]
>
> Let us move on to Rosemary and Nathan Herz. We have not seen them on their home ground, so let us join them…[4]
>
> You must not imagine me as speaking to you in my own person.[5]

Le lecteur fictif devient un narrataire[6] dont la fonction n'est pas passive: au fil du récit, il peut glisser d'auditeur à interlocuteur par le truchement des sollicitations du narrateur. Un rapport de familiarité se met en place entre le narrataire et le narrateur qui semble, en fait, chercher un subterfuge pour donner un avis ou se débarrasser de personnages secondaires, ou encore provoquer la curiosité. Le lecteur est pris par le narrateur dans un enchâssement de récits. Ce procédé a le double effet de jouer sur l'ambiguïté des rapports auteur/narrateur et de rompre l'illusion réaliste. Le roman s'ouvre alors sur d'autres perspectives.

La trace du roman gothique

Le roman gothique est, selon Ellen Moers, un véritable millefeuille narratif: "In Gothic writings fantasy predominates over reality, the strange over the commonplace, and the supernatural over the natural, with one definite auctorial intent: to scare."[7] Cependant, à première vue, les romans de Margaret Drabble apparaissent fort éloignés du genre gothique. En effet, les cimetières effrayants ou les châteaux hantés des romans d'Horace Walpole ou de Mrs Radcliffe ne font pas partie de l'univers des héroïnes de Drabble et le surnaturel ne se manifeste pas non plus dans leur quotidien. Pourtant, dans Jerusalem the Golden, l'angoisse ressentie par Clara, la jeune narratrice est marquée dans le récit. La maison familiale dont elle cherche absolument à s'échapper représente dans son imaginaire naïf, un lieu lugubre qui lui apparaît menaçant car elle craint de sombrer dans la médiocrité et l'anonymat:

---

[3] Margaret Drabble, *A Natural Curiosity* (London: Viking, 1989), 208.
[4] Margaret Drabble, *The Witch of Exmoor* (London: Viking, 1996), 121.
[5] Margaret Drabble, "Stepping Westward: A Topographical Tale," *A Day in the Life of a Smiling Woman* (New York: Houghton Mifflin Harcourt, 2011), 207.
[6] L'instance à laquelle s'adresse le narrateur.
[7] *Ellen Moers, Literary* Women (London: The Women's Press Ltd, 1978),90

> [...] she feared that if she fell asleep she might lose her determination and her faith, might wake up alone in her narrow bed, in the small back bedroom, overlooking the small square garden, backing onto the next small square garden, where for so many years she had lain and dreamed her subversive dreams.[8]

Dès son jeune âge, Clara s'évade fréquemment de cet espace qu'elle déteste, symbole de l'engrenage de sa destinée, pour se réfugier dans la littérature: "The world of the figurative was Clara's world of refuge. The literal world which she inhabited was so plainly hostile that she seized with ardour any references to any mode of being [...]."[9] Elle développe un goût prononcé pour tout ce qui est extérieur à son univers et rejette le quotidien qui ne lui apporte aucune satisfaction: "She liked the unknown, she liked to be familiar with the unknown."[10]

Dans ce paradoxe, elle se projette vers le futur et parvient à se construire un monde imaginaire qui lui donne le pouvoir de gommer la réalité médiocre de son environnement, une petite ville provinciale, banale et étouffante qu'elle veut fuir. La force qui anime Clara marque la dualité de la femme qui refuse, sans violence, l'espace qui l'entoure et parvient à s'en échapper toutefois grâce à son imaginaire. Ce trait de l'héroïne rapproche la romancière des auteurs de romans gothiques dans lesquels le contexte historique et géographique sert à mettre en place l'imaginaire, vecteur permettant à l'héroïne d'être transportée du monde réel contraignant vers un monde à la fois prometteur et illusoire.[11]

A plusieurs reprises, les premiers romans de Drabble mettent ainsi en situation des femmes qui se construisent un monde hors de la réalité sociale et s'enferment occasionnellement, dans le cocon de l'écriture: les images sensorielles dominent alors les descriptions de paysage avec une sensibilité particulière à la qualité de la lumière et l'envie de dépasser ce qui est perceptible pour aller au-delà de la simple vision des choses:

> The green hill slopes up behind them to the brilliant azure. Large pink lambs, surreal, tinted from the red earth, stand outlined on the hill against the blue. An extraordinary primal timeless brightness shimmers in the hot afternoon air. A slight breeze moves the grass like waves on the water.[12]

---

[8] Margaret Drabble, *Jerusalem the Golden* (Harmondsworth : Penguin Books, 1969), 27.
[9] Ibid., 31.
[10] *Ellen Moers, Literary Women* ., 155.
[11] Ibid. "Traveling Heroinism: Gothic for heroines", chapter 7.
[12] Margaret Drabble, *The Radiant Way* (Harmondsworth : Penguin Books, 1988).

La lumière radieuse évoque une possible puissance divine et la colline symbolise la longue ascension vers la liberté qu'accomplit la femme tout au long de sa vie, réminiscence de cette irrésistible envie d'horizons nouveaux chez les héroïnes gothiques. Le thème de « literary transport »[13] est récurrent dans la littérature gothique féminine. Cette lumière semble donner pouvoir et force intérieure aux narratrices avec toutefois quelques limites à la connaissance :

> They are not to know that the weather of England has changed utterly, and that the summer of '89 will bring them blue skies and unclouded sunshine, an unimagined and unchanging radiance. They will be rewarded, as they walk along the green ceiling of the limestone and by the singing river, with the glory of paradise. It will beat down upon them and reprieve them from mortality as they sit on the short starred sheep-cropped turf [...].[14]

Ce jeu sur le temps de la narration, comme une prophétie de l'instance narratrice omnisciente, illustre le combat de la femme avec le temps. Dans *The Radiant Way*, premier roman d'une trilogie[15] qui introduit la période de maturité dans l'œuvre de Margaret Drabble, la femme se soumet à la force du temps: "Time [...] is an illusion" (*TRW*, 394), alors que dans le dernier volet de la trilogie, *The Gates of Ivory* elle semble maîtriser ce temps, l'éternel ennemi; la narratrice devient désormais « héroïne », au sens propre du mot car elle n'hésite pas à affronter des situations parfois périlleuses et se distingue par sa bravoure.

Dans un souci de vraisemblance, la romancière tempère ce trait psychologique par la peur et la colère, sentiments humains qui, certes, suscitent le courage chez les femmes, mais surtout, déforment la perception et donnent l'illusion du réel. Jouant sur les contrastes, la romancière compense cette faillibilité des personnages en leur octroyant des pouvoirs nouveaux, une sagesse que seules les épreuves et l'expérience donnent à l'être humain : "[…] She has forgotten all about ageing and death. They all have forgotten. They may not be getting any younger, but they are not getting any younger either. Time is an illusion. They can arrest it." (*TGI*, 460)

Avec «The Merry Widow», une nouvelle parue en 1989, l'obsession du temps est également explorée dans un récit illustrant les rapports de la femme avec le temps. Ce dernier est représenté sous la forme allégorique

---

[13] Ellen Moers, *Literary Women* (London: The Women's Press Ltd, 1978), 128.

[14] Margaret Drabble, *The Gates of Ivory* (London: Viking, 1991), 460.

[15] *The Radiant Way* (1987), *A Natural Curiosity* (1989), *The Gates of Ivory* (1991). Hereafter abbreviated as *TRW, ANC* and *TGI.*

d'un vieillard[16], une faux à la main (en réalité un simple jardinier en train de faire son travail), menaçant et effrayant: "He was a terrible old man, gnarled, brown, toothless, with wild white hair."[17] En fait, la narratrice, veuve depuis peu, voit en lui l'image de la mort qu'elle redoute et elle découvre progressivement en elle, la force de l'écarter:

> And it came to her as she stood there in the early evening light that the old man was not Death, as she had feared, but Time. Old Father Time. He is the one with the scythe. She had feared that the old man was Death calling for her, as he had called for Philip, but no, he was only Time, Time friendly, Time continuing, Time healing.[18]

Il est trop tôt pour mourir, la lumière n'est pas encore tombée, cette conviction de la narratrice insuffle, à la fin du récit, une note d'espoir et d'optimisme. Le temps n'est pas un ennemi, il n'est qu'une épreuve que tout individu doit subir.

Outre les fantasmes et les craintes des personnages face à la réalité, d'autres éléments du roman gothique transparaissent dans les romans de Margaret Drabble: le surnaturel, par exemple, fait parfois des incursions dans le récit. Ainsi, dans *The Radiant Way* et *A Natural Curiosity*, Alix apparaît fréquemment impliquée dans des situations risquées, voire périlleuses, au-devant desquelles elle s'avance. Elle poursuit obstinément un objectif, celui de mieux comprendre le cheminement psychologique de Paul, jeune meurtrier pour lequel elle semble éprouver une véritable fascination et se trouve également confrontée physiquement et violemment à la mère de Paul, "the evil Angela" (*ANC*, 192) au cours d'une de ses expéditions solitaires.

The Gates of Ivory repousse les limites du réalisme et la romancière introduit désormais dans ses récits des espaces inhabituels et étranges. L'instance narrative prend ses distances avec le solipsisme des héroïnes antérieures pour s'intéresser davantage à l'histoire, la politique ou l'économie du monde et leur influence respective sur les individus. La première partie du roman entraine le lecteur dans un univers actuel, celui du Cambodge contemporain; la description est confiée d'une part à Liz, la narratrice, et d'autre part à une sorte de témoin-narrateur, parti d'Angleterre deux années auparavant et dont les récits de voyage sont empreints de

---

[16] Figure allégorique de Chronos, dieu du temps tel qu'il est représenté dans le portait d'Elizabeth I en 1610, par exemple :
http://en.wikipedia.org/wiki/File:Elizabeth-I-Allegorical-Po.jpg.

[17] *TMW*, 202.

[18] *TMW*, 204.

subjectivité. Cette rencontre épistolaire fortuite donne d'abord lieu à un voyage imaginaire, que Liz effectue à la manière des héroïnes de romans gothiques.

Au fil des romans, les personnages féminins se révèlent moins statiques en vieillissant et semblent parfois, à la manière des récits épiques, se jeter au-devant de l'inconnu, sortir de leur sphère habituelle, affronter volontairement des dangers, de Londres à Hanoï dans une surprenante obstination à aller jusqu'au bout des missions qu'elles se sont elles-mêmes attribuées.

L'intrusion de l'irrationnel

La distanciation du personnage par rapport à sa sphère immédiate et son besoin de découverte seront le point de départ, le prétexte (et le pré-texte) pour élargir la perspective du roman vers un espace moins restreint que celui de la société britannique, généralement dépeint par Drabble. Malgré son profond respect d'une longue tradition littéraire, Margaret Drabble s'est progressivement éloignée du réalisme classique et de récits centrés essentiellement sur l'étude de la conscience et de l'expérience propres à la femme.

Dans *The Gates of Ivory*, la structure d'enchâssement de plusieurs récits, un pluri-témoignage, une vision élargie du réel est le fondement même de l'énoncé, composé nettement de deux récits différents, l'un transmis par Stephen sous la forme d'un journal de voyage et l'autre, d'une narration présentée par plusieurs narrateurs de façon alternative. Pour la première fois dans son oeuvre, la romancière s'intéresse directement à un monde autre que le microcosme londonien dans lequel elle faisait jusqu'alors évoluer ses personnages. Les déchaînements de l'humanité contemporaine y occupent désormais une place prépondérante: les conflits internationaux sont présents et font ressortir l'immuabilité inhérente à la société britannique. Le roman présente deux mondes parallèles, surnommés métaphoriquement "Good Time" et "Bad Time", aux rapports souvent ambigus:

> Good Time and Bad Time coexist [...]. There is a relationship between Good Time and Bad Time. There are interpenetrations. Some cross the bridge into the Bad Time, into the Underworld, and return to tell the tale. Some go deliberately. Some step into Bad Time suddenly. It may be waiting, there, in the next room.[19]

Ces quelques lignes résument le récit et l'atmosphère inquiétante qui se dégage de la dernière phrase fait aussi figure de présage car la mort est au centre du récit; l'obsession du temps devient progressivement la crainte de la mort.

L'illusion est le fondement même du récit de *The Gates of Ivory*: au contexte réaliste se mêle une vision déformée de la réalité. Les premiers mots invitent le lecteur à éviter la confusion entre fiction et réalité tout en maintenant un doute dans son esprit: “This is a novel-if novel it be [...].” (*TGI*, 3).

Cette courte mise en abyme permet de poser l'ambiguïté narrative qui était annoncée, d'une part, dans la page des remerciements de l'auteur qui affirmait la réalité des faits traités: “I would like to thank all those who helped me with the background material for this book. I hope they will not disapprove of the fictional use to which I have put their facts.” et d'autre part, dans l'avant-propos, avec un extrait du Chant XIX de *L'Odyssée*:

> Dreams, said Penelope to the stranger, may puzzle and mislead. They do not always foretell the truth. They come to us through two gates: one is of horn, the other is of ivory. The dreams that come to us through the traitor ivory deceive us with false images of what will never come to pass: but those that appear to us through the polished horn speak plainly of what could be and will be.[20]

Le jeu de mots d'Homère sur la corne et l'ivoire exprime ici la dualité entre illusion et réalité. *The Gates of Ivory* fait écho à *The Gates of Horn*, un essai d'Harry Levin sur le réalisme traditionnel dont Margaret Drabble s'est progressivement éloignée. La référence à l'oeuvre d'Homère indique également l'importance dans l'étude du personnage de la mythologie comme élément essentiel de la construction de soi. L'irrationnel n'épargne pas non plus les personnages secondaires ; ainsi, Claudio, personnage mineur mais inquiétant de *The Radiant Way*, sur son lit de mort, se révèle d'une nature étrange et peu rationnelle:

---

[19] *TGI*, 3-4.

[20] *TGI*, avant-propos.

> Claudio was [...] still encouraging himself with dreams of hidden power and forbidden knowledge. He spoke of Gorgon and the Medusa and Géricault and Demigorgon and Salome and the Bessi and Thrace. He spoke of a witch he had known in Sofia.[21]

L'apparition de ce personnage, fréquemment qualifié de "wicked", de "satanic", de "werewolf", coïncide avec une abondance de références à la mythologie qui ponctuent les récits de la trilogie.

L'empreinte de la mythologie

L'exploration du passé individuel, la recherche de correspondances entre l'histoire individuelle et le comportement sont des thèmes privilégiés de Margaret Drabble. Cependant, elle apporte une nouvelle dimension à l'étude de la psyché avec l'introduction de plusieurs héroïnes aux points de vue différents. Ainsi, la romancière élargit-elle son analyse de l'être humain (et essentiellement de la femme), en puisant dans l'histoire collective et en s'appuyant sur les théories de la psychologie freudienne. L'ouverture sur l'irrationnel de *The Radiant Way*, contrebalance ainsi la logique scientifique incarnée par Liz, une psychiatre de renom attirée par l'inexplicable:

> [...] she had felt reality to be revealed at times in flashes beyond even the possibility of rational calculation: had felt danger [...] of too much knowledge, of a kind of powerlessness and sadness that is born of knowledge: for these reasons, perhaps, was it that she had decided to multiply the possibilities so recklessly, to construct a situation beyond her own grasping? [...] To be defeated, honourably, by the multiplicity of the unpredictable, instead of living with the power of her knowingness? With the limits of the known?[22]

D'autres personnages éprouvent ce désir de comprendre l'inconnu : Alix, par exemple, qui cherche sans répit une explication aux faits et gestes de personnages en marge de la société, que ce soit Paul, un meurtrier dont les décapitations font figure de rites ou encore Jilly, une jeune détenue amenée à devenir par la suite, à sa sortie de prison, la victime de Paul.

Alix se sert de Paul pour poursuivre sa recherche sur le comportement humain et se trouve malgré, elle, entraînée vers la mythologie celte, source

---

[21] *TRW*, 340.
[22] *TRW*, 9.

d'intérêt pour le jeune meurtrier: “Alix Bowen has got Roman Britain and severed heads on the brain.” (*ANC*, 194)

Sans offrir de réponse rationnelle au comportement du jeune homme, la mythologie propose toute une imagerie de têtes coupées qui a probablement servi de modèle implicite à ses actes:

> [...] Alix described what she had learned of Paul's childhood and background. The father was a butcher, the mother a hairdresser, in a small town in the north Midlands [...]. He had been taken on a school outing at the age of eleven to see the Bog Man of Buller. He had become obsessed by death and human sacrifice. He had devoured books on the Druids and Stonehenge, on the Celts and the Romans, on the old gods.[23]

Ainsi, *A Natural Curiosity* aurait pu être intitulé *A Severed Head*, non par ses ressemblances avec le roman de Iris Murdoch mais plutôt par la multitude de têtes coupées qui jalonnent le récit. En plus des malheureuses victimes du jeune Paul, s'ajoutent des références au mythe grec de la Méduse: la mère de Paul est qualifiée de “monster woman” (*ANC*, 40) ou encore “a fury, a harpy, a gorgon” (*ANC*, 41). Liz imagine même un traité dont le titre serait « *Medusa: Our Hidden Knowledge* » (*ANC*, 276). En outre, la narration se tisse autour d'un enchevêtrement de faits divers, d'éléments historiques, mythologiques et littéraires dans les nombreuses références à une tête coupée ou une gorge tranchée: une belle-mère décapitée, Louis XVI, Marie-Antoinette, la reine Mary, l'empereur Tacite, Titus Andronicus, Hamlet, etc.

Les animaux ne sont pas non plus épargnés dans cette macabre imagerie où on trouve des têtes de porc sur un étal, une tête de cheval décomposée, suspendue au milieu d'un charnier, (peut-être une réminiscence d'un ancien sacrifice romain au dieu Mars?) ou encore un marteau de porte en forme de tête de femme: “The knocker, a female tête coupée, smiled gravely and impassively at her. It was a common design [...]. Alix had never before thought of seeing it as a Medusa, as a Celtic offering.” (*ANC*, 197)

L'allusion récurrente à la Méduse, “the dreadful Medusan image” (*ANC*, 99) et “a sort of bland, mask-like Medusa defiance” (*The Middle Ground*, 8), fait écho à la représentation allégorique du mythe de la Gorgone, sous les traits de Jilly Fox, la femme aux “snake-like tresses” (*TRW*, 339) qui signe ses tableaux par “a twining serpent with a beaked cock's head” (*TRW*, 326) et hante la vie d'Alix. La tête de Jilly,

---

[23] *ANC*, 7.

soigneusement décapitée, se trouve d'ailleurs déposée dans la voiture d'Alix par le meurtrier.

Le réalisme se mêle au pathologique et à l'irrationnel. La proximité de ce cadavre suscite chez Alix, l'héroïne, une angoisse qui ira grandissant jusqu'au moment où, à la fin du récit, le meurtrier sera arrêté à proximité de sa maison. Ce fait divers poussera Alix, mue par une force intérieure, à pénétrer dans les méandres de la vie du jeune psychopathe: "The murderer had come Alix's way, without much intervention on her part. He had followed her, as it were. He and Alix were inextricably, mystically linked." (*ANC*, 2) Quant à la funeste destinée de Jilly, elle était annoncée implicitement dans le texte. En effet, elle écrivait peu avant sa mort: " [...] I shall be with you for ever. There is no death. There is death only for the body [...]." (*TRW*, 275). Elle traduisait sa folie dans des peintures macabres: "Daggers, pierced hearts, severed heads, dripping blood, gaping wounds, severed limbs, floating eyes." (*TRW*, 326)

Ces images psychotiques sont également illustrées par de multiples références textuelles à des oeuvres d'art appartenant à la tradition religieuse chrétienne: Sainte-Agathe, martyre aux seins mutilés, Saint Jean-Baptiste, décapité, un hymne évoquant la crucifixion et le martyre ou encore des renvois à la mythologie avec Persée et la Méduse. Comme dans le mythe de la Méduse, Jilly, la Gorgone, sera finalement vaincue et perdra ses pouvoirs sur Alix: "The eyes were open, and stared. Alix gazed at them, at Jilly's livid face, at the dishevelled hair [...]. Alix stared, and was not turned into stone." (*TRW*, 334) En revanche, le personnage de Jilly réapparaîtra, dans une résurrection littéraire que lui offre *The Gates of Ivory*, sous les traits de Hattie, narratrice bavarde, exubérante et actrice ratée, "with a great headful of snaky falsely frizzed Medusa curls." (*TGI*, 35-36)

Le mythe de la Méduse est présent dans les romans de Margaret Drabble, y compris pour les personnages mineurs où il témoigne d'une fragilité et "symbolise l'image déformée de soi"[24]:

> [...] it was Polly who appeared to have been petrified [...] The Jilly Fox episode put an end to Mike Gitting's spell as a mortuary assistant [...]. Mike Gitting's dreams were to be haunted for years to come. He had been educated at a Quaker school in York, and as a boy had been disturbed by descriptions of heads displayed above city gates, displayed and left to rot, the hair falling away, the teeth bravely grinning at the weather: now they turned to grin at him.[25]

---

[24] Paul Diel, *Le symbolisme dans la mythologie grecque* (Paris: Payot, 1966), 96.
[25] *TRW*, 340.

L'auteur introduit des éléments occultes qui modulent la réalité par une interprétation subjective où sont combinées les notions de réel et d'irréel, d'objectif et de subjectif, de conscient et d'inconscient, éléments déterminants du réalisme magique.

L'influence du réalisme magique

Margaret Drabble ne fait pas vraiment partie de ce petit groupe de romanciers qui dépassent les limites du réalisme au point de le rendre « magique». Elle propose sa propre définition du genre:

> Magic realist novels and stories have, typically, a strong narrative drive, in which the recognizably realistic mingles with the unexpected and the inexplicable, and in which elements of dream, fairy-story, or mythology combine with the everyday, often in a mosaic or kaleidoscopic pattern of refraction and recurrence. English Magic Realism also have some affinity with the Neo-Gothic[26]

Les romans de la maturité de Margaret Drabble s'orientent vers le réalisme documentaire et par leurs thèmes mêmes, dénoncent les maux de notre société. A ces faits réels se superposent parfois, à un autre niveau, des éléments qui jouent sur la confusion entre réalité et fiction: personnalités inquiétantes collectionnant des "Satanic videos" (*TGI*, 318), rêves macabres, hallucinations, apparition intempestive de personnages apparemment venus de nulle part comme "the professional mourner [...] repairing her vampire lips." (Ibid.) Tous ces détails jalonnent le texte et entraînent progressivement le lecteur hors du monde sensible. Les personnages semblent en proie à des sentiments très violents, inhabituels chez Drabble: ainsi, l'amour passionné, par exemple, ressemble plus à la damnation de Faust qu'à un voyage édénique:

> [...] she has left the real world of real people, and entered the dark world of passion [...]. They have willed disaster, and they set out on its dark salted flood, burning, glorious, redeemed, transformed [...] she is [...] in another kingdom.[27]

L'amitié féminine prend une importance nouvelle et la force des sentiments de Liz et d'Alix, pour leur amie Esther, confère à celle-ci de

[26] Margaret Drabble (ed.), *The Oxford Companion to English Literature* (Oxford: Oxford University Press, 1985), 606.
[27] Drabble, *ANC*, 260-261.

nouveaux pouvoirs: " '[...] it won't be sunny anywhere else', said Esther [...] 'I can only arrange the weather for this valley. Everywhere else, it is still raining. There is a limit to my powers.'" (*TRW*, 394) La voix narrative omnisciente fait écho à cette affirmation, renforçant ainsi l'illusion de réel: "Alix and Liz beg leave to doubt this. Her powers, they claim with admiration, are limitless, supernatural [...]." (ibid.)

Pourtant, dans *The Gates of Ivory*, Esther est déchue de son rôle de magicienne et rentre dans le rang des personnages ordinaires, sans doute par son mariage tardif et inopiné avec un homme raisonnable et différent du diabolique Claudio (déjà connu dans *The Radiant Way*) qui avait été longtemps son amant et dont Margaret Drabble avoue: "He's a satanist anthropologist. This was my little joke about magic realism, a sort of response to it. I'm interested in the way the irrational fits into the rational world."[28]

Dans ce roman, la romancière joue sur l'ambiguïté de la frontière qui sépare le réalisme magique du fantastique. En effet, le surnaturel, propre à la littérature fantastique, s'immisce lentement autour du personnage d'Alix. Celle-ci apparaissait comme une femme convaincue de la bonté de l'homme, du bien-fondé du progrès social. Petit à petit, elle évolue vers un vague mysticisme qui la pousse à se réciter, telle une incantation, des hymnes religieux pour conjurer la peur et à croire à la grâce et à la rédemption. Ce trait, apparemment inoffensif, évolue en tendance schizoïde lorsqu'elle avoue être obsédée par des visions de personnes mortes après, paradoxalement, avoir lu un livre philosophique traitant de "matter and neurons and quantum particles" (*TGI*, 286). Elle s'enfonce alors dans un obscur paradoxe: "[...] I have come round to the view that not only is there no God, there is no Spirit either [...]. But the odd thing is, the more I convince myself there is no Spirit, the more apparitions I see." (Ibid.)

Aucune explication rationnelle ou scientifique ne sera suggérée par son amie Liz, psychiatre de profession. Le lecteur est laissé dans l'ambiguïté et ce n'est qu'à la fin du roman que la réalité reprend le dessus: "She will not turn into that old witch, that mad recluse, that inspired sybil." (*TGI*, 432) L'abondance du vocabulaire spécifique de la magie avec la récurrence de mots tels que "evil", "witch", "omen", "vampire", "blessed", "mystic", ou encore "magician"annonce déjà le roman suivant, *The Witch of Exmoor*, où la narration gravite autour de Frieda, personnage central, anticonformiste et excentrique qui a décidé de vivre reclus dans une grande

[28] Olga Kenyon, *Women Writers Talk: Interviews with 10 Women Writers* (Oxford: Lennard Publishing, 1989), 37.

maison isolée et délabrée. Aux yeux de sa famille et de la société, l'ordre naturel des choses est ébranlé ; la vieille femme, aux allures de sorcière, inquiète et dérange, ses apparitions dans le récit donnent au roman une atmosphère inquiétante.

Le réalisme magique constitue une autre perception du réel: il présente une vision individuelle de la réalité à travers le filtre des mythes de l'humanité ; il souligne les rapports complexes des personnages avec l'espace réel et l'espace imaginaire et dépeint les relations entre les deux niveaux de la conscience humaine. Le rôle de l'écrivain réaliste se trouve donc modifié; il ne se contente plus d'être le simple témoin de son temps: au contraire, dans un jeu narratif, il établit une confusion entre sujet et objet, entre réel et irréel avec la volonté de mettre en évidence des réalités qui s'opposent à une vision stéréotypée du réel.

Margaret Drabble dépasse les frontières du réalisme traditionnel avec la volonté d'accentuer le côté sombre du contexte historique réel. Les personnages, pris dans l'étau du quotidien, le transforment, le détournent en laissant leur imaginaire l'emporter sur la raison.

Dans son cheminement littéraire de ces cinquante dernières années, la romancière s'est ainsi ouverte à des formes d'expression qui s'affranchissent progressivement du réalisme dogmatique. Ses incursions dans des genres où l'atmosphère devient sombre et pesante se fait par petites touches comme dans un puzzle volontairement inachevé.

Réalité, fiction. Tout s'entremêle dans les romans de Margaret Drabble: on fait l'écheveau au fil de l'écriture, on défait l'écheveau au fil de la lecture. "A quoi bon écrire des livres si on n'invente pas la vérité? Ou, encore mieux, la vraisemblance?" disait Jorge Semprun[29] et les derniers romans de l'écrivain britannique semblent remplir cet objectif. Elle affirme son appartenance, non à une tendance littéraire déterminée mais plutôt à un mode d'expression malléable et en constante évolution: "To me, the novel remains a superbly flexible and informal form, which can treat with good and evil, with social trends, with hem lines and holocausts."[30]

De fait, elle porte, dans ses derniers romans, un intérêt pour des sujets de plus en plus variés tels que les recherches sur l'ADN ou encore la biologie marine, mais aussi les thérapies, les problèmes environnementaux, des sujets qu'elle examine, non pas en surface, mais en se documentant avec minutie. A la manière de Zola, pour qui "le romancier est fait d'un observateur et d'un expérimentateur", Margaret

---

[29] Jorge Semprun, *Le mort qu'il faut* (Paris:Gallimard, 2001), 188.

[30] Margaret Drabble, "Some Thoughts about the Novel,"*Caliban* XXVIII (1990).

Drabble intègre dans ses récits des éléments scientifiques, ou pseudo-scientifiques pour se jouer du lecteur. Elle crée des atmosphères qui évoquent des facettes différentes du genre romanesque.

A quel moment s'arrête la vérité pour laisser place à l'imagination ? C'est au lecteur d'en décider.

Works Cited

Primary Sources

Drabble, Margaret. *Jerusalem theGolden.* Harmondsworth: Penguin Books, (1967) 1969.

—. *The Radiant Way*, Harmondsworth : Penguin Books, 1988, 1ère édition : 1987.

—. *A Natural Curiosity*. London: Viking, 1989.

—. *The Gates of Ivory*. London: Viking, 1991.

—. *The Witch of Exmoor*. London: Viking, 1996.

—. *The Pattern in the Carpet.* London: Atlantic Books, 2009.

—. "The Merry Widow." *Woman's Journal* (September 1989).

—. "Stepping Westward: A Topographical Tale." *A Day in the Life of a Smiling Woman.* New York: Houghton Mifflin Harcourt (2011).

—. *The Oxford Companion to English Literature*, ed. Oxford: Oxford University Press, 1985.

Secondary Sources

Barthes, Roland. *Le degré zéro de l'écriture*. Paris: Seuil, 1953.

Diel, Paul. Le Symbolisme dans la mythologie grecque. Paris: Payot, 1966.

Drabble, Margaret. "Some Thoughts about the Novel." *Caliban* XXVIII. Toulouse: Presses Universitaires du Mirail (1990).

Kenyon, Olga. Women Writers Talk: *Interviews with 10 Women Writers.* Oxford: Lennard Publishing, 1989.

Levin, Harry. *The Gates of Horn: A Study of Five French Realists*, Oxford University Press, 1963.

Moers, Ellen. *Literary Women*. London: The Women's Press Ltd, 1978.

Peyre, Claudine. "Three interviews with Margaret Drabble, 1989, 1992, 2011." *Cercles 21* (2011)., http://www.cercles.com/n21/peyre.pdf.

Semprun, Jorge. *Le mort qu'il faut*. Paris: Gallimard, 2001.

SELECT BIBLIOGRAPHY

In one of his books, the French philosopher and historian of science Michel Serres tells his readers that providing a bibliography these days is both a waste of paper and a waste of effort since any web search will provide a much wider range of references than can be gathered in a few pages. Although we tend to agree with that view, we nevertheless believe that a select bibliography can still prove useful as a starting-point for anyone embarking on further research. The items we list are meant to give a sense of the context out of which this volume on the interaction between science, fables and chimeras arose and reflect the diversity of authors and periods dealt with in the volume. Our list omits some of the material referred to by individual contributors, includes some additional material, and is arranged thematically so that the reader may recognize the overlaps, meeting points and blurred boundaries with which this volume as a whole is very much concerned.

History of Science and Literature

Aït-Touati, Frédérique. *Contes de la Lune.* Paris: Essais, Gallimard, 2011

Ameisen, Jean-Claude. *Sur les épaules de Darwin, les battements du temps*. Paris: France Inter, Les Liens qui Libèrent, 2012

Ameisen, Jean-Claude et Brohard Yvan. *Quand l'art rencontre la science*, Paris: La Martinière, 2009

Beer, Gillian. *Darwin's Plots: Evolutionary Narrative in Darwin, George Eliot and Nineteenth-Century Fiction*. London: Routledge & Kegan Paul, 1983

—. *Open Fields: Science in Cultural Encounter*. Oxford: Clarendon Press, 1996

Céard, Jean. *La nature et les prodiges. L'insolite au XVIe siècle*. Genève: Droz, 1996 (1977)

Canguilhem, Georges. *La connaissance de la vie, 2e éd.* Paris: Vrin, 1965

Conway Morris, Simon. *The Crucible of Creation: The Burgess Shale and the Rise of Animals*. Oxford: Oxford University Press,1998

Daston, Lorraine and Peter Galison. *Objectivity*. New York: Zone Books, 2007

Daston, Lorraine and Katharine Park. *Wonders and the order of nature: 1150 - 1750*. New York, NY: Zone Books, 1998
Dawkins, Richard. *The Ancestor's Tale: A Pilgrimage to the Dawn of Life*. London: Weidenfield & Nicolson, 2004.
—. *Unweaving the Rainbow: Science, Delusion and the Appetite for Wonder*. London: Allen Lane, 1998.
Fara, Patricia. *Science: A Four Thousand Year History*. Oxford: Oxford University Press, 2009Figuier, Louis. *Histoire du merveilleux*, vol.III. Paris: Hachette et Cie, 1860
Gould, Stephen Jay. *Wonderful Life, The Burgess shale and* the *nature of history*. New York: W. W. Norton & Company, 1989
—. *La vie est belle, Les surprises de l'évolution*, traduction par Marcel Blanc. Paris: Seuil, coll. Points Science, 1991
—. *Rocks of Ages: Science and Religion in the Fullness of Life*. New York, Ballantine Publishing Group: 1999
—. *The structure of evolutionary theory*. Cambridge (Massachussetts) and London: The Belknap Press of Harvard University Press, 2002
Gould, Stephen Jay, and Rosamond Wolff Purcell. 2000. *Crossing Over: Where Art and Science Meet*. New York: Three Rivers Press.
Keller, Olivier. *Aux origines de la géométrie. Le Paléolithique et le monde des chasseurs-cueilleurs*. Paris: Vuibert, 2004
—. *La figure et le monde. Une archéologie de la géométrie. Peuples paysans sans écritures et premières civilisations*. Paris: Vuibert, 2006
Latour, Bruno. *Pandora's Hope. Essays on the Reality of Science Studies*. Cambridge, Mass: Harvard University Press, 1999
Levine, George. *One Culture: Essays on Science and Literature*. University of Wisconsin Press, 1987
Moisseeff, Marika. 2005 « La procréation dans les mythes contemporains : une histoire de science-fiction », *Revue Anthropologie et sociétés*, vol. 29, n° 2 : 69-94
Levy-Leblond, Jean-Marc. *La pierre de touche. La science à l'épreuve*. Paris: Folio, 1996
—. *Impasciences*. Paris: Bayard, 2000
—. *La science n'est pas l'art: brèves rencontres*. Paris: Hermann, 2010
—. *Le Grand écart : la science entre technique et culture*. Paris: Manucius, 2013
Nicolson, Marjorie. *Science and Imagination*. Ithaca: Cornell University Press, 1956
O'Connor, Ralph. *The Earth on Show: Fossils and the Poetics of Popular Science, 1802-1856*. Chicago: University of Chicago Press, 2007

Rudwick, Martin J.S. *Worlds Before Adam: The Reconstruction of Geohistory in the Age of Reform.* Chicago and London: The University of Chicago Press, 2008

Secord, James A. *Victorian Sensation: The Extraordinary Publication, Reception, and Secret Authorship of* Vestiges of the Natural History of Creation. Chicago: University of Chicago Press, 2000

—. *Charles Darwin. Evolutionary Writings*. Oxford: Oxford University Press, 2008

Serres, Michel. *Hermès V: Le passage du Nord-Ouest*. Paris: Minuit, 1980

—, ed. *Eléments d'histoire des sciences*. Paris: Bordas, 1989

Wilson, Edward O. *On Human Nature*. Cambridge, Mass: Harvard University Press, 1978

—. *Consilience. The Unity* of *Knowledge*. New York: Alfred A. Knopf, Inc., 1998

Science and Religion

Bowler, Peter J. *Reconciling science and religion: The debate in early twentieth century Britain.* Chicago: University of Chicago Press, 2001

Brooke, John H. and Ronald L. Numbers, eds. *Science and religion around the world.* New York: Oxford University Press, 2010

Brooke, John H., and E. Ihsanoglu, eds. *Religious values and the rise of science in Europe.* Istanbul: Research Centre for Islamic History, Art and Culture, 2005

Brooke, John H., and Ian Maclean, eds. *Heterodoxy in early modern science and religion.* Oxford: Oxford University Press, 2005

Clayton, Philip, and Zachary Simpson, eds. *The Oxford handbook of religion and science*. Oxford: Oxford University Press, 2006

Desmond, Adrian and James R. Moore. *Darwin's sacred cause: Race, slavery and the quest for human origins*. London: Allen Lane, 2009

Dixon, Thomas. *Science and religion: A very short introduction*. Oxford: Oxford University Press, 2008

—. *From passions to emotions: The creation of a secular psychological category*. Cambridge: Cambridge University Press, 2003

Dupuy, Jean-Pierre. *La marque du sacré*. Paris: Carnets nord, 2008

Durant, John R., ed. *Darwinism and divinity: Essays on evolution and religious Belief.* Oxford: Blackwell, 1985

Ferngren, Gary B., ed. *Science and religion: A historical introduction*. Baltimore, MD: Johns Hopkins University Press, 2002

—, Edward. J. Larson, Darrel D. W. Amundsen, eds. *The history of science and religion in the western tradition: An encyclopedia*, New York and London: Garland, 2000

Gilbert, James. *Redeeming culture: American religion in an age of science*. Chicago: University of Chicago Press, 1997

Hart, D., D. Livingstone and M. Noll, eds. *Evangelicals and science in historical Perspective*. Oxford: Oxford University Press, 1999

Hooykaas, R. *Religion and the rise of science*, Grand Rapids, MI: Eerdmans, 1972

Knight, David and Matthew Eddy, eds. *Science and beliefs: From natural philosophy to natural science*. Aldershot: Ashgate, 2005

Latour, Bruno. *On the modern Cult of the Factish Gods*. Duke University Press, 2009

Lindberg, David C. and Ronald L. Numbers, eds. *God and nature: Historical essays on the encounter between Christianity and science*. Berkeley, CA: University of California Press, 1986

—. *When science and Christianity meet*. Chicago: University of Chicago Press, 2003

Livingstone, David N. *Adam's ancestors: Race, religion, and the politics of human Origins*. Baltimore, MD: Johns Hopkins University Press, 2008

McGrath, Alister E. *Science and religion: An introduction*. Oxford: Blackwell, 1999

Numbers, Ronald L. *Science and Christianity in pulpit and pew*. New York: Oxford University Press, 2007

—. ed. *Galileo goes to jail and other myths about science and religion*. Cambridge, MA: Harvard University Press, 2009

— and John Stenhouse, eds. *Disseminating Darwinism: The role of place, race, religion, and gender*. Cambridge: Cambridge University Press, 1999

Osler, Margaret. *Rethinking the scientific revolution*. Cambridge: Cambridge University Press, 2000

— and J. van der Meer, eds. *Science in theistic contexts: Cognitive Dimensions*. Chicago: University of Chicago Press, 2001

Olson, Richard. *Science and religion, 1450–1900: From Copernicus to Darwin*. Westport, CT: Greenwood Press, 1994

Proctor, James D. ed. *Science, religion, and the human experience*. New York: Oxford University Press, 2005

Richardson, W. Mark, and Wesley J. Wildman, eds. *Religion and science: History, method and dialogue*. London: Routledge, 1996

Roob, Alexander. *Le musée hermétique. Alchimie et Mystique*. Taschen, 2006
Russell, Colin A. *Cross-currents: Interactions between science and faith*. Leicester: Inter-Varsity, 1985
Shapin, Steven. *The scientific revolution*, Chicago: University of Chicago Press, 1996
Stanley, Matthew. *Practical mystic: Religion, science, and A. S. Eddington*. Chicago: University of Chicago Press, 2007
Taylor, Charles. *A Secular Age.* Cambridge, MA: Belknap Press of Harvard University Press, 2007
Van der Meer, Jitse, ed. *Facets of faith and science*, 4 vols. New York: University Press of America, 1996
Van Huyssteen, J. Wentzel, ed. *Encyclopedia of science and religion,* 2 vols. New York: Macmillan, 2003
Westfall, Richard S. *Science and religion in seventeenth-century England.* New Haven, CT: Yale University Press, 1958
Wood, Paul, ed. *Science and dissent in England 1688–1945*. Aldershot: Ashgate, 2004

Theory on Myth and Imagination

Aristote, *Poétique.* Paris: les belles lettres, [1932], 1969
Baltrusaitis, Jurgis. *Réveils et prodiges. Le gothique fantastique*. Paris: Armand Colin, 1960
Brunel, Pierre. *Le mythe de la métamorphose*. Paris: José Corti, 2004
Bachelard, Gaston. *L'air* et *les songes.* Paris: José Corti, 1943
— . *La terre et les rêveries du repos*. Paris: José Corti, 1939
Barthes, Roland. *Mythologies*. Paris: Seuil, [1957]1970
Bergson, Henri. *Les deux sources de la morale et de la religion.* Paris:Alcan, 1932
Campbell, Joseph. *The Power of Myth*. New York: Doubleday, 1988
Caillois, Roger. *Œuvres*. Paris: Gallimard, coll. « Quarto », 2008
Cassirer, Ernst. *La philosophie des formes symboliques.* (1923-1929), 3 tomes. Paris: Éditions de Minuit, 1972
Clair, Jean. *Eloge du visible : fondements imaginaires de la science*. Paris: Gallimard, 1996
Diel, Paul. *Le Symbolisme dans la mythologie grecque*. Paris: Payot, 1966.
Durand, Gilbert. *L'imagination symbolique*. Paris: Presses Universitaires de France, coll. « Quadrige Grands Textes », [1964], 2008

—. *Les structures anthropologiques de l'imaginaire*. Paris: Dunod, [1969], 1992
Eliade, Mircea. *Aspects du mythe.* Paris:Gallimard, 1963
—. *Mythes, rêves et mystères.* Paris: Idées, NRF, 1972
Foucault, Michel. *Les mots et les choses.* Paris: Gallimard, 1966
Freud, Sigmund. *L'inquiétante étrangeté et autres essais*. Paris: Gallimard, 1991
Frazer, James George. *The Gorgon's Head and other literary pieces.* London: Macmillan, 1927
—. *The Golden Bough, A study of magic and religion.* New York:Macmillan, 1922
Gaillard, Aurélia et Jean-René Valette, ed. *La beauté du merveilleux.* Bordeaux: Presses Universitaires de Bordeaux, 2011
Jung, Carl Gustav. *Métamorphoses de l'âme et ses symboles.* Paris: Poche/references, 1993 (1953)
Levi-Strauss, Claude. *Anthropologie structurale.* Paris: Plon, 1962
Secord, James A. *Myth: A Very Short Introduction*. Oxford: Oxford University Press, 2004
Segal, Robert. *Myth: Critical Concepts in Literary and Cultural Studies*. London and New York: Routledge, 2007
Serres, Michel. *Récits d'humanisme*. Paris: Le Pommier, 2009

Monsters and Chimeras

Absalon, Patrick. *Dragons, entre science et fiction*. Paris: De Monza, 2006
— et François Canard. *Les dragons, des monstres au pays des hommes*. Paris: découvertes Gallimard, 2006
Aldrovandi, Ulysse (et Ambrosino, Bartholomeo). *Monstrorum historia cum Paralipomenis historiae omnium animalium.* Bologne, N. Tebaldin, 1642
Allain , Yves-Marie et Guy Prouveur, Guy. *Monstruosités et chimères du monde végétal*, Paris: Ellipses, 2009
Ancet, Pierre. *Phénoménologie du corps monstrueux*. Paris: PUF, 2006
Baltrusaïtis, *Le Moyen Age fantastique. Antiquités et exotismes dans l'art gothique*, Paris: Flammarion, 1981
Benton, Janetta, Rebold. *Saintes Terreurs, Les Gargouilles dans l'architecture médiéval*. Paris: Editions Abbeville, 1997
—. *Medieval Monsters: Dragons and Fantastic Creatures*. Katonak, N.Y.: Katonah Museum of Art, 1995
Boaistuau, Pierre. *Histoires prodigieuses.* Paris, 1566

Borges, Luis. *Le livre des êtres imaginaires.* Paris: Gallimard, 1987

Charlier, Philippe. *Les monstres humains dans l'Antiquité.* Paris: Fayard, 2008

Chevrier, Jean-François. *L'hallucination artistique de William Blake à Sigmar Polke.* Paris: L'Arachnéen, 2012

Clair, Jean. *Méduse : Contribution à une anthropologie des arts du visuel.* Paris: Gallimard, 1989

Clair, Jean, ed. *L'âme au corps. Arts et sciences, 1793-1993, sous la direction de J. Clair.*, cat. Exp. Paris: Gallimard, RMN/Electa, 1993

—. *Hubris : la fabrique du monstre dans l'art moderne.* Paris: Editions Gallimard, 2012

Delacampagne, Ariane et Christian. *Animaux étranges et fabuleux ; Un bestiaire fantastique dans l'art.* Paris: Citadelles & Mazenod, 2003

Dayun, Yue and Alain Le Pichon, Alain eds. *La licorne et le dragon. Les malentendus dans la recherche de l'universel.* Peking: Peking U.P. 1995

Ellis, Richard. *Monsters of the Sea.* New York: Alfred A. Knopf, Inc., 1994

Gesner, Conrad. *Historiae Animalium.* 4 vols, Paris: 1551-58

—. *Curious Woodcuts of Fanciful and Real Beasts.* New York: Dover Publications, Inc, 1971

Geoffroy Saint-Hilaire, Isidore. Traité de tératologie, Histoire des anomalies de l'organisation. Paris, 1836

—. *Philosophie anatomique.* Paris, 1822

—. *Discours d'introduction à l'ouvrage, monstruosités humaines formant le deuxième tome, de la Philosophie anatomique.* Paris: 1826

—. *Considérations générales sur les monstres, comprenant une théorie des phénomènes de la monstruosité.* Paris: J. Tastu, 1826

Gould, Charles. *Mythical Monsters, with Ninety-three Illustrations.* London: W.H.Allen & Co, 1886

Guédron, Martial, Annika Petersen, Klara Tuszynski et Coline Valdenaire. *Monstres, merveilles et créatures fantastiques.* Paris: Editions Hazan, 2011

Kappler, Claude. *Monstres, démons et merveilles à la fin du Moyen-Age.* Paris: Payot, 1980

Lavers, Chris. *The Natural History of Unicorns.* New York: Harper Collins, 2009

Lascaux, Gilbert. *Le monstre dans l'art occidental.* Paris: Klincksieck, 2004

Le Douarin, Nicole. *Des chimères, des clones et des gènes.* Paris: Odile Jacob, 2000

Leeming, David Adams. *Mythical and fabulous Creatures. A Source Book and Research Guide,* ed. M. South, New York: Greenwood Press, 1987

Liceti, Fortunio. *De Montris*, 1665

Lycosthènes, Conrad. *Prodigiorum ac ostentorum chronicon*, Bâle: H. Petrus, 1557

Martin, Ernest. *Histoire des Monstres depuis l'Antiquité jusqu'à nos jours (1880)*, Paris: Jérome Million, 2002

Meurger, Michel. *Histoire naturelle des dragons*. Rennes: Terre de Brume, 2001

Mitchell, W. J. T. *The Last dinosaur Book: The Life and Times of a Cultural Icon*. Chicago : Chicago University Press, 1998.

Ottinger, Didier. *Chimères*. Arles: Actes Sud, 2003

Paré, Ambroise. *Des monstres et prodiges*, Genève: Fleuron, 1996

—. *Discours de la Momie et de la Licorne*. Paris: Gallimard, 2011

Pomet, M. *Traité des drogues*, 1695

Roux, Olivier. *Monstres Une histoire générale de la tératologie des origines à nos jours*.

Roussillon, Laurence. *Méduse au Miroir. Esthétique romantique de Dante Gabriel Rossetti*. Grenoble :ELLUG, 2008.

Roussillon-Constanty, Laurence. Des dragons de Ruskin aux dinosaures de Darwin : art, science et la peur de l'origine. *Interfaces*, vol.31, Spring 2011, 13-25

Rudwick, Martin J.S. *Scenes from Deep Time: Early Pictorial Representations of the Prehistoric World*. Chicago: University of Chicago Press, 1992

Stead, Evangelina. *Le monstre, le singe et le foetus:tératogonie et decadence dans l'Europe fin-de-siècle*. Genève: Droz, 2004

Tort, Patrick. *L'Ordre et les Monstre: le débat sur l'origine des déviations anatomiques au XVIIIème siècle*. Paris: Le Sycomore, 1980

Shepard, Odell. *The Lore of the Unicorn.* New York: Dover Publications, Inc, 1993

CONTRIBUTORS

EGIL ASPREM holds research positions at the Centre for History of Hermetic Philosophy and Related Currents at the University of Amsterdam, and the Department of Archaeology and Religious Studies at the Norwegian University of Science and Technology (NTNU), Trondheim. His research interests include the relationship between science and esotericism, the history of parapsychology, the role of science and technology for religious change, contemporary esotericism, and theoretical and methodological questions in the study of religion, esotericism, and the history of science. Asprem is the author of *Arguing with Angels: Enochian Magic and Modern Occulture* (State University of New York Press, 2012), and co-editor (with Kennet Granholm) of *Contemporary Esotericism* (Equinox Publishing Ltd. / Acumen, 2013).

ALEXANDRA BORSARI holds a Ph.D. in Political Science from the University Paris-Est and is currently holding a research position at Télécom ParisTech & Rennes 2 ("Chaire d'Enseignement et de recherche Modélisations des Imaginaires"). She is also an associate Member of the Centre de Recherche sur l'Imaginaire (Université Stendhal, Grenoble 3) and her research field is the Imaginary (fundamental structures of representation and epistemology theories on the imaginary). Her main research focuses on the construction of the Imaginary in nature and the environment and on the connections between the theory of evolution and theories of the Imaginary. She has a special interest in the Neolithic turn and the representation of man's place in the living world.

CÉLINE CADAUREILLE holds a Ph.D. in Fine Arts from the Toulouse Art School (LLA-Créatis Lab/ Université Toulouse 2-Mirail) and is currently engaged in both theoretical and artistic research. In her doctoral thesis entitled "L'obscénité et les limites du voir" (seeing and the obscene) she explores the various aspects of the obscene through individual case studies of scandalous works that challenge the body's integrity either sexually or morbidly. Her main area of interest is the exploration of borderlines and the limits and representation of desire and fear and she has published several articles on related issues.

MICHEL CAFFAREL (IRSAMC and CNRS-Lab. Chimie et Physique Quantiques) is a CNRS (French National Center for Scientific Research) senior research scientist at the University of Toulouse. His field of expertise is quantum physics and his work can be broadly classified as contributions to the understanding of the quantum N-body problem both in condensed matter physics and theoretical chemistry. Dr. Caffarel's research career brought him to explore various aspects of the quantum world such as the fascinating behavior of electrons in a wide range of contexts: superconductors, transition metal systems, electronic structure of molecules, new materials, etc. He is also Head of the doctoral school in Physics, Chemistry, and Material Sciences of the University of Toulouse.

HERVE DELAMARE is an artist based in the north of France. Born in 1969 in Le Havre, he graduated from the Ecole des Beaux-Arts in Paris in 1991 and has exhibited his works in various places in France (Montpellier, Nantes, Toulouse, Paris, Marseille) and abroad (Japon and Germany). In his work, Delamare enjoys working with mixed media, ranging from drawing, painting, sculpture to drawing as he explores metamorphosis in all its guises. From 1991 to 1996, he has worked more specifically on the transformations of cityscapes and urban space. From 1998 to 2010 his work focused on animal metamorphoses and hybrids. By contrast, his most recent work looks at the metamorphoses of modern man and the evolution of humankind. In his artistic practice, Hervé Delamare deliberately chooses to exhibit in unusual places, in the heart of the city and in public places (such as in libraries, in industrial business companies or in natural history museums or schools). Via his website and through his various interactive projects with the public, Hervé Delamare can truly be called an “artiste engagé”. To find out more, please visit his websites: http://h.delamare.free.fr and http://nuagedencre.free.fr/

MARGARET DRABBLE is a novelist. She was born in Sheffield, Yorkshire, England. Her father, John Frederick Drabble, was a barrister, a county court judge and a novelist. Author A.S. Byatt is her older sister. She attended the Mount School, York, a Quaker boarding-school and was awarded a major scholarship to Newnham College, Cambridge, where she read English and received double honors (a "starred first"). After being graduated from Cambridge University, she joined the Royal Shakespeare Company at Stratford during which time she understudied for Vanessa Redgrave. In 1960 she married her first husband, actor Clive Swift, who is best known for his role in the 1990 BBC television comedy “Keeping Up Appearances.” They had three children in the 1960s and divorced in 1975.

She subsequently married the biographer Michael Holroyd in 1982. They live in London and Somerset. Her novel The Millstone won the John Llewelyn Rhys Prize and she was the recipient of a Society of Author's Travelling Fellowship in the mid-1960s. She also received the James Tait Black and the E.M. Forster awards, was awarded the CBE in 1980 and the DBE in 2008. Her long and flourishing career includes many novels, among which *The Peppered Moth* (2000) and *The Sea Lady* (2006). She also wrote non-fiction works, literary biographies and edited two editions of the reference book *Oxford Companion to English Literature* (1985, 2000). She is often described as being the author one should read to get a clear view of what it's like to live in England. "A Writer's Britain" describes the literary history of the British landscape, and her novels, through the decades, have also reflected the dramatic political, economic and social changes that have taken place in Great Britain.

SEBASTIEN GREGOV is agrégé en études germaniques and a doctoral student. He holds a Master's Degree in Science of Religion from Sorbonne (Paris IV, Sorbonne Nouvelle, EPHE, Ecole Pratique des Hautes Etudes) and currently teaches German at the Institut d'Etudes Politiques de Rennes.

SABINE GRUFFAT (CRRLPM - Centre de Recherche sur les Relations entre Littérature, Philosophie et Morale, ENS Ulm) holds a Ph.D. in 17th Century French Literature and her research focuses on the relations between morals and aesthetics in La Fontaine's Fables. Her main field of expertise is the poetics of fables, moral literature and Classical Age drama. Her main publications include an edition of La Fontaine's *Fables* in the "Livre de Poche" series (collection "Classiques"), co-edited with J.C. Darmon in 2002. She has also annotated and discussed Racine's *Iphigénie* in the "Classiques Bordas" series in 2004 and has edited *Le Théâtre français du XVIIème siècle*, published by Ellipses ("Thèmes et Études" series). She has written about thirty articles dealing with fables (La Fontaine, Houdar de La Motte, Furetière), the moralists (La Rochefoucaud) and 17th century drama (Racine, Corneille, the lyrical tragedy). After teaching 17th century literature at the University of Lyon 3 (Jean Moulin) from 1996 to 2011, Sabine Gruffat is currently teaching French literature in "classes préparatoires" at the Lycée du Parc in Lyon.

MATTHEW JARRON is Curator of Museum Services at the University of Dundee, which includes the D'Arcy Thompson Zoology Museum. He is the co-author (with Cathy Caudwell) of D'Arcy Thompson and his

Zoology Museum in Dundee (University of Dundee, 2010) and has recently written the introduction to a special issue of Interdisciplinary Science Reviews devoted to D'Arcy Thompson (to be published in 2013). He is Chair of the Scottish Society for Art History and is currently writing a book on art in Dundee in the late-nineteenth and early-twentieth centuries.

OLIVIER KELLER is an independent scholar agrégé de Mathématiques and holds a Ph.D. in History of Science from the EHESS in Paris (Ecole des Hautes Etudes en Sciences Sociales). His main field of research is the production of key-concepts in the prehistoric period and early civilizations. He is the author of two books, *Aux origines de la géométrie. Le Paléolithique et le monde des chasseurs-cueilleurs* (Paris : Vuibert, 2004) and *La figure et le monde. Une archéologie de la géométrie. Peuples paysans sans écritures et premières civilisations* (Paris: Vuibert, 2006). He has also published several articles such as « Les fables d'Ishango, ou l'irrésisitible tentation de la mathématique-fiction », (2010) : http://www.bibnum.education.fr/mathématiques/algébre/l'os-d'ishango, «La géométrie des Sulbasutras. Exemple de géométrie rituelle de l'Inde védique : l'agrandissement de l'autel en forme de faucon. » *Repères-Irem* 40 (2000): 115-124. http://www.univ-irem.fr/commissions/reperes/consulter/40keller.pdf and « Eléments pour une préhistoire de la géométrie. » *L'Anthropologie* 105 (2001) : 327-349.

ALICE LAMY is Agrégée de lettres classiques and holds a Ph.D. in Philosophy from the University of Tours. She currently teaches classical literature in the Lycée Hélène Boucher in Paris. Her published works mainly deal with Aristotelian commentaries of the medieval period in European universities and scientific medieval writings on matter and measure. She has published several articles on the author of her thesis, Gauthier Burley and a book on the *Physics* of Burley, *La grandeur de l'être au XIVe siècle, une lecture introductive aux commentaires de la 'Physique' de Gauthier Burley*, Paris, Hermann, 2012. She is also interested in the relations between literature and philosophy and the role of creation and imagination the writings of scholars such as Pierre d'Ailly or Jean Gerson.

DANIEL S. LARANGÉ holds a Ph.D. in French Language and Literature from the Sorbonne (Université Paris 3-Sorbonne Nouvelle) and teaches at the Swedish University of Åbo Akademi (Finland). He also holds an M.A. in Fundamental theology from Martin Luther's Faculty (Erlangen

University) and the Evangelical Theological Faculty of Charles University in Prague. His field of expertise is French romanticism and the connections between socialism and esotericism in the 19th century. He is the author of several books and articles on F.M. Dostoevsky, Kahlil Gibran, John Amos Comenius and the representation of spirituality in the French literature of the 19th and 20th centuries. He's currently carrying out his research on romanticism in the French Department of McGill University (Montreal) and is an affiliated member of the CRIST (Center for Interdisciplinary Research on Socio-criticism in Texts).

MONIQUE MARTINEZ THOMAS is Professor of Spanish at the University of Toulouse (LLA/CREATIS, Toulouse 2-Mirail). She has published several articles and essays on contemporary Spanish drama and has played a major part in the renewal of Spanish studies through projects promoting contemporary Spanish fiction (from translation work, new editions through to the staging of plays in the original language with overhead translations in French and other related research projects), an initiative that was relayed and later applied in the "Universcènes" international drama Festival for German, English and Italian. Monique Martinez also heads the DRAMA research programme which associates drama and computer science, and has successfully set up a network of academics and people interested in the interactions between art and science (in collaboration with the business cluster STARS). She is Head of the Doctoral School of Humanities, ALLPH@ (Arts, Lettres, Langues, Philosophie et Info-communication), and Head of the Doctoral School of the University of Toulouse (governing all fifteen graduate schools, or "écoles doctorales").

FRANCOIS MIGLIO was born in Montpellier, in August 1958 and has been a resident of Beziers (Languedoc-Roussillon) for the past 30 years. He first started his professional career in the agriculture and pharmaceutical industries in France, after which he has devoted full time to his art, raising two children, and restoring his historic 18th century house along the Canal du Midi in Beziers. From 1990 to 1993, he achieved "Cosmogony", his first significant piece which led him to develop his pictorial concept, "Fractal Syncretism". François Miglio first travelled to the United States to in 1998 and he found that the Bay Area – being is a vast "open system" receptive to new ideas – was the place to be. For the past 12 years, he has spent at least 5 months a year, painting, exhibiting and promoting his art in the San Francisco Bay Area and New Mexico. He has participated in two solo exhibitions and lectures (about Fractal

Syncretism) at the International House of the University of California, Berkeley, and also held two solo continuing exhibitions for two years running at the department of Economics in UCL(2002-2003). He has also exhibited also at Stanford Art Space, Center for Integrated Systems, Stanford Universit (2006). He also gave a lecture at the conference of International Society of Systems Sciences in Sonoma State University of California (2006). François Miglio has also participated in various solo and group exhibitions in Vorpal Gallery (San Francisco) and Mc Carthy Gallery (Santa Fe New Mexico). He did a solo exhibition in the Zen Center and in a New College of San Francisco. During few years, he also exhibited with Art Ship Foundation in Oakland and in Somarts Cultural Center of San Francisco. He has also participated to the Biennale of San Francisco with Border Zone Art Foundation, at the Art Fair of Nîmes in France and at a Sony Tower Salon in Osaka in Japan. In 2011 he has participated to the Art fair of Beijin and he has exhibited at the Li Xinhuin of Nova Art Coordinates Gallery also in Beijin. François Miglio is actually full time present in the West Coast in USA and his two latest major exhibitions were held at the Institute for Computational and Mathematical Engineering of the Department Huang Engineering Center of Stanford University (20012/2013) and at Google Head Quarters, Mountain View, Silicon Valley California (2013). Whis son Hadrien François Miglio is currently developing an Art Project called "Miglio Art" in order to promote his pictorial concept of "Fractal Syncretism". To get a comprehensive view of François Miglio's art work, please visit his website: http://www.miglioart.com

PHILIPPE MURILLO (Université Toulouse 3-Paul Sabatier) holds a Ph.D in American Studies from Université Paris 3, Sorbonne Nouvelle. He also holds a M.A in Religious Studies from Université Bordeaux 3. He is a research fellow with CAS research center (Cultures Anglo-Saxonnes, Université Toulouse 2-Mirail) and specialized in New Religious Movements in the United States, with a particular focus on American metaphysical religion. As a member of CESNUR (Center for the Study of New Religious Movements), he has published many articles in his field of expertise and conducted several workshops at Cesnur annual conferences. He is currently writing a monograph on New Thought: *Transatlantic Encounters: From New Thought to Process New Thought*. He teaches English and American Studies at Université Toulouse 3 – Paul Sabatier, where he co-chairs the Department of Modern Languages. He is also a visiting professor with University of Central Florida, Orlando, USA.

MARIE-CHRISTINE NOAILLES-PIZZALOTTO is Senior Lecturer at the University of Toulouse (Université Paul Sabatier, Toulouse 3). She holds a Ph.D. in American literature and teaches English to life science students. She is particularly interested in the links between American literature and life sciences, especially medicine, and has written several articles on the topic.

RALPH O'CONNOR is Professor in the Literature and Culture of Britain, Ireland, and Iceland at the University of Aberdeen, where he teaches in the departments of Celtic, History, and English. He has published widely on mediaeval Irish and Icelandic and modern British literature. He is the author of *Icelandic Histories and Romances* (Tempus, 2002), *The Earth on Show: Fossils and the Poetics of Popular Science* (U of Chicago P, 2007), which won two international book prizes in 2008, *Science as Romance* (Pickering & Chatto, 2012), and *The Destruction of Da Derga's Hostel: Kingship and Narrative Artistry in a Mediaeval Irish Saga* (Oxford UP, 2013).

MARCO PASI is Associate Professor in the History of Hermetic philosophy and related currents at the University of Amsterdam (UvA). He has written extensively on the history of modern Western esotericism, especially in relation to magic, art, and politics. He is the author of *Aleister Crowley e la tentazione della politica* (1999), translated into several languages. He has edited *Peintures inconnues d'Aleister Crowley: La collection de Palerme* (2008), and co-edited *Kabbalah and Modernity: Interpretations, Transformations, Adaptations* (2010).

ISABELLE PERCEBOIS (Université Paris 4) is Agrégée de Lettres Modernes. She holds a Ph.D. in Comparative Literature from Paris IV-Sorbonne University where she successfully defended her thesis entitled *Science writings in the European fantastic imaginary (1816-1894)*. Her main publications are: "Vienne la scientifique", in *Germanica* 43 (2008, 75-84) ; "Du conte merveilleux au conte fantastique : la forêt dans *Lokis* de Mérimée" in *Otrante* 27-28 (2010, 113-120) ; "Le cadavre dans la littérature fantastique : la place du corps dans *Frankenstein* de M. Shelley et *Le Rêve du docteur Mišić* de K.S. Gjalski" in *Frontières* 23-2 (2011, 7-13) ; and "Mary Shelley: la jeune fille et la science ; empreinte du discours scientifique dans *Frankenstein*" in *Panthéons Littéraires et Savants XIX*e-*XX*e *siècles* (2012, 143-156). She has also written a paper entitled, "Humorous Duplicity: Ironic Distance and Fantastic Tension in Villiers de l'Isle-Adam's *Claire Lenoir*", due to be published in *Fastitocalon III* in 2013.

CLAUDINE PEYRE is Senior Lecturer at Paul Sabatier University in Toulouse (LAIRDIL, Toulouse 3) and co-Chair of the English Department. Her Ph.D. work focused on Margaret Drabble's literary work and women's fiction-writing. Her current research ranges from language-teaching and learning to contemporary fiction. Her most recent publications explore such key-notions as space and time in Margaret Drabble's work.

MATHIEU QUET is a postdoctoral student at the Institut de Recherche pour le Développement (IRD/IFRIS, UMR 201). His main research focuses on the interaction between communication, power and protest movements in science and medicine. His early research dealt with anti-nano activism and radical science movements in the 1970s. His most recent research projects examine controversies over the international regulation of pharmaceutical markets and look more specifically at the transnational movement for healthcare access between Europe, India and Kenya. He is the author of *Politiques du savoir. Sciences, technologies et participation dans les années 68* (Editions des Archives Contemporaines, forthcoming, 2013).

LAURENCE ROUSSILLON-CONSTANTY is Senior Lecturer at Paul Sabatier University in Toulouse (CAS, EA 801, Université Toulouse 2-Mirail). Her main research field is Victorian literature and painting and she has published many articles on Dante Gabriel Rossetti and John Ruskin. She is the author of *Méduse au miroir. Esthétique romantique de Dante Gabriel Rossetti* (Grenoble: ELLUG, 2007) and co-editor of *The Rossettis then and Now: Cosmopolitans in Victorian London* (London: Anthem Press, 2003). She also co-authored a translation into French of a selection of texts from John Ruskin's *Modern Painters* (Pau: PUP, 2006). She is the co-editor (with Anuradha Chatterjee) of the online academic journal, *The Eighth Lamp, Ruskin Studies To-day.* Her most recent articles focus on the interaction between art and science and museum culture.

A firm believer in interdisciplany research, she initiated the *Science, Fables and Chimeras* conference project, commissioned and coordinated the related exhibitions and events in Toulouse 3 University Library and at the Natural History Museum in Toulouse. In the wake of the Chimeras conference, she prompted the idea for another international interdisciplinary conference on "imprints, impregnation, impression", which she co-organised with Michael Parsons and Dominique Vaugeois at the University of Pau (UPPA) in June 2012. A selection of articles presented at the conference will be published in January 2014.

INDEX